BIOTECHNOLOGY and Society

HALLAM STEVENS

BIOTECHNOLOGY and Society

AN INTRODUCTION

UNIVERSITY OF CHICAGO PRESS

Chicago and London

The University of Chicago Press, Chicago 60637
The University of Chicago Press, Ltd., London
© 2016 by The University of Chicago
All rights reserved. Published 2016.
Printed in the United States of America

25 24 23 22 21 20 19 18 17 16   1 2 3 4 5

ISBN-13: 978-0-226-04596-2 (cloth)
ISBN-13: 978-0-226-04601-3 (paper)
ISBN-13: 978-0-226-04615-0 (e-book)
DOI: 10.7208/chicago/9780226046013.001.0001

Library of Congress Cataloging-in-Publication Data
Names: Stevens, Hallam, author.
Title: Biotechnology and society : an introduction /
Hallam Stevens.
Description: Chicago : University of Chicago Press, 2016. |
Includes index.
Identifiers: LCCN 2015044526 | ISBN 9780226045962
(cloth : alk. paper) | ISBN 9780226046013 (pbk. : alk. paper) |
ISBN 9780226046150 (e-book)
Subjects: LCSH: Biotechnology. | Biotechnology—Social
aspects. | Biotechnology—Moral and ethical aspects.
Classifications: LCC TP248.23 .S74 2014 | DDC 660.6–dc23
LC record available at http://lccn.loc.gov/2015044526

♾ This paper meets the requirements of ANSI/NISO
Z39.48–1992 (Permanence of Paper).

CONTENTS

INTRODUCTION

SCIENCE AND TECHNOLOGY IN CONTEXT

Christopher Reeve—best known for his role as Superman in the series of 1970s and 80s films—passed away in 2004. The causes were complications arising from a spinal injury he had suffered while horseback riding in 1995 and which had left him a quadriplegic. In the final years of his life, Reeve became a controversial figure. He argued forcefully that stem cell research had the potential to cure his paralysis and that of thousands of others. The restrictions on US federal government funding for stem cell research, enacted by George W. Bush in 2001, were delaying progress in this crucial area of medical research, Reeves said.

The supporters of the restrictions, many belonging to the Religious Right, argued that collecting human embryos for this research involved the destruction of life. Setting up a foundation to fund this research, Reeves became involved in a highly politicized and bitter struggle involving biotechnology and religion. Here, as elsewhere, the significance of biotechnology extends far beyond the walls of the laboratory or the hospital. "Making Superman walk again" linked stem cells to powerful cultural, religious, economic, and political issues that divide society. Biotechnologies have become objects in debates about the costs of health care, the appropriate roles of government regulation and funding, international scientific and economic competition, and our rights over our own bodies. On Wall Street, the fate of biotechnology and pharmaceutical companies influences the global economy. Questions central to biotechnology—such as those about the ownership of genes or cells—are hotly contested in the courts. Within popular culture, Hollywood movie plots center on cyborgs, deadly viruses, and artificial organs. More and more of the food that most of us buy in the supermarket contains genetically modified ingredients. And most everyone knows Dolly the sheep. So, understanding biotechnology—what it is, and where it came from—is undoubtedly important. But this means not merely understanding biotechnology as a technical phenomenon, but also seeing how it fits into our society. Such a perspective will shed light on some of the most pressing and controversial issues of our time. Biotechnology reveals much about present-day relationships between nature and culture, biology and technology, living and nonliving, human and

1

nonhuman. By studying and analyzing biotechnology we can come to see our own place in the world a little more clearly.

: : :

This book is intended to do two things. First, it is a history of biotechnology. It provides a narrative of how these technologies (and the industry that produces them) came into being, tracing some of the important historical transformations that biotechnology has engendered. Second, the book uses the tools of the social sciences (history, sociology, anthropology, philosophy, cultural studies) to analyze biotechnology. It critically examines biotechnology in a wide variety of contexts. Many of the arguments in this book work from the assumption that the technical and the social, the technical and the political, and the technical and the economic are always intertwined. It is impossible to understand the origins or significance of technical things without understanding their social, political, and economic context. There have been many books that examine the economic effects of biotechnology, or explore some of its ethical ramifications, or speculate about its possible consequences for the future. Rather than taking any one of these approaches, *Biotechnology and Society* synthesizes these different perspectives on biotechnology.

In the first chapter, a case is made that biotechnology should be understood as a *sociotechnical system*: a complex and interacting set of elements that includes both technical (test tubes, gene sequencing machines) and social things (laws, institutions, science fiction movies). Really understanding biotechnology means seeing the big picture, seeing the connections and interactions between all these things. One straightforward but powerful framework for analyzing technological constructs in this way is called *coproduction*. This idea suggests that the outcomes of science and technology are always coproduced by social and technical circumstances—we cannot understand stem cells or climate change or quantum mechanics without paying attention to both technical and social circumstances *at the same time*. One might even want to go so far as to say that the technical is inseparable or inextricable from the social—there is nothing that is technical that is not always also social. The Apollo spacecraft, for example, was no doubt a feat of science, engineering, and technical prowess; but its existence and design can be fully explained only by referring to the social, political, and economic conditions of the Cold War, the space race, nuclear fear, US prosperity, and peculiarly American conceptions of the "final frontier." All kinds of things in the world are coproduced by technological and social circumstances.

How might we discover the social or political aspects of things around us? Historians of technology have shown how even seemingly mundane objects have "politics" built into them. Take a highway overpass, for instance. This seems like a fairly boring, neutral sort of object. Many such overpass bridges have been built all over the United States since the 1950s. In New York City, many such bridges were planned and designed by Robert Moses (1888–1981), one of the most influential urban planners of the twentieth century. Head of numerous public authorities in New York, Moses wielded wide-ranging power over the construction of public facilities around the city. Unfortunately, Moses was also a racist and publicly espoused racial segregation. In Brooklyn, this had an unusual consequence for road overpass bridges: Moses built them especially low. So low in fact that, while passenger cars could pass under them without issue, public buses could not use the roadways. Since it was mostly poor African-Americans who used the buses as their primary means of transportation, Moses knew that his low bridges limited the access of this minority to the beaches on Long Island. Moses did not want racial mixing on Jones Beach, Lido Beach, and Rockaway Beach, and built this prejudice into the design of his highways. The mundane highway overpass became an object that reflected a particular ideological agenda, expressing a political view in concrete and rebar. This is an extreme example, but it suggests how unpacking the history of an object reveals cultural and social meanings. This book will do a similar thing for genetically modified foods, stem cells, and personal genomics—it will open up their history and their politics.

These kinds of analyses—using coproduction and looking for the politics inside technologies—will help to expose the connections between biotech's technical side and its political and cultural dimensions. They will show the importance of thinking of biotechnology as a system that affects many aspects of our lives.

AUDIENCE AND SCOPE

Biotechnology and Society has several intended audiences. For the general reader, I hope that this book provides an introduction to the subject of biotechnology: what it is, where it came from, what its significance is, and where it might be leading us. For this audience the book can be read as a straightforward historical narrative. For students of "science studies," "science, technology, and society," or the history of science, the book provides a broad overview of the history of biotechnology. This is a starting point for further reading and research in this field; it introduces not only the main topics, but also most of the major theoretical and methodological approaches to this material. I have provided annotated bibliographies at the end of each chapter that suggest works that provide more detailed accounts and analyses of the subject matter. Finally, the book is aimed at biomedical scientists who probably already know a lot about biotechnology as a technical field but wish to know more about its history. For these readers, I hope that connecting biotechnology to its various contexts helps them to see (and perhaps even do) their scientific work from new and broader perspectives.

Despite the emphasis on the social and political aspects of biotech, this book does not shy away from technical details. It is simply not possible to understand or analyze the social implications of biotechnology without some understanding of the science and technology itself. This book does not attempt to gloss over this detail: where necessary, scientific and technical concepts are explained from first principles, in plain language. In some cases, simplification is necessary. However, the aim is that the reader will come away with some technical knowledge about how particular aspects of biotechnology work, as well as an understanding of its historical and social contexts.

: : :

The definition of *biotechnology* I will offer in chapter 1 is wide: it applies to vitamin supplements and genetic counseling as well as much in between. Such a definition reminds us how much of our modern lives are intertwined with the biotechnological in some way or another. However, some selectivity is always necessary and it is worthwhile to point out some specific domains that this book does *not* examine in detail.

First, the book will have relatively little to say about "older" biotechnology, except in chapter 2 and except where it is useful to highlight similarities and differences with newer biotech objects and techniques. Other books have done an excellent job of tracing the prehistory of biotechnology, and the references to those works can be found in the Further Reading sections at the end of each chapter.

Second, this is not a book about the development of the biotech industry per se. The approach here is to understand biotechnology as a sociotechnical system. This system includes companies and economic elements (venture capitalists, stock markets). These are vital to the story of biotechnology. However, the emphasis here is on trying to understand how these financial and economic elements fit together with biological objects, social institutions, laws, and so forth to form the complex we call biotech. Once again, there are many books that have described the rise and development of the biotechnology industry, and I refer the reader to these works where appropriate.

Third, the chapters do not address many consequential developments in biology itself or in medical research and practice. Of course, biology and medicine do enter the story at particular points—especially where they have immediate and transformative social, political, and economic consequences (for instance, the development of the contraceptive pill and psychoactive pharmaceuticals warrant particular attention). This is not to suggest that developments in biology and medicine are unimportant—on the contrary, they form the background to much of the history described here. However, by necessity, *Biotechnology and Society* focuses its attention on instances where contact between the biological and the social is at its sharpest.

Finally, although the book does pay attention to the relationships between biotechnology and culture, it is not about the public understanding or public reception or responses to biotechnology. Films, science fiction, and other technological "imaginaries" certainly play a role in shaping technologies and their patterns of use and acceptance; several examples of this will be discussed in detail. However, the emphasis is on understanding the interplay between culture and technology: it is not just a question of determining the impact of technology on culture, or the influence of culture on technologies. Rather, we need to examine the complex feedback loops (ramifying through politics, economics, law, medicine) through which each constantly remakes the other.

As a whole, *Biotechnology and Society* is not just about understanding the technological details of biotech, nor is it about understanding the effects of biotechnology on the economy, or the effects on medicine, or the effects on the law. Rather, it is about trying to comprehend the complex interconnections between all these elements. That is, it is about how biotech is transforming our society and culture as a whole. *Biotechnology and Society* is not just about technology, but also about *what we are becoming through this technology*.

: : :

Writing a book about such a diverse range of topics entails a further challenge. Biotechnology is a rapidly changing field. No doubt, by the time this book goes to press, some important new developments will have emerged. Does this mean that this book is already out of date? Fortunately, no. For one thing, the historical accounts will remain relevant, even if our perspective on the past changes on account of new discoveries or inventions. But the frameworks for analyzing biotechnology described here will continue to be applicable, too. These frameworks will continue to be relevant to new biotechnologies and the chapters here will continue to serve as examples of how we might think about these new discoveries and inventions critically from social, political, and economic viewpoints.

This book is composed of twenty-four short chapters. They tell the story of biotechnology in roughly chronological order. Each of these chapters can be approached on its own. However, more than chronology ties the chapters together. Studying the history of biotechnology reveals some recurring themes. These themes summarize some of the lessons we might learn from a critical analysis of biotechnology and suggest how we might move forward in resolving some of the problems and debates that biotech raises.

FIVE THEMES

THEME 1: BIOTECH IS NOT NEW

There are a broad range of things that might count as biotechnology. When most people think of biotech, they think of recent techniques (such as direct manipulation of cells or DNA, for instance). Most of these technologies have been around only since the 1970s. But limiting our view of biotech to just DNA and stem cells seems, on closer examination, quite arbitrary. After all, genetically modifying corn (a recent practice) and selectively cross-pollinating corn to create larger ears (an ancient practice) both aim at the same end (increasing food yield). And, they're modifying the same object (that is, the maize genes). Admittedly, the genetic modification techniques are more direct and no doubt quicker, but there is something fundamentally similar going on. In chapter 2, I examine the history of beer as an extended example of how humans have been using biology (in this case yeast, a microorganism) for their own ends for many centuries. So at least in agriculture, baking, brewing, cheese-making, dye-making, and significant parts of the chemical industry, biotechnology has a long history. Parts of this history will enter into this book, although the book does not trace the long history of biotech in detail (if you are interested in this story, I recommend Robert Bud's *The Uses of Life*).

This long history is important for two reasons. First, showing that biotechnology has a history shows how some things that we think are radical and new about biotech are actually quite old. Showing that problems, debates, and controversies have a history can sometimes help us to resolve, or at least better understand, them. Contemporary debates about genetically modified foods, for instance, have parallels in debates about the Green Revolution in the 1960s and '70s and in British concerns about class and food resources in the nineteenth century. Putting these debates in this context suggests that some of the anxiety about genetic modification may have more to do with concern over the planet's diminishing resources than with new techniques, per se.

The second important reason for examining biotech's long history is that there are some special and interesting things about post-1970s biotechnology. The hype of new biotechnologies often makes everything seem novel. The longer history can present a point of comparison—it can help us pin down what is really new and interesting about biotech in the twenty-first century. Such an analysis is important because of the political and economic stakes involved. Proponents of biotechnologies often have an interest in showing that many of its practices have a long history. Portraying it in this way may make biotechnologies seem less dangerous or frightening. Opponents of biotechnologies, too, may draw on historical examples intended to warn us about the dangers of meddling with nature. Examining the history of biotechnology in more detail will allow us to critically evaluate both kinds of claims.

THEME 2: PLASTICITY

In 1895, the renowned science fiction author H. G. Wells wrote an article for the London weekly, *The Saturday Review*, titled "The Limits of Individual Plasticity." Wells wrote:

> A living being may also be regarded as raw material, as something plastic, something that may be shaped and altered, that this, possibly, may be added and that eliminated, and the organism as a whole developed far beyond its apparent possibilities . . . a living thing might be taken in hand and so moulded and modified that at best it would retain scarcely anything of its inherent form and disposition; that the thread of life might be preserved unimpaired while shape and mental superstructure were so extensively recast as even to justify our regarding our result as a new variety of being.[1]

Although Wells was a novelist, he was also active in the science and politics of late Victorian Britain. Influenced by socialism, Wells was deeply interested in scientific questions about the molding or shaping of humans. *The Island of Dr. Moreau* (1896) is an extended consideration of this subject—a rogue scientist uses the techniques of vivisection to gradually transform animals into human-like forms. The novel is in part a warning about the dangers of unchecked scientific curiosity, but it also speculates about the "limits" of life's plasticity and raises the question, "How much could life be changed and still remain life?"

Much of contemporary biotechnology is an exploration of this question. What are the limits of life? To what degree can it be manipulated and shaped? What are the consequences of this shaping? The history of biotechnology is largely a history in which we have discovered that life is far more plastic than

1. H. G. Wells, "The Limits of Individual Plasticity," *Saturday Review*, January 19, 1895. Also reproduced in H. G. Wells, *H. G. Wells: Early Writings in Science and Science Fiction* (Berkeley: University of California Press, 1975), 36–39.

we thought. Much of the controversy about biotech is about coming to terms with this ability to intervene, manipulate, and reshape at will (biologists are often criticized for "playing God"). Following Wells, we might consider biotechnology as a set of ways in which living beings might be used as raw material. This also suggests a connection between controlling life in labs and controlling the lives of populations. Since scientists and others in the past have often thought of these two problems as connected, this book also considers them as inseparable parts of a whole that must be discussed together (genetics and eugenics, for instance).

THEME 3: PROMISE

Biotechnology is often portrayed as the science and technology of the future. When biotechnology is discussed in the media or in public forums, we hear mostly about the biotechnology that is to come: the breakthrough just about to be made, the life-saving drug just around the corner, the possibilities for extending life into distant, but not unimaginably far off, years to come. Whether or not biotechnology actually lives up to such promises, of course, remains to be seen. However, this book argues that these kinds of promises are constitutive of biotechnology—it is always oriented towards the future, the science of "never quite there." This is partly because of economics. As an industry, biotechnology needs to make promises to its investors that they will receive high returns. Those start-ups that can make big promises are likely to see the financial returns of the big investments. Here the economics of speculation crosses with the creativity of the laboratory to create what might be called *promissory science*.

But this hype associated with biotech also has another important source. Popular culture—especially science fiction in books, movies, and video games—also generates expectations and foreshadows the biotech developments of the future. These scientific and technical imaginaries exert an influence on how biotech develops, influencing the direction of scientists' own work. But they also, perhaps more importantly, influence how we understand biotechnologies: what we expect, what we desire, and what we fear. *Jurassic Park* (1993), *Outbreak* (1995), *Gattaca* (1997), *In Time* (2011), and even the X-Men movies (2000, 2003, 2006, 2009, 2011, 2014) shape our ideas about cloning, bioterrorism, genetic privacy, immortality, and genetic mutation. So analyzing biotechnology means paying careful attention to how cultural sources have shaped and are continuing to shape biotechnology, especially through hype, promise, and fear.

One of the most serious consequences of promissory science is that it makes biotech especially difficult to assess for the purposes of any kind of debate. Biotechnology almost always requires weighing the value of present "knowns" against the promise of potential future benefits. Synthetic biology, for instance, creates a risk that some specially engineered organism could escape from its laboratory and run amok in the environment. Does this mean we should cease all research in synthetic biology? To make such a decision we need to balance these risks against the great promise, speculation, and hype about the potential future gains from this line of work (new therapies, new energy sources, etc.). Analyzing biotech as a promissory science means, as much as analyzing the science and technology itself, finding ways of analyzing the hype, speculation, and expectations that surround it.

THEME 4: BEYOND CONTROVERSY

Much of the popular attention devoted to biotechnology concerns a set of high-profile controversies. They include debates about the safety of genetically modified foods, worries about the privacy of genetic information, the possibilities of bioterrorism, biotech's contribution to the rising costs of health care, and the politically divisive debate about stem cell research. These are accompanied by smaller-scale battles including those concerning cloning (what forms are acceptable, for what purposes?), assisted reproductive therapies (who should have the right to use them?), the patentability of genetic information, and laboratory safety. These controversies are all too often framed around the question of "*should we or shouldn't we?*": Should we allow tax dollars to be spent on stem cell research or shouldn't we? Should we require genetically modified foods to be labeled in supermarkets or shouldn't we? Should we allow personal genomics companies access to our genomes or shouldn't we? And so on. Asking "should we or shouldn't we?" about various technologies is often important, and usually it is the most pressing question for immediate public policy purposes.

But this book encourages the reader to step back from "should we or shouldn't we?" This question all too often obscures or obfuscates a number of other more fundamental and sometimes more important questions that need to be asked about biotechnologies. First, asking questions about the history of biotechnologies can often reveal much about why a debate is taking place at all. Showing why an issue emerges at *this time* and in *that place* can tell us much about what is really at stake. Second, putting biotechnology in context can show how some controversies about technology are actually manifestations of more deeply rooted and long-term cultural conflicts. Third, often the "should we or shouldn't we?" debate proceeds without a clear analysis of who stands to gain and who stands to lose from particular developments. We need to ask questions that sort out the interwoven strands that connect biotechnologies to the interests of governments, institutions, and corporations—often in obscure ways. The social sciences offer a range of tools and frameworks that go beyond the obvious ethical or moral questions in order to interrogate biotech from a variety of perspectives. Again, this is not to suggest that the ethical dimensions of biotechnology are not important, but rather to make sure we are fully equipped to provide solutions to moral quandaries that encompass more than just the technical dimensions of biotech.

THEME 5: RISK

Finally, biotechnology demands that we engage with the problem of risk. The biotech breakthroughs that offer great hope for curing cancer or cleaning up pollution also have the potential for great harms. These harms may be environmental or medical (posing a threat to human health), or they may be legal, social, and economic (inventions that are so disruptive that they destabilize). They may be immediate and predictable, or they may emerge only in the very long term (and be largely unforeseeable). The problem here is that most of the frameworks for assessing and understanding risk are unable to deal with the challenges posed by biotech. One issue is that risk management frameworks tend to try to measure costs versus benefits for a specific technology. In the case of biotechnologies, long-term costs and benefits may be hard to foresee. In addition, cost-benefit analyses tend to require some form of quantification—risks and rewards that can be measured in probabilities and dollars. The kinds of "social risks" that biotech often presents are hard to quantify and hence are often ignored by risk rubrics.

In the 1970s and '80s, there was a large public opposition to nuclear power in the Western world. Scientific experts (physicists, government officials) repeatedly and consistently assured the public that nuclear power was "safe" and that the risks were small. Yet the opposition continued, or even intensified. How and why did the public perceive the risk of nuclear power in this way? In the public imagination, nuclear power was tied to Cold War anxieties and the risks of nuclear war. Moreover, the public perceived the nuclear power industry as a hegemonic technological system that was associated with state control and increasing state power over people's lives. It was not that the public "didn't understand" nuclear power or were acting irrationally; in fact, in some sense, they understood nuclear power only too well. Public opposition reflected not just a technical assessment of risk, but a more holistic sociopolitical assessment of the cultural and political meanings of a nuclear power industry.

Brian Wynne, a sociologist who studies the public understanding of science, argues that we need to take far greater account of how "personal understandings" affect perceptions of risk. "Specific publics," he writes, "are likely to be skeptical, critical, or simply hostile to scientific statements — often because such statements seem to emerge from an idealized and inappropriate model of real world conditions."[2] This book argues that new, and more sophisticated, methods for understanding risk and the perception of risk (such as Wynne's) are required in the case of biotechnology. We need to understand, in particular, the complex relationships between "experts" and the "public," how knowledge circulates between them, and even how a clear distinction between these two domains might be breaking down (through movements such as citizen science and Do-It-Yourself Bio, for instance). Using biotech wisely, carefully, and productively requires better ways of understanding and measuring risk.

: : :

These five themes suggest how this book can also be read as an intervention into the debates about biotechnology. They show how and where we can make progress towards creative, productive, safe, and socially responsible uses of biotechnology.

FURTHER READING

There are several other books that have attempted to give more or less comprehensive accounts of biotechnology. The most thorough historical overview of biotechnology is Robert Bud, *The Uses of Life: A History of Biotechnology* (Cambridge: Cambridge University Press, 1994). This work focuses on manifestations of biotechnology before the 1970s, although there is some attention to more recent developments too. For an introduction to the science of biotechnology, the best places to start are W. T. Godbey, *An Introduction to Biotechnology: The Science, Technology, and Medical Applications* (Cambridge: Woodhead Publishing, 2014) and Ashim K. Chakravarty, *Introduction to Biotechnology* (New Delhi: Oxford University Press, 2013). These provide many more technical details than will be given here. Biotechnology is changing rapidly so technical books, in particular, may quickly be out of date.

There are several monographs that take up cultural and philosophical approaches to biotechnology: Jon Turney, *Frankenstein's Footsteps: Science, Genetics, and Popular Culture* (New Haven, CT: Yale University Press, 1998), Gregory Stock, *Redesigning Humans: Choosing Our Genes, Changing Our Future* (Boston: Mariner Books, 2003), and Robert Carlson, *Biology Is Technology: The Promise, Peril, and New Business of Engineering Life* (Cambridge,

---

2. Alan Irwin and Brian Wynne, eds., *Misunderstanding Science: The Public Reconstruction of Science and Technology* (Cambridge: Cambridge University Press, 1996), 9.

MA: Harvard University Press, 2011). These provide accounts of the relationship between biotechnology and culture, the impact of genetically engineering humans, and biological engineering, respectively. This is by no means a complete list of books about biotechnology, but these are good places to start if you are looking for a broad scope.

The various frameworks mentioned in this introduction derive from important scholars in the field of science and technology studies. The notion of coproduction comes from the work of Sheila Jasanoff, *States of Knowledge: The Co-Production of Science and the Social Order* (New York: Routledge, 2004). The story of Robert Moses and his bridges can be found in Langdon Winner, "Do Artifacts Have Politics?" *Daedalus* 109, no. 1 (1980): 121–136. Winner is arguing that technologies (including seemingly mundane, everyday objects) are not politically neutral—they are designed in particular ways and used in particular ways that have social and cultural effects and that therefore they are political. However, see also the arguments of Bernward Joerges, "Do Politics Have Artefacts?" *Social Studies of Science* 29, no. 3 (1999): 411–431.

The notion of plasticity in biotech has been discussed by Hannah Landecker, "Living Differently in Time: Plasticity, Temporality, and Cellular Biotechnologies," *Culture Machine* 7 (2005). There is a significant literature that addresses the theme of promise in biotech in various ways. For anthropology this is summarized in Karen Sue-Taussig, Klaus Hoeyer, and Stefan Helmreich, "The Anthropology of Potentiality in Biomedicine," *Current Anthropology* 54, no. S7 (2013): S3–S14. Particularly worthy of mention are Nik Brown, Alison Kraft, and Paul Martin, "The Promissory Pasts of Blood Stem Cells," *Biosocieties* 1, no. 3 (2006): 329–348, and Michael Fortun, *Promising Genomics: Iceland and deCODE Genetics in a World of Speculation* (Berkeley: University of California Press, 2008).

Finally, the arguments of Brian Wynne are laid out in numerous places, but the best place to begin is Brian Wynne and Alan Irwin, eds., *Misunderstanding Science? The Public Reconstruction of Science and Technology* (Cambridge: Cambridge University Press, 2004). Another, related, approach to the public understanding of science can be found in Helga Nowotny, Peter Scott, and Michael T. Gibbons, *Re-thinking Science: Knowledge and the Public in an Age of Uncertainty* (Cambridge: Polity, 2001).

PART I
THE LIMITS OF
BIOTECHNOLOGY

1 WHAT IS BIOTECHNOLOGY?

WHAT IS TECHNOLOGY?

To talk about *biotechnology* we need to know more precisely what this word means. At least for the purposes of this book, we need a definition to guide our discussions about what is new, or different, or interesting, or controversial about biotech. And to do that, it is helpful to begin with a discussion of technology itself. We should sort out what we might mean by *technology* before we can agree on what *biotechnology* might be.

Human beings have been making and using tools for a very long time: sharpened sticks and rocks, hollowed tree trunks, dried grasses, animal hides, and so on. Some philosophers have even argued that making tools to control the environment should be considered the definitive characteristic of humans. Rather than *Homo sapiens* (wise man), perhaps we should be called *Homo faber* (man the creator). On this view, technologies are integrally part of human existence and human interactions with the world. The word *technology*, then, doesn't just refer to objects. When we say *technology* we also mean the knowledge required to use those objects. Knowing how to start a fire using specially shaped sticks, or how to surf the web, is part of the technology of fire, or computer networks, respectively. Recipes, know-how, and rules of thumb might all be considered technologies, or parts of technologies. This extends to the processes of creating the objects themselves. The actual activity of carving a tool or manufacturing a computer is surely part of the idea of technology too. So technologies are not only objects, but also the knowledge, activities, processes, methodologies, and uses that are necessarily attached to them. Technologies are all around us, and include many of the things we do in our lives.

One way to talk about this complex set of things is to use the concept of a *sociotechnical system*. Rather than speaking about a technology as it if it were an isolated object, we can speak of a sociotechnical system that includes all the elements just mentioned. This concept reminds us that when we pick up our cellular phones to make a call, we are really depending on the operation of a large system: cell phone towers for mounting radio transmitters, radio transmitters to capture and re-transmit the signal, software to encode and decode the signal, computers to run this software, standards that determine how the encoding and decoding should work, laws that prevent inter-ference in particular parts of the radio spectrum, companies that maintain the towers, transmitters, and computers, engineers who design the specifi-cations of the standards, databases that keep track of call times for billing, managers and accountants to run the phone companies, and so on. The list could be expanded further. What is crucial about it, however, is that it in-cludes both objects (what we ordinarily identify as technologies) and "social" elements: people, institutions, conventions, and laws. And all these things are multiply connected to one another in complicated ways, in a sort of network or system that includes both technical objects and social elements. This is a sociotechnical system.

All but the most elementary tools can be understood as being embedded in such a system. Even a prehistoric stone tool depended on know-how about how to make such tools being passed down from generation to generation; and on norms and conventions for its proper or appropriate use; or perhaps on systems of barter and exchange that allowed groups of people to get the right kinds of rocks. Without any of these elements, the technology would fail. The stone tool could not be made, or no one would know how to use it. The cellular phone would be useless without all of the elements being in place. It would cease to be a technology at all.

At least for the purposes of this book, then, it will be useful to understand technology not just as objects or sets of objects, but as networks of social and technical things linked together in complicated and interdependent ways.

BIOTECHNOLOGY AS SOCIOTECHNICAL SYSTEM

The definition of technology that I have just offered presents some problems for the notion of *bio*technology. First, to distinguish something as a tool usually means distinguishing something *made by humans*. What makes *Homo faber Homo faber* (that is to say, what makes us us) is the ability to cre-ate new things not already present in the world. Usually we think of things already present in the world as belonging to nature, as natural things. Things that we make, tools and technologies, are human-made or "cultural" things. But the *bio* in biotechnology refers to things already in nature, living things not made by humans. So, at first glance, it seems like biotechnology presents us with a contradiction: bio refers to things made by nature, but technology refers to things made by humans. What could such a term possibly mean?

But there's yet another problem. If we take the idea of a sociotechnical sys-tem seriously, then such networks of elements should include "natural" stuff. Couldn't your hand (a natural object) for holding your cellular phone or carv-ing the stone tool also be considered part of the system? And isn't the rock itself (or the silicon in your phone) something made by nature? Or consider the technology of a car: an internal combustion engine runs on gasoline that was once living matter (very biological stuff). Or a power grid: if it's hydroelectric then should we include the (again quite biological) river as part of our sociotechnical system? Mapping out sociotechnical systems might lead us to the conclusion that almost all technologies are in some sense biotechnologies since they include elements of nature.

We seem to have two possibilities: biotechnology is an impossible contradiction or biotechnology includes all technologies. Neither of these can be right. Most people would agree that biotechnology is a real thing that we can point to in the world around us, and that it is really distinct from things like the technologies of silicon chips, combustion engines, and electric power. Genetically modified food, cloning, stem cells, and other things we will discuss in this book all rightly belong to the category of biotechnology. iPods, ceramics, and metalworking, for instance, do not.

The definition of biotechnology that is proposed here, and which will be used throughout this book, includes the most familiar biotechnologies such as genetically modified foods, stem cells, and cloning. It also attempts to resolve the dilemmas outlined above. The definition has two parts. The first part is this:

1. Biotechnology is a sociotechnical system in which some of the elements are active biological processes.

What is an active biological process? Bacteria reproducing, plants growing, DNA replicating, yeasts metabolizing, and animals respiring, would all count. Using dead plants for gasoline, or rocks, or a flowing river, does not involve any *active* biological process. This definition attempts to distinguish between just making tools with objects found in nature and those tools or techniques that depend on the continuous and active functioning of growth, reproduction, respiration, locomotion, metabolism, and other biological processes. Genetic engineering depends on bacteria reproducing, cloning and stem cells depend on animal reproductive processes, genetic modification of foods still depends on ordinary plant growth, and so on. Power plants, cellular phone networks, computers, metalworking, and building bridges cannot be said to be dependent on biological processes in this same way.

This definition also suggests that biotechnology is not a mere collection of objects, but rather a whole complex of social and technical elements, only some of which need be strictly "biological." This accords well with usages of the term *biotechnology* that refer not just to genetically modified mice or cell lines, but also to laboratories, institutions, companies, methodologies, and laws. Biotechnology is a whole system of animate and inanimate elements that must function together. Considering biotech as a sociotechnical system resolves the contradiction of bio+technology noted above: biotechnology involves taking biological processes and placing them within a human-made system, within a complex of natural and social things. These natural things are not broken down or destroyed but rather embedded within a system that makes them into useful tools. Unlike many other tools, biotechnology does not take a natural object (a stone) and make it into something human-made (a blade). Rather, natural processes become part of a system. This system is "technology" made by humans, but it also includes elements of the "bio" made by nature. All this suggests that biotechnology *mixes* the categories of the social and the natural—the complex systems into which biological processes become enlisted make them both social and natural at the same time. We will see many examples later in the book of how biotechnologies seem to do exactly that.

BIOTECHNOLOGY AS MOLECULAR CONTROL

The definition that I have offered remains rather broad. For example, it includes all kinds of agriculture and food production. The kinds of settled agriculture that have been going on for many millennia should certainly be considered a sociotechnical system: they depend not only on technological objects like plows, but also on whole systems of social organization (feudalism, for example). And since they depend on plant growth, they are certainly making use of a biological process. Animal husbandry, selective breeding of cattle or crops, fish farming, and so on are all sociotechnical systems that make use of biological processes and therefore fall under the first part of the definition offered here.

In fact, there are many other practices that fall under this definition—the brewing of beer, the fermentation of wine, and the leavening of bread—that have a long history. In the next chapter we will learn about zymotechnology, the study and industrial application of fermentation that stretches back at least to the seventeenth century. There are good reasons for including these topics in any discussion of biotechnology, particularly in order to highlight some of the continuities and differences between current biotechnology and older practices. We should continue to bear in mind that not everything biotechnological is new and that humans have been trying to turn nature to their own purposes for a long time.

However, during the twentieth century, advances in our understanding of biological processes increased the specificity with which we can intervene in them. Cloning Dolly the sheep using somatic cell nuclear transfer is a substantially more powerful intervention then selectively breeding sheep in order to alter the offspring over many generations. We need a definition of biotechnology that captures something of our increasing power over biology. So a second part is needed:

2. Biotechnology is directed towards control over biological processes at the molecular level.

Particularly since the 1970s, most things that we would want to call *biotechnology* are part of a sociotechnical system that aims to manipulate biological molecules. They read, edit, and rewrite the molecular mechanisms of biology. This applies not only to the manipulation of DNA and RNA, but also to proteins, immunoglobulins, lipids, polysaccharides, vitamins, hormones, neurotransmitters, sterols, metabolites, and the many other small molecules that play a role in biological processes. Even where biotechnology deals with meso- and macroscopic objects such as cells, blood, and whole organs—such as in blood banking or xenotransplantation—these depend, to varying degrees, on understanding and controlling processes on the molecular level.

It is important to note that old biotechnologies such as animal husbandry ultimately depended on hereditary processes that are molecular (DNA recombination, and so on). But the practitioners in these cases suspected nothing of such molecular mechanisms—so their work could not possibly be said to be *directed* towards molecular control. This does not mean that there is a completely sharp and clear break between old (let's say pre-1970s) and new (1970s onward) biotechnology. From the early years of the twentieth century, for instance, vitamins and hormones were used as molecular interventions in human and animal bodies. But the definition does capture how biotechnology (and biology more generally) has generally moved towards manipulation of biological objects on a smaller and smaller scale.

This definition also helps us understand the relationship between biotechnology and biology. In conceptual, institutional, economic, and cultural terms, biotechnology is very closely related to the study of living things. Traditionally, however, biology has been conceived as directed toward *understanding* living organisms. This second part of the definition makes it clear that biotechnology is directed towards *control* rather than understanding. But the history of biotechnology has blurred this distinction considerably: biotech inventions have proved crucial for basic biology, more and more "biological" research has become oriented towards "useful" ends, and synthetic biology raises the question of whether the best way to understand organisms isn't just to go ahead and build (or rebuild) them (see chapter 23). This suggests that biotechnology is helping to redraw the boundaries between science and technology. One way of putting it might be to say that basic biology is becoming a node within the sociotechnical system of biotechnology.

: : :

The definition offered here will not be the last word on the meaning of biotechnology. After all, biotechnology itself is changing all the time and our ideas about its meanings must also evolve. Others have offered (and will continue to offer) different definitions that place more or less emphasis on continuity with the past, or on biotech's novelty, or on its commercial orientation, or on its connections to medicine. No matter which definition we adopt, it is important to think critically about what it includes, what it excludes, and why. The controversy, hype, and money surrounding biotech mean that there is often much at stake in its definition.

FURTHER READING

On the idea of human as tool-maker (*Homo faber*) see Henri Bergson, *Creative Evolution* (New York: Henry Holt, 1911). Bergson considers the ability to make artificial objects and tools as a key part of the definition of intelligence. Hannah Arendt, in *The Human Condition* (Chicago: University of Chicago Press, 1958), uses the concept of *Homo faber* to describe the separation of the human and natural worlds.

There is a large literature that addresses the question "What is technology?" A good place to start is Stephen J. Kline, "What Is Technology?" *Bulletin of Science, Technology, & Society* 1 (1985): 215–218. More advanced approaches include Andrew Feenberg, *Critical Theory of Technology* (Oxford: Oxford University Press, 1991), and Martin Heidegger's essay "The Question Concerning Technology," originally published 1954 and now widely available including in *The Question Concerning Technology and Other Essays* (New York: Harper Torchbooks, 1982).

The literature on sociotechnical systems comes from the history and sociology of technology, especially the school of "social construction of technology" or SCOT. Some of the most important texts in this field are Thomas P. Hughes, Wiebe E. Bijker, and Trevor J. Pinch, eds., *The Social Construction of Technological Systems: New Directions in the Sociology and History of Technology* (Cambridge, MA: MIT Press, 1987), Weibe E. Bijker, *Of Bicycles, Bakelites, and Bulbs: Towards a Theory of Sociotechnical Change* (Cambridge, MA: MIT Press, 1997), and Thomas P. Hughes (*Human Built World: How to Think about Technology and Culture* (Chicago: University of Chicago Press, 2004).

2   THE LONG HISTORY OF BIOTECHNOLOGY

WHEN DID BIOTECH BEGIN?

Biotechnology brings to mind images of a futuristic world inhabited by cyborgs and clones. The very idea of biotechnology is associated with the future. Very seldom is the *past* of biotechnology thought to be interesting or significant. To where and when can we actually trace that past? The answer depends, of course, on the definition of biotechnology that we choose to use. In chapter 1, we adopted a definition of biotechnology that limited it to technologies (1) that involve active biological processes, and (2) that rely on controlling biological processes at the molecular level. According to this definition, the most obvious place to look for the beginnings of biotech is in the development of genetic engineering in the 1970s. These breakthroughs led to the founding of a biotech industry and prompted widespread debates about the safety of genetic research. We will discuss these events in chapters 3 and 4.

This chapter, however, examines some examples of biotech that existed before the 1970s. Some of them will satisfy part 1 of our definition and not part 2. Others will (arguably) satisfy both. This complicates the answer to the question "When did biotech begin?" Nevertheless, the aim of this chapter is to establish that biotech does have a past and that this past is important for understanding biotechnology today. For one thing, it can help to sort out what is really *new* about modern biotechnology as opposed to what is just hype. A historical perspective shows not only that humans have been trying to put nature to use for a long time, but also that this control has very often been deployed to serve commercial and political ends.

Another reason to examine the history of biotechnology is to shed more light on the relationship between technology and society. It is hard to escape present-day ways of thinking and doing. Hindsight is often clearer. It is easy for us to believe, for example, that nineteenth-century biology was deeply influenced by prevailing ideas about race. Victorian accounts of evolution must be understood in the context of fears about racial competition and racial degeneration. It is often harder to see the connections between our own scientific ideas and our own culture and politics. How might the current debates about stem cells and cloning, for instance, be connected to our political and cultural concerns with abortion and reproductive choice? Historical examples can show us where and how to look for these sorts of connections.

This chapter does not cover all possible biotechs of the past. Those selected here emphasize the different sorts of similarities and parallels between older and contemporary biotech.

ZYMOTECHNOLOGY (OR THE HISTORY OF BEER)

The art of brewing beer was known to the ancient Egyptians. Carvings on the walls of tombs show that some time between 5500 BCE and 3100 BCE the Egyptians discovered how to ferment grain and water into beer. Archeological evidence suggests that beer drinking was widespread amongst the various levels of society and that beer formed an important part of religious rituals. We also know that some of the earliest agricultural societies in Mesopotamia, including the Sumerians, the Babylonians, and the Akkadians, also brewed, probably even before the Egyptians.

The rise of agriculture and the discovery of beer are closely connected. The seeds of grasses (such as wheat and barley) are not especially tasty or nutritious. Processing the grain to make bread and beer improves both the taste and nutritional qualities of these plants. For both, grain is first germinated (called *malting*) and then fermented. The fermentation is done by single-celled micro-organisms called yeast (figure 2.1). The yeast feed on the starches and sugars in the grain and break them down into alcohol and carbon dioxide gas. For beer, the alcohol is retained and the carbon dioxide is a by-product (creating the fizziness of beer). For bread, the alcohol is a by-product but the carbon dioxide causes the bread to rise (making it tastier and easier to digest).[1]

The ancient Egyptians and Mesopotamians knew nothing of micro-organisms. The first beer was probably produced by airborne yeast floating into a container of grain and water. The beer they produced was likely to have been a frothy, cloudy beverage with a low alcohol content (0.5%–1%). It probably didn't taste much at all like a modern-day lager. But this is not to say that beer-making wasn't a highly developed skill and a highly technical process. Again, archeological evidence suggests that Egyptians knew how to make many varieties of beer. Some had a long shelf life and some were used

1. In beer brewing, the mashing process makes much more starch available to the yeast and fermentation is allowed to go on much longer, producing more alcohol. In bread, very little alcohol is produced and this is mostly evaporated during baking. Dough that has been left to rise for too long will often smell of alcohol.

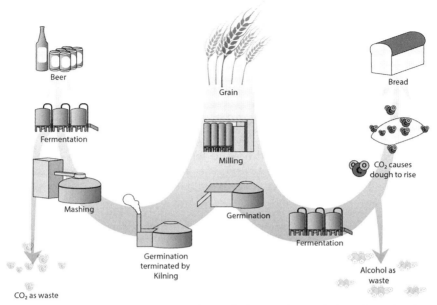

Beer

Grain

Bread

Fermentation

Milling

CO_2 causes
dough to rise

Mashing

Germination

Germination
terminated by
Kilning

Fermentation

CO_2 as waste

Alcohol as
waste

2.1 Beer and bread. The processes used to make beer and to make bread are closely related. Both require germination and fermentation of grains using yeast. In beer, ethanol (alcohol) is retained in the liquid, while in bread ethanol is cooked off during baking. Early agricultural societies had sophisticated means of producing both beer and bread. Source: Illustration by Jerry Teo.

for special purposes such as herbal remedies. Carvings show elaborate diagrams specifying methods for milling, malting, heating, filtering, and capping the beer. *Zymourghos* (yeast makers) knew how to maintain a yeast culture, "seeding" a new batch of beer with some of the previous brew. In ancient Egypt and elsewhere, beer brewing was a well-developed craft and part of an elaborate commercial process.

The techniques and traditions of beer brewing were perpetuated in Europe first by the Roman Empire, and then by Christian monks. In the Middle Ages, monasteries were often equipped with sophisticated brewing facilities. Beer was considered to have both spiritual and medicinal functions. For many people, drinking beer was probably healthier than drinking water since the heating of the wort (the mixture of water and malted grain) helped to kill bacteria. By the twelfth and thirteenth centuries, increasing urbanization increased demand for beer and led to the creation of the first commercial breweries. These operations became increasingly sophisticated, using copper kettles for heating as well as special containers for cooling and fermentation, and adding new ingredients (especially hops) for flavor.

Beer brewing in the Middle Ages and the Renaissance was a lucrative commercial craft. But the scientific revolution of the seventeenth century generated philosophical interest in processes such as fermentation and putrefaction. In Prussia (part of modern-day Germany), Georg Stahl (1659–1734) inaugurated the field of zymotechnology (from the Greek word "to leaven") for the study of fermentation. In 1697, Stahl's *Zymotechnia Fundamentalis* outlined a mechanical theory of fermentation. For Stahl, fermentation consisted of a rapid motion of particles that caused materials to recombine, producing alcohol and expelling gases. Although Stahl did not realize that fermentation depended on living organisms, his zymotechnology marked the beginning of philosophical and scientific efforts to understand and control fermentation.

In the eighteenth and nineteenth centuries, zymotechnology became central to scientific and philosophical enquiry and, at the same time, played a crucial economic and strategic role within Europe. In the 1780s, Antoine Lavoisier (1743–1794) showed that sugars were made up of carbon, hydrogen, and oxygen and that the mass of sugar was equal to the combined masses of the gas, alcohol, and acid that was produced by fermentation. Such experiments laid the foundations for modern chemistry. But they also provided a basis for a practical science of brewing. Especially in England and in German-speaking lands, brewing was becoming increasingly big business. By the late eighteenth century, breweries near London were installing steam engines (a new and expensive technology) for grinding, stirring, and pumping. But brewing was a complex process—results were often inconsistent and sometimes even poisonous. In 1816, brewers in Bohemia asked the university in Prague to establish courses in brewing so as to bring greater knowledge, precision, and rigor to their work. Elsewhere too, brewers began to be trained in the use of thermometers, hydrometers (for measuring density), and saccharometers (for measuring sugar content). Scientific and systematic brewing brought more consistent results and better flavor, leading to higher sales.

By the late nineteenth century, the governments of Europe saw zymotechnology as a science of national importance. In 1876, Louis Pasteur published "Studies on Beer: Its Diseases, Their Causes, and Processes for Making the Beer Constant—With the New Theory of Fermentation." This was part of Pasteur's broader program to establish that seemingly "spontaneous" processes such as fermentation, putrefaction, and human disease were all in fact caused by microscopic organisms. But Pasteur also recognized the industrial and commercial importance of beer. His "Studies" showed that yeasts were responsible for fermentation and that the spoiling of beer was caused by contamination by bacteria. Moreover, he showed that different kinds of yeast were involved in different stages of the fermentation process.

Pasteur's findings provoked a revolution in brewing science. Scientists began to study the effects of temperature, acidity, and other conditions on yeasts. In England, breweries set up an Institute for Brewing in 1886 for the scientific study of brewing processes. In Germany, the *Verein der Spiritus Fabrikanten* (Federation of Spirit Producers) set up in 1872 evolved into the *Institut für Gärungsgewerbe* (Institute for Fermentation Technology), established in Berlin in 1897. This was a large state-sponsored institute, employing eighty scientists and engineers. As important as beer was to the economies of Britain and Germany, zymotechnology took on even greater significance towards the end of the nineteenth century. Understanding fermentation could improve the quality and durability of other food and drink (wine, vinegar, bread, cheese, meats). And fermentation could be used to produce industrially important substances such as acetone and glycerol. Zymotechnology even came to have immense strategic importance during World War I. (England used fermentation-produced acetone in explosives and Germany used large quantities of industrially produced yeast for animal feed.)

Zymotechnology exhibits many of the features of modern biotechnology: specialized and detailed knowledge of brewing was deployed as part of a technical system that had great economic and political significance.

PEACHES AND PLUMS

If we're looking for other early examples of biotechnology, food is a good place to start. After all, food production and consumption is a realm in which humans necessarily come into direct contact with the products of nature. Brewing and zymotechnology take advantage of microscopic organisms, but all food—at least since the beginning of settled agriculture—is based on the manipulation and control of (larger) plants and animals. Archeologists and biologists who have examined the diet of ancient civilizations report that modern foods look nothing much like their wild ancestors. Wheat, corn, tomatoes, apples, all have been modified from their original forms to become hardier, more edible, tastier, larger, and so on. The same thing goes for livestock such as cows and sheep.

Humans have achieved these modifications by careful selective breeding over many centuries. Individual plants or animals that have desirable qualities are selected for breeding; these pass their characteristics on to the next generation. This is not perhaps as precise or dramatic as genetic engineering—it may take many generations and many years of breeding for the changes to stick. But, it is effective: look at the differences between varieties of heirloom tomatoes or breeds of dogs, for instance. It can also be a highly technical process. Selecting the right animals or plants for breeding, growing and managing the stock, and controlling the reproductive process itself (including through hand pollination or artificial insemination) all call for high levels of skill and knowledge.

In the plant world, this skill and knowledge reached its highest peak in the work of a Californian amateur horticulturalist. With no formal scientific education, Luther Burbank (1849–1926) became the most widely known and widely respected plant breeder of the early twentieth century. Working from his land in Santa Rosa, California, Burbank experimented with a variety of techniques including grafting (attaching branches from one plant to the stem of another), interspecies and intraspecies hybridization (breeding two closely related plants), and backcrossing (breeding a hybrid with one of its own parents to achieve an offspring more like that parent). This was painstaking and technically difficult work.

Burbank's output was prolific. Apart from the Burbank potato (which later became the Russet Burbank), the Shasta daisy, the Freestone peach, the plumcot (plum + apricot), and spineless cactus, Burbank created hundreds of other varieties of fruits, nuts, grasses, grains, vegetables, and ornamental flowers. Although the scientific community never accepted him as a peer, Burbank's plants were spectacularly successful, giving him celebrity status. Apart from their popularity, some of the new varieties also had economic importance. The spineless cactus, for instance, was cultivated as a new form of cattle feed in dry regions; and Burbank's potato remains one of the world's dominant varieties.

By the early twentieth century, many scientists were interested in manipulating plants to produce new varieties of crops. The rediscovery of Mendel's laws of heredity in 1900, in particular, promised new possibilities for agricultural science. Before World War II, the US government (especially the US Department of Agriculture) invested significant time and money in developing new varieties of crops, especially by cross-breeding stocks from different parts of the globe. The most dramatic result of these experiments and studies was the development of hybrid corn. Hybrid corn varieties were developed through an arduous process in which corn had to be inbred to produce pure-line strains and then recrossed to produce first-generation hybrids. This idea had originated in the work of George H. Shull (working at the Station for Experimental Evolution at Cold Spring Harbor) in 1908, yet it took almost thirty years of further work before hybrid corn was available for commercial sale.

Hybrid corn revolutionized US agriculture, rapidly increasing corn yields. But it also made farmers more dependent on science and on commercial seed producers. Hybrid corn was difficult and time-consuming to produce and, crucially, the offspring of hybrid corn varieties no longer had the same properties as their parents. This meant farmers could not replant corn seed, but rather had to repurchase seed from producers. In the 1920s, companies such as Hi-Bred Corn (forerunner of Pioneer Hi-Bred) were established to take advantage of this new market. By the 1940s and '50s these companies had established hybrid varieties of other crops such as sugar beet, sorghum, and onion.

Domestic animals have also been subjected to extensive study and modification through breeding. We will examine some of these efforts in later chapters in relation to discussions of assisted reproduction. Plant and animal breeders may not have been able to directly modify genes, but they succeeded in exerting a high level of control over their creations.

SEX, HORMONES, AND REJUVENATION

In the early twentieth century, humans too became the targets of such interventions. It was a heady time for biology: new knowledge about evolution, heredity, and genes inspired biologists to believe that huge advances in shaping living things might soon be possible. The European émigré biologist and physiologist Jacques Loeb (1859–1924) had succeeded in transforming simple animals, creating multiple-headed worms and other hybrid forms. In 1907, Burbank wrote "The Training of the Human Plant," in which he attempted to apply the lessons of his successful plant-breeding techniques to the improvement of children. Although this might seem far-fetched, the achievements of Burbank, Loeb, and others suggested that the manipulation of higher animals was entirely possible.

One promising avenue of research was in hormones. In 1889, the physiologist Charles-Édouard Brown-Séquard (1817–1894) injected himself with a solution containing fluid extracted from the testes of dogs and guinea pigs. The 72-year-old doctor claimed that the serum had an immediate rejuvenating effect, alleviating his frailty. Such claims were initially greeted with skepticism in the medical community. But by the turn of the century, the use of extracts from various bodily tissues was becoming distinctly popular. The rise of *organo-therapy*, as it was called, also prompted serious experimental work on the "ductless glands" in humans and animals (including work leading to the isolation and purification of adrenalin [1901] and insulin [1921]).[2] The testes, the ovaries, the thyroid, the pineal gland, the pituitary gland, the pancreas, and the adrenal cortex all secreted substances that had a powerful effect on the body.

By 1920, there was medical support for the popular view that hormones could control human growth, development, sexuality, behavior, and even intelligence. "We know definitely now," a popular medical work reported, "that the abnormal functioning of these ductless glands may change a saint into a satyr; a beauty into a hag; a giant into a pitiful travesty of a human being; a hero into a coward, and an optimist into a misanthrope."[3] In the United States, the leading advocate of hormone therapy was a Columbia Medical School physician, Louis Berman (1893–1946). Berman published a series of widely read books including *The Glands Regulating Personality* (1921), *Study of the Relation of Ductless Glands to Homosexuality* (1933), *Food and Character* (1933), and *New Creations in Human Beings* (1938). The titles alone make clear Berman's view that hormones provided a means of controlling human nature through chemicals.

What was especially attractive was the fact that hormones promised "rejuvenation": the possibility of making the old young again. For men this particularly meant the restoration of sexual function and virility. In Europe and the United States, large numbers of men underwent "Steinach operations"—vasectomies that came with the promise of physical, mental, and sexual makeovers. Eugen Steinach (1861–1944), director of the Physiological Section of the Institute of Experimental Biology in Vienna, theorized that suppressing sperm production through vasectomy might revitalize hormone-producing cells in the testes. Experiments with rats seemed to confirm this hypothesis, and soon the technique was extended to human patients. Many reported spectacular results.

In an era before widespread understanding of genes and DNA, hormones could determine your fate. Especially in Europe, this had important political

2. In physiology, the "ductless" or endocrine glands are distinguished from the "ducted" or exocrine glands, such as the tear glands, sweat glands, mucous glands, and sebaceous glands.

3. H. H. Rubin quoted in Chendak Sengoopta, "'Dr. Steinach Is Coming to Make the Old Young': Sex Glands, Vasectomy, and the Quest for Rejuvenation in the Roaring Twenties," *Endeavour* 27, no. 3 (2003): 122–126. Quotation p. 122. The name of Rubin's book was *Your Mysterious Glands: How Your Glands Control Your Mental and Physical Development and Moral Welfare* (1925).

implications in the aftermath of World War I: nations needed to be restored and replenished (especially their stock of young men). Controlling the glands was the key to regulating human bodies and rebuilding human society. Such ideas have echoes in more recent attempts to use genetic therapies and genetic manipulation to shape individuals and populations.

VITAMINS AND WAR

Vitamins were another important discovery of the early twentieth century. Indeed, both hormones and vitamins were part of a broader shift towards understanding life on the *molecular* level. At the turn of the century, experiments with animals showed that it was not possible for them to live on a diet of carbohydrates, proteins, fats, and minerals alone. Certain "accessory food factors" were also required. Further research showed that deficiencies in these same factors seemed to be linked to certain human diseases such as rickets, scurvy, and beriberi.

With the onset of World War I, scientific interest in accessory food factors increased. Particularly in Britain, the government and the military became concerned not only with the health of the men at the front, but the diet of the civilian population at home. War raised the possibility of critical food shortages. If the basic components of nutrition were known, it might be possible to minimize the effects of a dwindling food supply. The Lister Institute in London, as well as the Medical Research Committee and the Food (War) Committee, all set researchers to work on this problem.

These studies focused on showing that deficiencies in accessory food factors caused certain diseases, determining the minimum requirements of accessory food factors in the diet, and determining the chemical and structural properties of accessory food factors. This was difficult work: the substances occurred in only minute quantities in food and appeared to be highly unstable when subjected to heating, drying, and other chemical processes. In 1912, Casimir Funk at the Lister Institute identified amines that appeared to be linked to beriberi and scurvy. He called these "vital amines" or "vitamines." By 1920, further work differentiated and isolated vitamins A and D.

After the war, vitamins remained a subject of immense importance. Evidence showed that fresh foods were high in vitamins and milled rice, white bread, dried milk, and canned foods lacked vitamins. Not wanting to lose sales, manufacturers set up their own laboratories and began to add vitamins to their foods. In the 1920s, manufacturers produced foods "enriched" with vitamins, claiming that they were superior to fresh equivalents. Glaxo, a company that sold milk powder intended to "build bonnie babies," fortified their product with vitamin D (figure 2.2). This led Glaxo into the business of

2.2 Builds Bonnie Babies. Advertisement displayed in London opposite King's Cross Station in 1921. Glaxo milk formula was enriched with vitamin D in order to "Build Bonnie Babies." Source: GlaxoSmithKline. Used by permission.

selling vitamin preparations (in liquid and tablet form) as diet supplements and pharmaceutical products. The market for vitamins has remained strong ever since.

In both World War II and the Cold War, vitamins again played their part in the struggle. Vitamins could keep the fighting forces and the workers, mothers, and children at home in the best possible shape. In the United States, pharmaceutical and food companies created "vitaminized" products including candy, chewing gum, cosmetics, and even tobacco. Vitamins came to have an almost mythic power in the public imagination. They could not only stave off illness but also decrease fatigue, depression, poor appetite, and irritability. During wartime, both health and morale were of critical importance: "V is for vitamins as well as victory." Consuming vitamins meant maintaining health and vitality and health and vitality were good for the nation. Vitamins became patriotic.

CONCLUSIONS

The kinds of biotechnologies described here persisted through the twentieth century and into the twenty-first. Zymotechnic processes were used to produce penicillin in industrial quantities during World War II and other drugs such as cortisone in the postwar period. In the early 1970s, Cetus Corporation began operations producing chemicals and antibiotics through fermentation and other microbial processes. Numerous efforts have also been made to harness bacteria and yeast for producing food and fuels. And, of course, beer brewing, vitamins, hormones, and the production of new agricultural varieties all remain active areas of research and important industries.

In all these examples, we have seen how the understanding and manipulation of nature was closely linked to the interests of private enterprise or governments or both. Especially in the nineteenth century, beer brewing and zymotechnology became closely linked to the economic interests of European states. They supported scientific institutions to ensure its continued development. By World War I, it even had strategic significance. The same applied to vitamins, which were closely tied to state interests in maintaining healthy populations of soldiers and workers during wartime. Hormones were associated with *national* (as well as personal) rejuvenation. Burbank's horticultural experiments yielded important results for commercial agriculture. But they also had implications for "training the human plant."

The historians who have turned a critical eye to these episodes have shown how study of organisms was rarely a disinterested enterprise—it was often closely aligned with commercial or political interests and the control of people or populations. Whether we classify agricultural breeding, hormones, vitamins, and zymotechnology as biotechnology or not, these examples show that the association between science, technology, governments, and commerce has existed for a long time. Like more recent biotechnologies, too, they attempted to explore the plasticity of living things and were accompanied by hype, promise, and controversy. Some of the techniques of modern biotech may be new, but many of the questions and concerns it raises are quite old.

FURTHER READING

The long history of biotechnology is explored in much greater detail in Robert Bud, *The Uses of Life: A History of Biotechnology* (Cambridge: Cambridge University Press, 1994), although Bud emphasizes the development of zymotechnology in particular. The history of beer and brewing draws on Ian S. Hornsey, *A History of Beer and Brewing* (Cambridge: Royal Society of Chemistry, 2004), L. F. Hartman and A. L. Oppenheim, *On Beer and Brewing Techniques in Ancient Mesopotamia* (Baltimore: American Oriental Society, 1950), and Richard W. Unger, *Beer in the Middle Ages and the Renaissance* (Philadelphia: University of Pennsylvania Press, 2007).

The story of Luther Burbank is discussed in the context of vernacular or popular science by Katherine Pandora, "Knowledge Held in Common: Tales of Luther Burbank and Science in the American Vernacular," *Isis* 92 (2001): 484–516. A full biography of Burbank has been written by Jane S. Smith, *The Garden of Invention: Luther Burbank and the Business of Breeding Plants* (New York: Penguin, 2009). Interested readers should also refer to Burbank's own works, including "The Training of the Human Plant," *Century Magazine*, May 1907. Philip J. Pauly, *Fruits and Plains: The Horticultural Transformation of America* (Cambridge, MA: Harvard University Press, 2008) puts Burbank's work in a much wider context, describing a longer history of plant manipulation in the United States. The more scientific sides of plant breeding are described in Jack Ralph Kloppenberg Jr., *The First Seed: The Political Economy of Plant Biotechnology*, 2nd ed. (Madison: University of Wisconsin

Press, 2004), while Noel Kingsbury, *Hybrid: The History and Science of Plant Breeding* (Chicago: University of Chicago Press, 2011) begins his history of plant manipulation with the birth of agriculture (and also provides a more global story).

Philip J. Pauly, *Controlling Life: Jacques Loeb and the Engineering Ideal in Biology* (New York: Oxford University Press, 1987) is the source for Jacques Loeb's life and work. The history of hormones and especially sex hormones is covered in Chendak Sengoopta, *The Most Secret Quintessence of Life: Sex, Glands, and Hormones, 1850-1950* (Chicago: University of Chicago Press, 2006) and Christer Nordlund, *Hormones of Life: Endocrinology, the Pharmaceutical Industry, and the Dream of a Remedy for Sterility, 1930-1970* (Sagamore Beach, MA: Watson Publishing International / Science History Publications, 2011). More specific treatments of Eugen Steinach and Louis Berman can be found in Chendak Sengoopta, "'Dr. Steinach Is Coming to Make the Old Young': Sex Glands, Vasectomy, and the Quest for Rejuvenation in the Roaring Twenties," *Endeavour* 27, no. 3 (2003): 122–126; and Christer Nordlund, "Endocrinology and Expectations in 1930s America: Louis Berman's Ideas on New Creations in Human Beings," *British Journal for the History of Science* 40, no. 144, Part 1 (2007): 83–104. For a longer overview of ideas about impotence there is Angus McLaren, *Impotence: A Cultural History* (Chicago: University of Chicago Press, 2007).

For an overview of the history of vitamins in the American context, the best source is Rima D. Apple, *Vitamania: Vitamins in American Culture* (New Brunswick: Rutgers University Press, 1996). The role of vitamins in World War II is explored in Rima D. Apple, "Vitamins Win the War: Nutrition, Commerce, and Patriotism in the United States during the Second World War," in *Food Science, Policy, and Regulation in the Twentieth Century: International and Comparative Perspectives*, ed. D. F. Smith and J. Phillips (London: Routledge, 2001), 135–149. For a more British-centered perspective see Harmke Kamminga, "Vitamins and the Dynamics of Molecularization: Biochemistry, Policy, and Industry in Britain, 1914–1939," in *Molecularizing Biology and Medicine: New Practices and Alliances, 1910s-1970s*, ed. S. de Chadarevian and H. Kamminga (Reading, UK: Harwood Academic, 1998), 83–105; and the collection of essays in Harmke Kamminga and Andrew Cunningham, eds., *Science and Culture of Nutrition, 1840-1940* (Amsterdam: Editions Rodopi, 1995). Funk's seminal 1912 paper is Casimir Funk, "The Etiology of the Deficiency Diseases," *Journal of State Medicine* 20 (1912): 341–368.

PART II
GENETIC ENGINEERING

3 INVENTING GENETIC ENGINEERING

In 1932, the English writer Aldous Huxley (1894–1963) imagined a world in which human beings were manufactured rather than born. Huxley describes how "Podsnap's technique"—which speeds up the maturation of eggs within an ovary—is combined with "Bokanovsky's process"—that causes fertilized eggs to divide into identical copies—to produce large numbers of identical humans in a "Hatchery and Conditioning Centre." Of course, Huxley made up all of this. But it was not just wild speculation: Huxley was well versed in the latest biology of the 1930s. Aldous' brother Julian was a well-known evolutionary biologist who had taught at King's College London in the 1920s. Given Julian's enthusiasm for eugenics, it is likely that the brothers had discussed the biological possibilities of manipulating life through the application of chemicals, heat, hormones, and selective breeding.

Brave New World is most widely known as a satirical critique of totalitarian society. With the Soviet Union consolidating its power in the wake of the Russian Revolution and fascism on the rise in Italy and Germany, Huxley's book was a warning about where centralized planning of society and the economy could lead. In this *Brave New World*, planning and control extends even to reproduction. Huxley's vision shocks us because it depicts the government reaching even into the most intimate aspects of our lives. Humans have become automata, to be programmed and reprogrammed according to the needs of those in charge. But they have also become commodities, mass-produced in factories according to the principles of Fordist efficiency.

Genetic engineering still conjures up Huxley-inspired visions of mad scientists creating babies inside test tubes. It is variously understood as a triumph of biomedical progress, as a symbol of scientific hubris, and as scientists "playing God." As we try to understand the social, political, and economic significance of genetic engineering, it is worthwhile to keep Huxley's vision in mind. This scientific imaginary—in which manipulating life is associated with the totalitarian manipulation of society—influences the way society thinks about biotechnology and its consequences.

This chapter provides a background for understanding the so-called revolution that took place in biology in the 1970s. The term *genetic engineering* is now often used in a loose way to refer to many techniques in biotechnology. But in the 1970s, genetic engineering came to be associated with a specific technique for making copies of DNA that was invented in 1972.[1] Why was this discovery so important?

LIFE AS CODE

The idea that life depends on molecules is relatively new. The subdiscipline of *molecular* biology coalesced only after World War II as biologists developed the tools to investigate life on the smallest scale. One of the factors that caused this development was the influx of physicists into biology right around this time. Physics had enjoyed enormous success by examining smaller and smaller bits of the world: atoms, electrons, protons, neutrons, and photons. Some physicists imagined they would have equal success by applying the same techniques to biology—that is, by trying to understand the very smallest constituents of a system.

An influential figure in this respect was Erwin Schrödinger. In 1925, Schrödinger had played a major role in the formulation of quantum mechanics, the key theory that underpinned physicists' understanding of matter on the subatomic level. In 1944, Schrödinger wrote a short book called *What is Life?* based on some lectures he had given the previous year in Dublin. His premise—provocative to physicists and biologists of the 1940s—was that life should be understood at the level of physics and chemistry. Schrödinger speculated on ways in which similarities between parents and offspring (heredity) could be explained in molecular terms:

> It has often been asked how this tiny speck of material, nucleus of the fertilized egg, could contain an elaborate code-script involving all the future development of the organism . . . For illustration, think of the Morse code. The two different signs of dot and dash in well-ordered groups of not more than four allow thirty different specifications. Now, if you allowed yourself the use of a third sign, in addition to dot and dash, and used groups of not more than ten, you could form 88,572 different "letters."[2]

Schrödinger was suggesting that the fertilized egg contained a set of molecular symbols, like Morse code, that could specify how to build an organ-

---

1. This technique is also known as *molecular cloning* since it is used to make many copies (clones) of a DNA molecule. This is not be confused with the popular use of *cloning*, which now usually refers to reproductive cloning or therapeutic cloning (see chapter 16).

2. Erwin Schrödinger, *What Is Life?* (Cambridge: Cambridge University Press, 1967 [original publication 1944]). Quotation p. 61.

ism. Schrödinger didn't know about DNA so he just called this an "aperiodic crystal."

Schrödinger's book did not announce any new discoveries or new theories about biology. But it was important for a different reason. As other scientists began to do experiments on the molecules in cell nuclei (proteins, DNA, and RNA) they began to adopt Schrödinger's code-script idea. They began to talk about molecules containing codes and passing information from one to another. In 1953, James Watson and Francis Crick discovered the structure of DNA. This achievement was widely celebrated because the structure immediately suggested the means by which a molecule might carry a code (see figure 3.1 and box 3.1).

All this was happening at the same time that information and communication sciences were making their first appearance. The first electronic computers were developed during World War II and were soon being put to use in a range of scientific fields. The mathematician Claude Shannon published "A Mathematical Theory of Communication" in 1948, laying the groundwork for a new discipline called information theory.

Historians have documented how "information" and "code" came to be powerful metaphors in molecular biology in the 1950s and 1960s. Biologists thought of DNA as acting like a computer program: it contained a code that was read out by the machinery of the cell in order to build proteins. Of course, DNA was not literally a piece of software and the cell did not literally act like a computer. But the language of "codes" and "information" played a crucial role in shaping how biologists thought about molecules and organisms.

For one thing, the metaphor meant that the most urgent problem for molecular biology was "cracking the code"—that is, discovering exactly how DNA built the proteins that made up living things. A single strand of DNA comprises a chemical "backbone" plus a chain of "nucleotides" or "bases"—adenine (A), guanine (G), cytosine (C), and thymine (T). This is why a DNA molecule is often represented by just a string of letters: AAGGATGCC, for example. The nucleotides can be strung along the backbone in any order: thymine-thymine-cytosine (TTC); adenine-cytosine-guanine-guanine (ACGG); and so on (see box 3.1). Molecular biologists suspected that particular strings of As, Gs, Ts, and Cs made up a code that provided instructions for building a protein molecule. But how did this work? Between 1961 and 1965 the painstaking laboratory work of Marshall Nirenberg, Heinrich Matthaei, and Har Ghobind Khorana eventually solved this problem. This made it possible to read the "code of life" (see table 3.1). This only encouraged biologists to take their metaphors more seriously. If DNA was a code, then it also represented a kind of "text" and if it was a text then the whole collection of DNA was a

DNA Replication

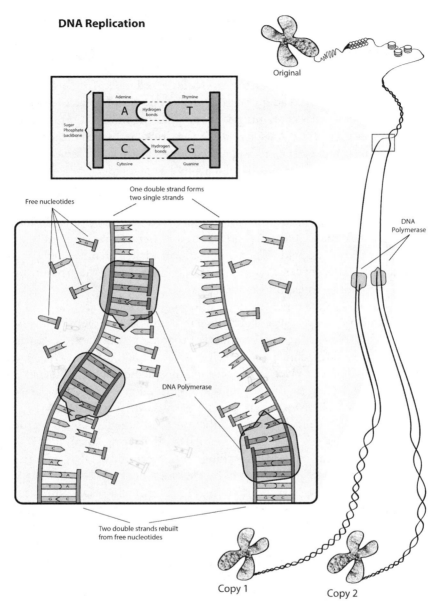

3.1 DNA replication. Refer to box 3.1. DNA is a complex double-stranded molecule twisted into a double helix shape. The order of the nucleotide bases forms a "code" that can be used to make proteins. The separation of the two strands allows each strand to be used as a template for copying the entire DNA molecule. Source: Illustration by Jerry Teo.

Box 3.1 A DNA Primer

Deoxyribonucleic acid is a very large molecule that consist of millions or even tens of millions of atoms (carbon, hydrogen, oxygen, nitrogen, and phosphorus). The molecule consists of two strands—the strands are each twirled into a helix and so the molecule overall is shaped like a double helix with the two strands twisting together in parallel.

Each strand has a "backbone" that is made of a kind of sugar (this is deoxyribose) and stretches along the entire length of the molecule. At regular intervals along the backbone, another kind of molecule (called *nucleotide* or *base*) is attached. These nucleotides can be one of four types: adenine, guanine, cytosine, and thymine (usually abbreviated A, G, C, and T). These are complex molecules in and of themselves, made up of one or two rings of carbon atoms. The nucleotides can be attached along the backbone in any order, forming a distinct pattern—AAGGATCCA, for instance. DNA molecules are very long, so in fact there can even be millions of nucleotides in a row.

This sequence of letters is referred to as the DNA sequence. Some parts of the DNA sequence are genes. This means that they can act as a template for building proteins. Within a gene, each triplet of nucleotides corresponds to an amino acid. For example, AAG corresponds to the amino acid called lysine; and GAT corresponds to aspartic acid (see table 3.1). When a gene is "expressed" this means that the DNA is being "read out" by the cellular machinery and made into a chain of amino acids. This chain of amino acids folds into a protein.

The nucleotides from each strand are also arranged to stick out towards each other so that they almost touch in the middle. If we were to flatten the double helix out onto a two-dimensional surface, it would look something like a ladder, with each rung made up of two nucleotides, one from each strand. But only specific combinations of nucleotides will join together to form rungs. As will only join with Ts and Gs will only join with Cs. So if the nucleotides on one strand are AAGGATCCA (from bottom to top), then the nucleotides on the other strand must be TTCCTGGT. If the nucleotides don't match up in this way, the rungs will not join and the two strands will split apart.

This double-strand system provides the means of copying DNA molecules. When DNA is to be copied inside the cell, the two strands are pulled apart and separated by a special enzyme. Since the nucleotides on one strand must match the nucleotides on the other (A with T and G with C), it is possible to rebuild two double strands from two single strands (figure 3.1). This is the job of a molecule called DNA polymerase. DNA polymerase moves along each single strand and rebuilds a double strand: where it senses an A on one strand, it places a matching T on the other strand; where it senses a G, it builds a C, etc. Eventually two complete and identical DNA strands can be reconstructed.

TABLE 3.1 THE GENETIC CODE

FIRST NUCLEOTIDE		SECOND NUCLEOTIDE				THIRD NUCLEOTIDE
		T	C	A	G	
	T	TTT—Phe	TCT—Ser	TAT—Tyr	TGT—Cys	T
		TTC—Phe	TCC—Ser	TAC—Tyr	TGC—Cys	C
		TTA—Leu	TCA—Ser	TAA-STOP	TGA—STOP	A
		TTG—Leu	TCG—Ser	TAG—STOP	TGG—Trp	G
	C	CTT—Leu	CCT—Pro	CAT—His	CGT—Arg	T
		CTC—Leu	CCC—Pro	CAC—His	CGC—Arg	C
		CTA—Leu	CCA—Pro	CAA—Gln	CGA—Arg	A
		CTG—Leu	CCG—Pro	CAG—Gln	CGG—Arg	G
	A	ATT—Ile	ACT—Thr	AAT—Asn	AGT—Ser	T
		ATC—Ile	ACC—Thr	AAC—Asn	AGC—Ser	C
		ATA—Ile	ACA—Thr	AAA—Lys	AGA—Arg	A
		ATG—Met	ACG—Thr	AAG—Lys	AGG—Arg	G
	G	GTT—Val	GCT—Ala	GAT—Asp	GGT—Gly	T
		GTC—Val	GCC—Ala	GAC—Asp	GGC—Gly	C
		GTA—Val	GCA—Ala	GAA—Glu	GGA—Gly	A
		GTG—Val	GCG—Ala	GAG—Glu	GGG—Gly	G

3-letter code	*Amino acid name*
Ala	Alanine
Arg	Argenine
Asn	Aspargine
Asp	Aspartic acid
Cys	Cysteine
Gln	Glutamine
Glu	Glutamic Acid
Gly	Glycine
His	Histidine
Ile	Isoleucine
Leu	Leucine
Lys	Lysine
Met	Methionine
Phe	Phenylalanine
Pro	Proline
Ser	Serine
Thr	Threonine
Trp	Tryptophan
Tyr	Tyrosine
Val	Valine

"book of life." Some biologists even imagined DNA as a script or a language that had biblical resonances.

Molecular biologists still describe biology in terms of information and codes. This is the way it is taught in classrooms and textbooks. It is hard to imagine it any other way. Can you describe the relationship between DNA and protein without using the words *code* or *information*? This suggests the deep influence this has had on our way of understanding life. But we should not make the mistake of thinking that code and information are the *only* way of describing genetics. After all, the As, Gs, Ts, and Cs are not like English or Japanese—they are not really a language. Nor are they really a code. Morse code, for instance, takes an alphabetic language and represents it as a series of dots and dashes. But the DNA code doesn't represent any other underlying language.

With enough thought it might be possible to imagine describing biology using different metaphors: templates or molecules acting as locks and keys, perhaps. In any case, it is important to remember that information and code are *metaphors* rather than literal descriptions of how biology works on a molecular level.

WHAT IS GENETIC ENGINEERING?

What does all this code talk have to do with genetic engineering? It is only really possible to understand why genetic engineering was considered to be so important if we understand that molecular biologists saw DNA as a piece of *text*. To be fluent in a language, you need not only to be able to *read* it, but also to be able to *write* it. By cracking the code molecular biologists had figured out how to read DNA, but they had not yet figured out how to write in this language, or even how to edit it. This is what genetic engineering is all about.

In 1972, Herbert Boyer was a thirty-six-year-old biochemist and molecular biologist working at the University of California in San Francisco. Relaxed and gregarious, he usually wore jeans, running shoes, and a leather vest. He ran his lab in a casual style too, often gambling on new ideas that emerged from brainstorming sessions over a beer. Boyer's subject of research was restriction enzymes—special proteins that occur naturally within organisms and which are used to cut or cleave DNA at particular sites. These enzymes are designed to recognize specific sequences of DNA—AAGGAT, for instance—and make a cut only at this site (for instance, it could cut between the two Gs in this example).

Significantly, Boyer found that these molecular scissors did not make a straight cut across a double-stranded piece of DNA. Figure 3.2 shows how

Recombinant DNA with Sticky Ends

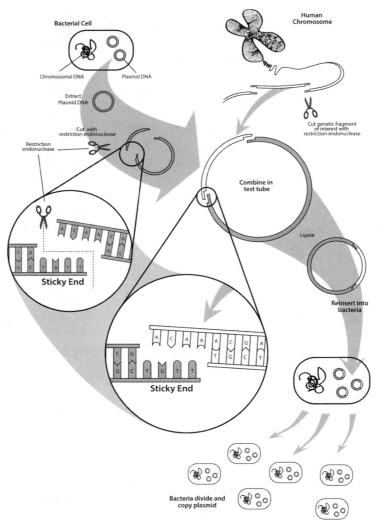

3.2 Recombinant DNA with sticky ends. A vector or plasmid ring is spliced open with restriction endonucleases to form an open ring. The donor DNA to be inserted into the ring is spliced with the same endonucleases. Mixing the open rings and the donor DNA and adding the enzyme ligase allows the rings to incorporate the foreign DNA. The now complete rings are reinserted into bacterial cells. The bacterial cells divide and copy, making millions of copies of the donor DNA. This DNA can then be extracted from the bacterial cells. The bacterial cells can also be used to make the protein associated with the donor DNA. Source: Illustration by Jerry Teo.

the restriction enzyme could cut at an angle across the double strand, leaving overhanging pieces of single-stranded DNA on both sides of the cut. These overhangs were called "sticky ends" since they could be used to re-stick pieces of DNA back together. In imagining how this works it is useful to think of a carpenter trying to join together two long pieces of wood end-to-end. Just sticking the pieces together would not make a very effective join (or it would require some very strong glue). A stronger join would be formed by removing half the thickness of each piece of wood and then overlapping them.[3]

Stanley Cohen, a geneticist at nearby Stanford University, had little in common with Boyer. Although they were almost the same age, Cohen was meticulous, private, and circumspect. He was the consummate professor: beard, baldness, glasses, sports jacket, and serious demeanor. Cohen's interest too, was not particularly similar to Boyer's: he studied little rings of double-stranded DNA that existed inside bacteria. These plasmids, as they were called, were separate from the bacteria's main DNA chromosomes and seemed to provide the organisms resistance against antibiotics. Cohen wanted to understand how this worked. In order to study the plasmids effectively Cohen often needed to break the DNA up into small pieces. He could do this in a blender, but this left him with random fragments that were difficult to study. He wanted a more systematic approach.

In November 1972, Cohen planned a conference in Honolulu on the topic of plasmid research. Happening to have just heard about Boyer's work on restriction enzymes, Cohen invited him, almost as an afterthought. The two men had never met and knew nothing of the details of each others' work. But as Boyer presented his research, Cohen realized that the precise cuts made by Boyer's restriction enzymes were exactly what he needed.

A walk along the beach after the day's proceedings allowed them to share their ideas in more detail. The stroll ended up at a deli near Waikiki Beach and over beer and sandwiches Boyer and Cohen realized that—more than just solving Cohen's problems—they might have hit upon a way to copy or clone pieces of DNA. Not just bacterial plasmid DNA, but *any* piece of DNA. They became "jazzed" with the idea, as Boyer later put it, immediately sensing its possible importance.

Boyer and Cohen's insight can be explained in five steps (figure 3.1). First, you needed to remove plasmids from their bacterial cells. Cohen's lab already knew how to do this. Second, you could use Boyer's special cutting enzyme to cut open the rings at a specific point. This would leave open rings with sticky

3. In carpentry this is known as a half-lap joint. Just sticking the pieces end to end is known as a butt joint and is the weakest of all joints.

ends. Third, you could take another piece of DNA (this could be more or less any piece of DNA) and again cut it with Boyer's restriction enzyme. Because it was cut with the same enzyme, it would have sticky ends that paired with those in the rings. Fourth, mix the open-ring plasmids with the cut DNA. Adding a special enzyme called ligase (which promotes the joining of DNA) would cause some of the plasmids to incorporate the foreign DNA, reforming a ring, but now including an extra piece. Fifth, reinsert these plasmids into the bacteria (again, Cohen's lab knew how to do this).

These bacteria containing the specially modified plasmids could then be grown in the lab. As they reproduced, not only would they copy their own DNA, but they would also make copies of the foreign DNA. One copy could be amplified into millions as the bacteria quickly divided.

At least, that was the plan. In Hawaii, Boyer and Cohen made an agreement to give the experiments a try. Beginning in January 1973, Cohen's lab isolated plasmids and Boyer's team began working with the enzymes. At first, the results were ambiguous, but by March it became clear that the "recombination" (between plasmid DNA and some foreign DNA) had worked. The plasmids had taken up the foreign DNA and copied it accurately. In just a couple of months—remarkably fast by the standard of most scientific work— they had succeeded in cloning DNA. Boyer was stunned:

> The [DNA] bands were lined up [on the gel] and you could just look at them and you knew . . . [that DNA recombination and cloning] had been successful . . . I was just ecstatic . . . I remember going home and showing a photograph [of the gel] to my wife . . . You know, I looked at that thing until early in the morning. . . . When I saw it . . . I knew that you could do just about anything . . . I was really moved by it.[4]

Suddenly Boyer and Cohen had in their hands a straightforward technique for taking any piece of DNA and copying it. They knew they could do "just about anything" (figure 3.3).

It wasn't quite as simple as this, of course. Boyer and Cohen's initial experiments had in fact used DNA from the same bacteria as the plasmid (a common laboratory organism, *Escherichia coli*). It still remained to be shown that the system would work for DNA from a different species of bacteria (let alone from a higher organism like a frog or a human). Ultimately, the answer was a resounding "yes," but it took Boyer and Cohen a year's more work to show it. Until this point in 1973, studying the DNA of higher organisms had

4. Herbert Boyer quoted in Sally Smith Hughes, *Genentech: The Beginnings of Biotech* (Chicago: University of Chicago Press, 2011), 16.

3.3 Herb Boyer. Boyer in his lab during the 1970s. Source: Getty Images. Used by permission.

hardly been possible—only the tiniest amounts, just one or two copies of a gene, could be extracted from a cell. But Boyer and Cohen's method provided a way to produce millions of copies—this opened up a wealth of new possibilities for research.

: : :

A small but growing group of researchers—biochemists and molecular biologists—quickly became aware of Boyer and Cohen's work. A handful of biologists and biochemists began to use and improve the recombinant techniques. Some suspected that the bacterial cells might even be able to *express* the foreign DNA that was inserted inside it—that is, the bacteria might be able to make the protein corresponding to a foreign gene inserted into it. Insert the gene for human insulin in the bacteria and it would produce human insulin. If that possibility could be realized, it would really seem that biologists had gained the ability to build and control organisms, to write in the language of DNA.

THE MULTIPLE MEANINGS OF GENETIC ENGINEERING

I want to interrupt this history of genetic engineering at this point to take stock of some of the long-term consequences of recombinant DNA (the story will be picked up in chapters 4 and 5). Why is this particular idea so important? I have already described some of the reasons for its *scientific* importance, but it also has a *symbolic* importance that needs to be explored. This symbolic importance can be divided into three kinds of meanings we might attribute to recombinant DNA: social meanings, political meanings, and economic meanings.

SOCIAL MEANINGS

We usually think of the world as neatly divided into two categories: natural things and social things. Natural things include inanimate objects like rocks and water, as well as living things like bacteria, trees, and human bodies. Social things include things made by us: bridges, schools, microscopes, synthetic rubber molecules, and even institutions like "science" or "government." If we take this as a reasonable description of our world, then recombinant DNA poses a big problem. Does it fit into the first category or the second? Is it "natural" or "social"? It is made up of things that we would usually think of as natural: bacteria, DNA, enzymes. It is also made using these same things as tools. But it is also constructed by humans—it is engineered like a bridge. It can even be used for human ends like making technologies or medicines.

Recombinant DNA appears to be neither entirely "natural" nor "social." It is perhaps both: a kind of *hybrid*. Why does this matter? First, it requires us to rethink justifications based on "naturalness." We hear these all the time: in arguments about everything from homosexuality to climate change one side or the other often invokes the idea that something is "natural" or "unnatural" to support their point. Such arguments certainly wouldn't make sense for a genetically engineered bacterium—they are neither natural nor unnatural. More generally, their very existence suggests that these categories are problematic—that what is natural versus what is social depends on your point of view. So recombinant DNA might call into question *all* arguments that rely on such categories.

Consequently, recombinant DNA is provocative or even disruptive. Its hybridity calls into question the kinds of categories and reasons that we use to make sense of the world around us and justify our beliefs.

POLITICAL MEANINGS

Genetic engineering also raises possibilities for radically altering ourselves. It may lead not only to cures for diseases, but also towards the capability to enhance our physical and mental characteristics. So far, these ideas have mostly remained in the realm of science fiction (although in later chapters we will see some examples of ways in which these and other techniques are becoming closer to reality).

As biologists began to extend and explore the ways in which they could manipulate and recombine DNA, some social scientists became concerned about the *political* implications that such changes might bring. If the possibility of remaking one's body or mind did become a reality, it was likely to be available only to a select few. The high cost of health care meant that it was likely to be available only to the wealthy; more likely still, it would be limited to North America and Europe, leaving poorer countries behind.

Disparities in wealth would lead not only to inequities in opportunities for education and employment, but now to unequal bodies and minds. Those with access to biotechnologies could be built better, faster, stronger, and smarter. The ultimate result would be an even more divided society, or an even more divided world.

The political scientist Francis Fukuyama has predicted even graver results. Many of our political institutions—especially in a democracy—are based on the idea that all humans are "created equal." This is the basis of liberal thought and the cornerstone of human rights. If biotechnological enhancements are available for some and not others, humans may become literally unequal (some people may have better eyesight, or be able to think faster, or to live longer). This fact would undermine our democracies and our human rights, Fukuyama argues.

It is possible that we will never face these problems in the dramatic way suggested here. Nevertheless, genetic engineering's promise to reconstruct humans makes it fundamentally challenging to ideas of human equality.

ECONOMIC MEANINGS

As we will see in chapters 5 and 6, genetic engineering quickly became an economic phenomenon. Cohen and Boyer's discovery formed the basis for what became the biotech industry. Globally, the combined revenue of this industry was over $250 billion in 2013.[5] To put this in perspective, this is comparable to (but smaller than) the size of the global software industry and more than half the size of the global market for seafood. But the industry continues to grow rapidly: it includes not only the sale of drugs and therapeutics, but also genetic tests, personalized medicine, the sale of technologies such as DNA chips and sequencing machines, trade in patents, the running of global clinical trials, and the use of biotechnologies in agriculture.

And all of this seems to play a more and more important role in our everyday lives. The anthropologist Kaushik Sunder Rajan has even argued that "the

5. Based on a 2013 report from Transparency Market Research. See http://www.prweb.com/releases/2013/6/prweb10848846.htm.

life sciences represent a new face, and a new phase, of capitalism."[6] There is no doubt that biotech is now immensely important for many national economies. But Rajan and others are also saying that biotechnologies have made living matter inextricably bound up with capitalism. Boyer and Cohen's discovery became a workable technology only through the workings of capitalism. This is true of most other biotechnology — it depends on capitalism. To understand our economy we need to pay attention to biotech and to understand biotech we need to pay attention to the economy. We have entered the era of "biocapital." Capitalism is no longer just about making and selling goods, but about harnessing living stuff (cells, proteins, DNA) in order to make money.

CONCLUSIONS

Genetic engineering has not created Huxley's *Brave New World*. But the issues that Huxley chose to explore in his fiction are exactly those that generate concerns around biotechnology. Huxley imagined a world of stark inequality between Alphas, Betas, Gammas, Deltas, and Epsilons — each social rank engineered to suit its appointed tasks. Genetic engineering of humans remains figuratively connected to concerns about a living in a radically unequal and divided society. Moreover, Huxley's book examines the consequences of the manufacture of human beings — in *Brave New World* humans are made on a production line, not born. Society today worries about the devaluation and commodification of life that accompanies biocapital. At the heart of both these issues — equality and commodification — is the problem of control. Ultimately, Huxley's society was intolerable because it enabled the exercise of total power. This too is perhaps what we fear most about genetic engineering: that it provides a means for a few to gain social, political, or economic power over others.

Huxley should receive some credit for his accurate depiction of the biotech future. But these convergences also suggests how our understanding of technology is influenced by imagination. Boyer and Cohen may never have read *Brave New World*, but the meanings we attach to the technology they invented are shaped by that book.

FURTHER READING

Huxley's famous book is Aldous Huxley, *Brave New World* (London: Chatto & Windus, 1932). There is a larger body of science fiction that

---

6. Kaushik Sunder Rajan, *Biocapital: The Constitution of Postgenomic Life* (Durham, NC: Duke University Press, 2006), 3.

deals with the possibilities of genetic engineering of humans, including Nancy Kress, *Beggars in Spain* (New York: William Morrow, 1993), Margaret Atwood, *Oryx and Crake* (Toronto: McClelland & Stewart, 2003), and Paul Di Filippo, *Ribofunk* (iBooks, 1996).

The relationship between molecular biology and writing, text, script, and communication theory is explored in Lily E. Kay, *Who Wrote the Book of Life? A History of the Genetic Code* (Palo Alto, CA: Stanford University Press, 2000). Some of the consequences of this code-based understanding of biology and genes are elaborated in Evelyn Fox Keller, *Century of the Gene* (Cambridge, MA: Harvard University Press, 2000); the best sources on the relationship between physics and biology in mid-twentieth century are Evelyn Fox Keller, "Physics and the Emergence of Molecular Biology: A History of Cognitive and Political Synergy," *Journal of the History of Biology* 23, no. 3 (1990): 389–409; and Horace Freeland Judson, *Eighth Day of Creation: Makers of the Revolution in Biology* (New York: Cold Spring Harbor Laboratory Press, 2006).

For the most comprehensive account of the invention of recombinant DNA see Sally Smith Hughes, *Genentech: The Beginnings of Biotech* (Chicago: University of Chicago Press, 2011). There is also an article-length version focusing on the patenting of the method rather than the company: Sally Smith Hughes, "Making Dollars out of DNA: The First Major Patent in Biotechnology and the Commercialization of Molecular Biology, 1974–1980," *Isis* 92 (2001): 541–575. On the social and political meanings of recombination I relied on Hans-Jörg Rheinberger, "Beyond Nature and Culture: Modes of Reasoning in the Age of Biotechnology and Medicine," in *Living and Working with the New Medical Technologies*, ed. Margaret Locke, Alan Young, and Alberto Cambrosio (Cambridge: Cambridge University Press, 2000), 19–30; and Francis Fukuyama, "The World's Most Dangerous Ideas: Transhumanism," *Foreign Policy*, September 1, 2004. The longer-term economic consequences of genetic engineering and biotechnology, and the notion of biocapital, come from Kaushik Sunder Rajan, *Biocapital: The Constitution of Postgenomic Life* (Durham, NC: Duke University Press, 2006).

INTRODUCTION: SCIENCE IN THE COUNTERCULTURE

Chapter 3 gave an account of the science that led to the development of recombinant DNA techniques. Here we need to begin by taking a step back in order to explore the social and political context in which these events occurred. This background is crucial for understanding how genetic engineering was understood and received in the 1970s. In particular, we need to examine why genetic engineering met with fierce opposition from local communities, politicians, and some scientists. Although debates over recombinant technology took place in other nations, the United States exhibited the conflict in its sharpest form (likely because genetic engineering was emerging there first and more rapidly).

The 1960s was a difficult time to be a scientist. In the immediate postwar period of the 1940s and '50s, scientists had been celebrated as heroes—they had created the atomic bomb, as well as a huge number of other technologies, that had helped win World War II. In the early 1950s, biologists and physicians had worked together to all but eliminate the crippling disease polio. Science seemed invincible. But beginning in the early 1960s, popular attitudes towards science and scientists began to change. In 1959, it was revealed that a drug called thalidomide that had been prescribed by doctors to thousands of pregnant women caused birth defects. The Surgeon General of the United States, in an abrupt change of course, now warned that smoking cigarettes was in fact dangerous to one's health. And, the early 1960s also saw the publication of several books—mostly notably Rachel Carson's *Silent Spring*—that demonstrated the damaging effects of chemicals on the environment. Science suddenly seemed not only fallible, but perhaps also culpable. At the very least, some scientists seemed to be spending taxpayers' money on obscure research that was of little or no benefit to the public.

Critics of science also pointed to its close association with the military. After twenty years of nuclear standoff with the Soviet Union, nuclear weapons no longer seemed such a great idea. America's involvement in the Vietnam War amplified many of these concerns. Those who opposed the war perceived the military and the government to be directing the might of US technology and industry against poor, largely defenseless peoples. Through military sponsorship of their work and through the military-industrial complex, scientists, many anti-war crusaders believed, were complicit in this injustice. Weren't scientists designing and perfecting the weapons (such as napalm and Agent Orange) that were allowing the United States to wage such a war? To the public, many scientists seemed to be guilty: they were receiving military money, helping to build weapons, or contributing to think tanks that devised military strategy. In 1960, as he left office, President Eisenhower had warned the nation of the dangers of a military-industrial complex that could corrupt democracy. For some, scientists seemed to be a central part of this potential problem.

This growing opposition to science was taken up within a set of cultural movements that historians have labeled the *counterculture*. This term is really a shorthand for a variety of different movements—including the free speech movement, the civil rights movement, the women's rights movement, the hippies, and the New Left—that were loosely related. Some parts of the counterculture were philosophically opposed to science. In 1969, Theodore Roszak—an intellectual leader of the counterculture—published *The Making of a Counterculture*. This book argued that the problem with American society was that it had become a technocracy: it was ruled by a scientific elite with a "technocratic mindset." In other words, science—as a way of approaching the world—was everything the counterculture was fighting against. These movements often perceived scientists to be colluding with the government and the military.

The counterculture was primarily a movement of young people. It sprang from university campuses, especially around San Francisco. These students organized protests, sit-ins, and teach-ins that criticized science and technology. Some condemned science absolutely, arguing that it only led to death and destruction. Other groups called for a redirection of science towards more "humanitarian" activities—science that could bring positive benefits to humanity. Many scientists in the late 1960s (particularly around San Francisco) were placed in a difficult position. Younger researchers, especially, had a good deal of sympathy with the counterculture and with the anti-war movement. But their own universities often received money as military contractors. And it was not a problem that could be ignored—scientists had to walk to their labs past or through the protests, sit-ins, and teach-ins that spearheaded many aspects of the counterculture. Scientists were caught in the middle.

Some scientists responded by joining in the protests, and even by organizing their own. On March 4, 1969, faculty at Massachusetts Institute of Technology (a science and engineering university) organized a teach-in to generate a "public discussion of problems and danger related to the present role of science and technology in the life of our nation."[1] Several faculty members spoke out in opposition to the Vietnam War and the military presence on campuses. The teach-in provoked a national debate and a spate of other sit-ins and teach-ins on other campuses. More and more scientists realized that they could not remain on the sidelines of these political upheavals.

ASILOMAR

But beyond just joining in, scientists also wanted to heed the calls for less abstract and more humanitarian research. Science, they hoped to show, could be responsible, could be beneficial for society. The biologist David Baltimore spoke of a "transformation" in his attitude towards activism:

> I guess I had kind of undergone a certain transformation over the period from probably '68 to '70, which a lot of people did, from being involved or trying to be involved in larger political issues . . . [I'd] been involved in the Left Wing in San Diego, and here [at MIT], to a certain extent, I'd been involved in the March 4th organization, that kind of thing—to the feeling that if I was going to do anything, it ought to be within the field I know best, because I'd been . . . ineffective outside of it. Like everybody else was, or almost everybody else. And so I was sensitized to issues that involved the biological community, and felt that if I was going to put in political time, it should be there rather than anywhere else.[2]

Politically active biologists realized that they could serve the cause best by addressing issues within biology. In particular, they could make a biology that was more useful to society.

The most immediate way to achieve this was to pursue research that had a direct bearing on human health. And any line of research that might result in an ability to manipulate human genes certainly seemed to hold great medical potential. In short, some younger biologists believed that genetic engineering held the promise of producing a kind of applied biology that could have the greatest positive impact on humanity. At Stanford, for instance, Paul Berg switched his research from bacteria to mammalian cells to pursue this line of work. One Stanford graduate student commented that working on genetic

1. Boris Magasanik, John Ross, and Victor Weisskopf, "No Research Strike at M.I.T.," *Science* 163 (1969): 517.

2. David Baltimore, interview, May 3, 1977. Transcript available in MC100, box 1, folder 6: 49. Recombinant DNA History Collection, MIT Institute Archives and Special Collections, Cambridge, MA.

engineering was "one of the few times a scientist really had an opportunity to do something for the general public."[3]

But the counterculture protests had also sensitized biologists to the potential dangers of their work. The new environmental movement had raised concerns about the health and environmental effects of chemicals (for instance, dichlorodiphenyltrichloroethane or DDT). In the 1960s, biologists were working with a range of potentially dangerous agents: bacteria, viruses, and carcinogens. Some people worried that these pathogens might escape from labs or somehow spread into the environment with disastrous consequences. Biologists like Baltimore could demonstrate their commitment to responsible science by ensuring that biologists took account of environmental and health concerns.

These were the issues on many biologists' minds when Boyer and Cohen announced their work on recombinant DNA. The participants at the Gordon Conference on nucleic acids in June 1973 who heard about Boyer and Cohen's work immediately raised concerns about the potential biohazards. The organizers of the meeting, Maxine Singer and Dieter Soll, sent a letter to *Science* pointing to both the benefits and the risks of recombinant DNA. They also asked the National Academy of Sciences (NAS) to establish a committee to study the issue and set guidelines for experimental work. The NAS asked Paul Berg, who had also been pursuing recombinant experiments, to head the panel. From 1966, Berg had been working with a cancer-causing animal virus (called SV40). Since the virus had the potential to transform into a human virus, Berg was especially sensitized to issues of biosafety.

In addition to Berg, the "Committee on Recombinant DNA Molecules, Assembly of Life Sciences" included James Watson, David Baltimore, and a number of other leading molecular biologists. The Committee published their conclusions in *Science* and *Nature*, stating the issue clearly: "There is serious concern that some of these artificial recombinant DNA molecules could prove biologically hazardous."[4] One of the main concerns was that recombinant DNA techniques used *E. coli* bacteria: "New DNA elements introduced into *E. coli* might possibly become widely disseminated among human, bacterial, plant, or animal populations with unpredictable effects." The Committee divided recombinant DNA experiments into three types in which:

---

3. Quoted in Eric Vettel, *Biotech: The Countercultural Origins of an Industry* (Philadelphia: University of Pennsylvania Press, 2006), 183.

4. Paul Berg, David Baltimore, Herbert W. Boyer, Stanley N. Cohen, R. W. Davis, D. S. Hogness, D. Nathans, R. Roblin, James D. Watson, S. Weissman, and Norton D. Zinder, "Biohazards of Recombinant DNA," *Science* 185 (1974): 3034.

1) genes for toxins or antibiotic resistance were spliced into bacteria; 2) DNA from cancer-causing viruses or other animal viruses was spliced into bacteria; and 3) other animal DNA was randomly spliced into bacteria. Berg and his colleagues proposed that biologists and biochemists cease work on the first two types of experiments until proper safety guidelines could be established. They also called for further study and a conference amongst researchers in the field in order to discuss the risks.

This conference was eventually held over four days in February 1975 at the Asilomar Conference Grounds in Pacific Grove, California (near Monterey and less than one hundred miles south of Stanford and Silicon Valley). The conference was well funded and attended by leading molecular biologists from both the United States and the rest of the world, as well as several lawyers, journalists, and science administrators. The basic problem the biologists faced was one of ignorance: they had no idea whether recombinant DNA techniques really even worked (could proteins actually be expressed?), let alone what its effects would be or how dangerous it was. So the Asilomar Conference was convened to discuss and study the issue, to try to assess the risks, and to come to decisions about the conditions under which recombinant DNA experiments could be conducted.

But Asilomar was also preemptive defense against criticism from outside. Given the anti-science climate, the last thing biologists wanted was the media or politicians exaggerating the dangers of their work, with all the potentially negative publicity this might entail. For this reason, the conference tried to focus on narrowly technical issues. Many biologists felt that they needed to contain the threat of outside interference, protecting their right to self-determination and freedom of inquiry.

But the issues that the Conference faced made this difficult. One of the tasks that the biologists set themselves at Asilomar was to evaluate possible scenarios that could arise in the course of recombinant DNA work. Examining a couple of these scenarios shows just what the participants were up against:

SCENARIO 1

(a)   Some plants are susceptible to specific toxins that can be traced to a genetic sequence;

(b)   Assume that the gene responsible for the toxin is implanted in bacteria;

(c)   Assume that the bacteria with the plant toxin spread to susceptible plants;

Conclusion: Under the above conditions, there is some likelihood that susceptible plants can be wiped out.

SCENARIO 2

(a) In the human diet, roughage is primarily unprocessed cellulose;
(b) Roughage passes through the bowels undigested or undegraded;
(c) Some evidence links low roughage diets to higher incidence of bowel cancer;
     Conclusion: If the resident bacteria in the human gut degraded cellulose in the diet, thereby eliminating the usefulness of roughage, there could be higher incidence of bowel-related diseases, including cancer.

The language used indicates the first major problem: "assume," "some evidence," "could be," "if," "some likelihood," etc. The scientists were dealing with highly opened-ended problems based largely on speculation. What made one scenario likely and another science fiction? Berg captured the problem succinctly: "Dr. [Robert] Sinsheimer has said 'what if?' and you can go on with 'what ifs' eternally, and there is no way to answer all possible 'what ifs.'" Possible risks might be easy to imagine, but extremely difficult to measure and assess.

Second, assessing the risks would involve more than technical knowledge or calculation. In the first scenario, whether and how a toxic bacteria spread into the environment would depend not only on the characteristics of the bacteria and the plant, but also on plant import regulations and agricultural practices. Likewise, in the second scenario, the actual incidence of bowel disease would depend on food intake, public health measures, and advances in medicine. In other words, a real assessment of the risk would have to extend to include a range of related social and political issues.

To sidestep these problems, Asilomar participants again tried to focus on technical issues—these, at least, could be managed. The major outcome of the Asilomar Conference was a set of recommendations. The most important of these were (1) a four-fold classification of experiment types along with safety standards for performing each; and (2) the development of "safe" vectors that could not (in theory at least) escape from labs.[5]

5. Asilomar also recommended a continuing moratorium on some kinds of experiments—those involving toxins, drug resistance, viruses, cancers, release of organisms into the environment, and large-scale experiments.

According to the first recommendation, different kinds of experiments with recombinant DNA would be assigned different kinds of risk levels (P1 = lowest risk, P4 = highest risk). Laboratories would have to conform to different safety and physical containment standards according to their work. These measures included special fume hoods, decontamination procedures, and safety training for workers. This classification system suggested that risks could be accurately assessed and that they could be balanced by appropriate safety technologies.

For the second recommendation, biologists would use recombinant techniques to engineer bacteria that required the uptake of certain specific and rare nutrients in order to survive. These bacteria would be kept alive in the lab by feeding them this nutrient but they would not be viable outside this environment. Biologists proposed that recombinant DNA itself could mitigate the complex risks posed by this new technology. The strategies of physical and biological containment provided a way to transform the uncertainties associated with recombinant DNA experimentation into seemingly measurable hazards and clear rules for action.

The Asilomar participants also recommended that the National Institutes of Health (NIH) draw up its own guidelines for regulating recombinant DNA work. Since the NIH funded much of this work (at least in academic labs), it would have the power to enforce its rules. The NIH guidelines emerged in July 1976, just over a year after the Conference. They relied heavily on Asilomar's recommendations. In the short term, the Asilomar Conference achieved its objective: it kept debate largely within the scientific community and allowed recombinant DNA work to proceed. Ironically though, it also drew more public attention to the possible dangers of recombinant DNA research.

FRANKENSTEIN AT HARVARD

For the general public, recombinant DNA research was never a narrow technical issue. Rather, it was an issue of broad concern that related to questions of political representation, global health, environmental protection, and scientific responsibility. Many people believed that the regulation of genetic engineering should be something decided by the courts or by legislatures rather than by scientists or bodies that represented them. During the second half of the 1970s, local activist groups in various cities across the United States attempted to open up the recombinant DNA issue for broader social and political debate. They enjoyed some short-term success, but in the end the biologists regained control of the issue.

One of the most visible activist efforts occurred in Cambridge, Massachusetts. Cambridge can be used as a case study to explore how these movements began, what was at stake, and why they ultimately faded away. Controversy began in early 1976 when Harvard University applied for permission to upgrade one of its biological laboratories. Harvard wanted to conduct work with viruses and recombinant DNA that would make its lab P3 (the second highest level of risk). Harvard's campus was scattered throughout the center of Cambridge, its buildings—including the biological labs—in close proximity to residences, schools, and parks. The stage was set for a clash between town and gown.

By mid-1976, the issue was gaining momentum. The mayor of Cambridge, Alfred Velucci, made a rousing speech in front of the Cambridge City Council in which he declared his opposition to the lab: "Biologists may even come up with a disease that can't be cured—even a monster! Is this the answer to Dr. Frankenstein's dream?"[6] Although comparing recombinant *E. coli* to Frankenstein's monster was certainly hyperbole, this quotation suggests how the public's view of the issue differed from that of scientists. For the public, genetic engineering conjured up fears associated with other widely known scientific failures: radium, asbestos, thalidomide, vinyl chloride, dieldrin, and Agent Orange.[7] From the point of view of many nonscientists, science had a bad track record when it came to identifying hazards (figure 4.1).

The Cambridge City Council appointed a committee to study the issues of recombinant DNA and make a recommendation regarding the new lab. Between mid-1976 and early 1977 the Cambridge Experimental Review Board heard over one hundred hours of testimony. Biologists weighed in on both sides. Some, like the Harvard biologist Ruth Hubbard, opposed the lab.

6. Quoted in Albert R. Jonsen, *The Birth of Bioethics* (Oxford: Oxford University Press, 1998), 185.

7. Radium was used as an ingredient in paint for watches, aircraft switches, dials, clocks, etc., to make them self-luminous (glow in the dark). Radium was also included in some toothpastes, hair creams, and foods but was later found to cause sores, anemia, and bone cancer. Asbestos is a naturally occurring substance often used as a building material before the 1950s. It is now known to cause deadly lung diseases including cancer and mesothelioma. Thalidomide was an anti–morning sickness drug prescribed to pregnant women between 1957 and 1962. It is now known to cause birth defects. Vinyl chloride is used for the production of plastics such as PVC (polyvinyl chloride). Although it is now considered a dangerous carcinogen, before the 1960s workers were widely exposed and vinyl chloride was also released into the atmosphere. Dieldrin was developed as an insecticide in the 1940s but persists and accumulates in the environment, resulting in extreme toxicity to animals and humans. Agent Orange was used as a defoliant in the Vietnam War—it is now believed to have a range of toxic effects including cancer and to cause birth defects in the children of veterans.

4.1 Protesters against recombinant DNA experiments in the balcony of Cambridge City Chambers. Cambridge, Massachusetts, 1976. Source: *Harvard Magazine* 79, no. 2 (1976): 19. Used by courtesy of the estate of Rick Stafford.

Others took it upon themselves to educate the Cambridge public about molecular biology and recombinant DNA techniques (figure 4.2). One of the most successful strategies for the biologists was to focus attention on medical benefits that came from biological research. Mark Ptashne argued, "The degree of risk involved in carefully regulated DNA experiments is less than that in maintaining a household pet." Pets and other animals, after all, were known carriers of human pathogens. But, more forcefully, he added, "If we were warranted in putting a stop to recombinant DNA experiments, then on the same logic we should put a stop to experiments involving animal viruses, animal cells, carcinogens and mutagens—this would signal the end of biomedical research."[8] Allowing recombinant DNA research might have risks, but banning it involved the risk of missing out on unknown diagnostic and therapeutic breakthroughs.

In the end, the Cambridge Experimental Review Board did not recom-

8. Mark Ptashne in "Hearings on Recombinant DNA Experiments, City of Cambridge," June 23, 1976. MC100, box 33, folder 458: 3, Recombinant DNA History Collection, MIT Institute Archives and Special Collections, Cambridge, MA: 73. See also: http://video .mit.edu/watch/hypothetical-risk-cambridge-city-councils-hearings-on-recombinant -dna-research-1976–7192/.

mend banning recombinant research. They insisted merely on some additional safety requirements and some monitoring by the City Council. Similar debates arose in other cities and communities around the US. Local and state laws governing recombinant DNA were enacted in various jurisdictions. Some activists attempted to make recombinant DNA a national issue. In 1977 and 1978, more than a dozen bills that aimed to regulate genetic engineering were put before Congress. None of them emerged from the committee stages.

By 1980, the debate seemed to have gradually faded away. No major legislation was passed restricting experiments. In other words, the biologists had regained control of the recombinant DNA issue. Scientists had succeeded in convincing most local, state, and federal legislators that they themselves could evaluate and manage the risks. In particular, they had argued that science should be left to make its own decisions—scientists were the ones most competent to make decisions about technical issues, they claimed. Any outside interference would be an intrusion of politics into science that could curtail scientific progress. This kind of argument became particularly important as the commercial possibilities of biotechnology became more apparent (more on this in chapters 5 and 6).

4.2 Educating the public about genetic engineering. Cambridge, Massachusetts, 1976. Biologists David Baltimore and Walter Gilbert speak to members of the public at a street fair. Source: *Harvard Magazine* 79, no. 2 (1976): 16. Used by courtesy of the estate of Rick Stafford.

Although the debate may have moved to the background, this did not mean the issue was dead. Although no Frankenstein monster emerged from the labs, many people remained profoundly uncomfortable with genetic engineering. They distrusted scientists' technical evaluation of risks and cost-benefit scenarios. The historian Alan Gross called the end of the recombinant DNA debates, "a seemingly permanent clash of purposes, an uneasy truce that left open the question of whether this particular conflict had been settled."[9] In retrospect, these controversies were just a first act in a longer drama that remains unresolved. Similar conflicts have continued to play out in public concern about genetically modified foods, cloning, stem cells, and synthetic biology.

CONCLUSIONS

Understanding the origins of these debates can shed some light on the reasons *why* they persist. At Asilomar, the biologists attempted to construct the recombinant DNA issue as a technical problem. This allowed them to reduce the complexity of the issues before them. But it also meant that they could portray the debate as one that fell within their own domain of expertise, thereby closing down the possibility for interference from outside the scientific community. The framing of both the problem of recombinant DNA (measurable risk and cost-benefit) and its solution (safe vectors, levels, training, containment) were cast in technical terms. Scientists claimed that more knowledge of recombinant DNA was required in order to fully assess the risks. Ironically, this meant that reduction of risk required *more*, not less, recombinant DNA experimentation.

As the clash in Cambridge showed, the public did not see things in the same narrow terms. For many, safeguarding risky technology with more technology was not a satisfactory solution. Moreover, the public had a wider conception of the dangers of science that linked technical concerns far more closely with social and political problems. Their fears drew on science's track record of failures, scientific imaginaries (e.g., Frankenstein's monster), countercultural suspicion toward science, and the environmental movement (which linked science and technology to global health and ecological problems). For the public it was not just the possibility of a dangerous bacteria escaping from the laboratory. It was also the possibility that their house prices might diminish (due to being located near a P3 lab), or that biotech break-

---

9. Alan Gross, "The Social Drama of Recombinant DNA," in *The Rhetoric of Science* (Cambridge, MA: Harvard University Press, 1996), 180–192. Quotation p. 190.

throughs would lead to less affordable health care, or that biotech might lead to new types of weapons.

In the short term the biologists were able to assert their technical expertise and define the terms of the debate. But since the 1980s, the language of risk assessment, cost-benefit, and containment has been less effective in controlling the debate or convincing citizens that technologies are "safe."

FURTHER READING

On the relationship between the emergence of biotechnology and the counterculture see Eric J. Vettel, *Biotech: The Countercultural Origins of an Industry* (Philadelphia: University of Pennsylvania Press, 2006). Theodore Roszak, *The Making of a Counterculture*: *Reflections on a Technocratic Society and its Youthful Opposition* (Berkeley: University of California Press, 1969) is useful for understanding countercultural attitudes towards science and technology. There are a number of excellent accounts of the March 4th movement at MIT. Jonathan Allen, ed., *March 4: Scientists, Students, and Society* (Cambridge, MA: MIT Press, 1970) is a record of the talks and panel discussions during the event itself. Dorothy Nelkin, *The University and Military Research: Moral Politics at MIT* (Ithaca: Cornell University Press, 1972) and particularly Stuart W. Leslie, "'Time of Troubles' for the Special Laboratories," in *Becoming MIT: Moments of Decision*, ed. David Kaiser (Cambridge, MA: MIT Press, 2010) provide more historical context.

Maxine F. Singer and Dieter Soll, "Guidelines for DNA Hybrid Molecules," *Science* 181, no. 4105 (1973): 1114, was the letter resulting from Boyer and Cohen's first announcement of their recombinant DNA work. The report of the "Committee on Recombinant DNA Molecules" can be found at Paul Berg, David Baltimore, Herbert W. Boyer, Stanley N. Cohen, R. W. Davis, D. S. Hogness, D. Nathans, R. Roblin, James D. Watson, S. Weissman, and Norton D. Zinder, "Potential Biohazards of Recombinant DNA Molecules," *Science* 185, no. 4148 (1974): 303. The discussion at the resulting Asilomar Conferences was not transcribed but the official report was published by Paul Berg, David Baltimore, Sydney Brenner, R. O. Roblin, Maxine F. Singer, "Summary Statement of the Asilomar Conference on Recombinant DNA Molecules," *Proceedings of the National Academy of Sciences USA* 72 (1975): 1981–1984. Historical accounts include Sheldon Krimsky, *Genetic Alchemy: The Social History of the Recombinant DNA Controversy* (Cambridge, MA: MIT Press, 1982); D. S. Fredrickson, "Asilomar and Recombinant DNA: The End of the Beginning," in *Biomedical Politics* (Washington, DC: National Academy Press, 1991), 258–292; and Alexander Morgan Capron and Renie

Schapiro, "Remember Asilomar? Reexamining Science's Ethical and Social Responsibility," *Perspectives in Biology and Medicine* 44, no. 2 (2001): 162–169. This topic is also covered in Errol Friedberg, *A Biography of Paul Berg: The Recombinant DNA Controversy Revisited* (Hackensack, NJ: World Scientific, 2014). The resulting guidelines for recombinant research were National Institutes of Health, "Guidelines for Research Involving Recombinant DNA Molecules," *Federal Register* 41, no. 131 (1976): 27911–27943.

The public debates about recombinant DNA that took place in the second half of the 1970s are also very well documented. For a contemporary report on the debates see Sheldon Krimsky, "A Citizen Court in the Recombinant DNA Debate," *Bulletin of the Atomic Scientists* 34, no. 8 (1978): 37–43; and D. A. Jackson and S. P. Stich, eds., *The Recombinant DNA Debate.* (Cambridge, MA: MIT Press, 1979). See also the memoir by D. S. Fredrickson, *The Recombinant DNA Controversy: A Memoir* (Washington, DC: ASM Press, 2001). For a historical account of the debates about Harvard's lab see Everett Mendelsohn, "'Frankenstein at Harvard': The Public Politics of Recombinant DNA Research," in *Tradition and Transformation in the Sciences,* ed. E. Mendelsohn (Cambridge: Cambridge University Press, 2003), 317–335. More sociological analyses can be found in Susan Wright, "Molecular Biology or Molecular Politics? The Production of Consensus on the Hazards of Recombinant DNA Technology," *Social Studies of Science* 16 (1986): 593–620; Susan Wright, "Recombinant DNA Technology and Its Social Transformation, 1972–1982," *Osiris* (1986): 303–360; and Alan Gross, "The Social Drama of Recombinant DNA," in *The Rhetoric of Science* (Cambridge, MA: Harvard University Press, 1996), 180–192. There are also two books that compare the development of the debates in the United States and the United Kingdom: Susan Wright, *Molecular Politics: Developing American and British Regulatory Policy for Genetic Engineering* (Chicago: University of Chicago Press, 1994) and Herbert Gottweis, *Governing Molecules: The Discursive Politics of Genetic Engineering in Europe and the United States* (Cambridge, MA: MIT Press, 1998). Other useful primary sources related to these topics are collected in James Watson and John Tooze, *The DNA Story: A Documentary History of Gene Cloning* (San Francisco: W. H. Freeman, 1983).

PART III
OWNING LIFE

5 BIOTECHNOLOGY AND BUSINESS

INTRODUCTION

Biotech companies now exist all over the world, from Iceland to Indonesia. In the United States, Canada, Europe, and Australia alone the collective earnings of these companies is around US$90 billion per year (2012).[1] In 2009, the pharmaceutical giant Roche paid $US47 billion to acquire one of the oldest and most successful biotech companies, Genentech.[2] The biotech industry has created hundreds of new medicines and treatments for a range of diseases, including diabetes, hepatitis B and C, various cancers, arthritis, hemophilia, and cardiovascular disorders. Biotech is big business and has entered the mainstream of medical practice.

However, this industry originated at a very specific time in a very particular place: namely, in the late 1970s in the area just south of San Francisco known as Silicon Valley. This small region, stretching roughly between Palo Alto and San Jose (see figure 5.1), is home to a vast number of high-tech corporations. It was the birthplace of Hewlett-Packard, Intel, Sun Microsystems, Cisco Systems, Logitech, and many other computing (and later Internet) companies. How did this concentration of intellectual and engineering talent and innovation arise? Why here and not elsewhere? Why did this place spur the creation of the biotechnology industry in the 1970s? And what can this history tell us about the present-day biotech industry? This is a story that turns on the intersection of entrepreneurialism, geography, and high technology.

SILICON VALLEY TO CELL VALLEY

The person who is generally acknowledged to have had the most influence on the creation of Silicon Valley is Frederick Terman. Terman graduated from the Massachusetts Institute of Technology in 1924 with a doctor-

---

1. See Ernst & Young, "Beyond Borders: Matters of Evidence," *Biotechnology Industry Report* 2013, p. 23. http://www.ey.com/Publication/vwLUAssets/Beyond_borders/$FILE /Beyond_borders.pdf.

2. On the Genentech acquisition see: http://www.nytimes.com/2009/03/13/business /worldbusiness/13drugs.html.

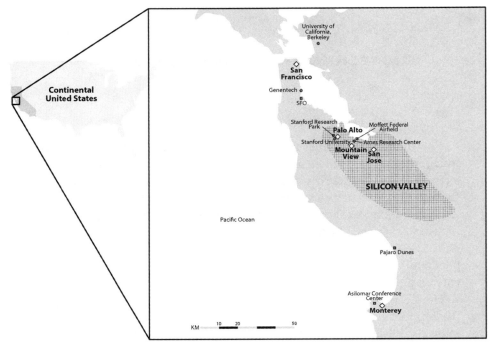

5.1 Map of Silicon Valley. Northern California around San Francisco showing Asilomar, Pajaro Dunes, original Genentech headquarters, University of California Berkeley, San Francisco International Airport, Ames Research Center, Moffett Field, Stanford University, and Stanford Research Park. Source: Illustration by Jerry Teo.

ate in electrical engineering. The following year, he joined the engineering faculty at Stanford University, aiming to build up Stanford's reputation as a top-class engineering school. By the middle of the 1930s, with the Great Depression at its peak, jobs for engineering graduates were hard to come by. Terman's solution was to encourage his students to set up their own companies, sometimes even providing them with small amounts of start-up capital himself. One beneficiary was Charles Litton, who in 1934 founded his own company to produce vacuum tubes for radios. In 1939, another pair of Terman's students, William Hewlett and David Packard, invented a novel way of manufacturing an audio oscillator (an instrument for testing sound equipment); with an investment of $538 from Terman, they set up Hewlett-Packard in Packard's garage in Palo Alto.

World War II (1939–1945) transformed science and engineering in the United States. Large numbers of scientists were recruited into big wartime efforts such as the Manhattan Project (for the atomic bomb) or the Radiation

Laboratory (a program devoted to radar research at the Massachusetts Institute of Technology). Terman himself was recruited to the Radio Research Laboratory at Harvard to study and build devices for counteracting radar. The results of this research and development were a range of new devices fit for commercialization and a large body of highly trained and practical-minded engineers. Radar research, in particular, spun off many kinds of new electronic devices which had potential uses (and markets) in audio, television, and computing.

Returning to Stanford as the dean of the School of Engineering after the war, Terman ensured that Palo Alto companies were well placed to take advantage. To this end, in 1951 Terman directed the creation of the Stanford Industrial Park (later called Stanford Research Park) as a site for attracting and supporting industries near Stanford (figure 5.2). There are now hundreds of industrial parks, business parks, office parks, and technology parks worldwide. But in the 1950s, Terman's notion was a novel one: the university-owned land was leased to high-tech companies in the hope of creating synergistic relationships both amongst the firms themselves, and between firms and the university. Companies would also have ready access to the talented pool

5.2 Aerial view of Stanford Research Park. The site of Fred Terman's Stanford Research Park in Silicon Valley in the 1950s. Source: Guy Miller Archives, Palo Alto Historical Association. Used by permission.

of young scientists and engineers graduating from the campus next door. Amongst the first tenants were Hewlett-Packard, Varian Associates, Eastman Kodak, General Electric, and Lockheed (Facebook is a current resident).

The area around Stanford also became an important hub of the military-industrial complex. Although military spending on scientific research and development decreased immediately following World War II, the exigencies of the Cold War caused a rapid rebound. The Department of Defense awarded lucrative contracts to universities and businesses, not just for the construction of military equipment but also for basic research that could be considered loosely connected to future military applications (high-energy physics, linguistics, and ecology were amongst the disciplines that received funding). Apart from any particular spin-offs, the aim was also to maintain a large group of highly qualified scientists and engineers ready for mobilization in the event of a major conflict. Both Stanford itself and the industries around it did particularly well in securing defense contracts. This was partly due to the efforts of Terman, but also because of the proximity of Moffett Federal Airfield (between Mountain View and Sunnyvale, CA) and the Ames Research Center that allowed researchers to build relationships with their military sponsors. Cold War patronage allowed local companies to flourish and attracted more technology-oriented firms to the region.

The ready sources of human and financial capital must have been in the mind of William Shockley when he decided in 1956 to commercialize his new invention in Palo Alto. In 1947 at Bell Telephone Laboratories in New Jersey, and along with John Bardeen and Walter Brattain, Shockley had invented a new kind of electronic amplifier made out of semiconductor silicon. Falling out with his bosses there, Shockley relocated to the West Coast (he had grown up in Palo Alto) to found Shockley Transistor Company. Transistors could be used as digital switches not only in computers, but also in communications, audiovisual, navigation, and telemetry equipment for planes and long-range ballistic missiles. The transistor generated a cascade of technologies (the integrated circuit, the microprocessor) and a cascade of companies (Fairchild Semiconductor, Signetics, Amelco Semiconductor, Intel, Advanced Micro Devices) that came to define Silicon Valley in the 1960s and '70s.

Although other factors have been suggested to account for the rise of Silicon Valley (the particularly pleasant climate and landscape, for instance), the best part of the explanation lies in the convergence of an ambitious, expanding engineering university, Cold War contracts, and the fortuitous location of a few key companies. Of course, once Terman had started the ball rolling by drawing a handful of high-tech companies around Stanford, any new company looking for a base was attracted to the concentration of talent, resources, and venture capital that already existed. This network effect was perhaps enhanced by what has been called a peculiar *West Coast style* that developed within Silicon Valley companies. This has been lucidly described by the author Tom Wolfe:

> Corporations in the East adopted a feudal approach to organization. . . . There were kings and lords, and there were vassals, soldiers, yeomen, and serfs, with layers of protocol and prerequisites, such as the car and driver, to symbolize superiority and establish the boundary lines. . . . [Robert] Noyce . . . rejected the idea of social hierarchy at Fairchild. . . . Everywhere the Fairchild engineers went, they took the Noyce approach with them. It wasn't enough to start up a company; you had to start a community, a community in which there were no social distinctions, and it was first come, first served in the parking lot, and everyone was supposed to internalize the common goals. The atmosphere of the new companies was so democratic, it startled businessmen from the East.[3]

Anna Lee Saxenian, for instance, argues that ultimately Silicon Valley was able to outcompete companies on the East Coast because it fostered informal social relationships that encouraged experimentation, entrepreneurship, and cooperative partnerships. Silicon Valley pioneered a flexible industrial system that quickly responded to change and in which specialist producers could work together effectively on complex but related technologies. The dense social network of the region encouraged companies learning from one another, collaborating, and communicating. Whether this theory is correct or not, it is certainly helpful for characterizing the "feeling" or social climate that pervaded Silicon Valley when biology entered this story in the early 1970s.

GENENTECH

The beginnings of the biotechnology industry in Silicon Valley can best be described by following the story of one of its first companies in some detail. Historians disagree about which organization should be considered the "first" biotech company. But the distinction is usually granted to Genentech. The other main contender for this title is Cetus Corporation, founded in 1971 (five years before Genentech) by the biochemist Ronald Cape, the physician Peter Farley, and the Nobel Prize–winning physicist Donald Glaser. Cetus' early research involved improving the yield of vitamin and antibiotic fermentations, and producing bioengineered products like xanthan gum for

---

3. Tom Wolfe, "Two Young Men Who Went West," in *Hooking Up* (New York: Farrar, Straus & Giroux, 2000), 17–65. Quotation p. 39.

oil extraction, and (in collaboration with Standard Oil of California) producing fructose from corn-derived glucose. So although Cetus was first, it was mostly using older forms of biotechnology (more similar to the zymotechnology described in chapter 2), and actively resisting the new recombinant technologies until the late 1970s (after Genentech had adopted them).

Both Genentech and Cetus have been the subject of whole books: *Genentech: The Beginnings of Biotech* (2011) by Sally Smith Hughes and *Making PCR: A Story of Biotechnology* (1996) by Paul Rabinow. These are important narratives not only because they depict the origins of the biotechnology industry, but also because these companies became models for how the biotechnology industry operated and continues to operate in the twenty-first century.

Chapter 4 described the invention of recombinant DNA techniques by Herb Boyer and Stan Cohen in the early 1970s. Although Boyer and Cohen realized their invention was important and had some practical applications, they did not immediately attempt to patent and commercialize it. Even after they had been convinced that commercialization was the best way to proceed, it was not clear *how* to turn their inventions into a technology, or whether such a strategy could have any success. The climate around Stanford was certainly a deeply entrepreneurial one. Engineers, chemists, and agricultural scientists routinely patented their inventions and often started companies in Silicon Valley or elsewhere to profit from their inventions. In biology and medicine, however, such commercialization was almost unheard of. This was partly due to the belief that biomedical inventions should remain in the public domain, to be freely used for the benefit of all. Perhaps more importantly, patent law generally excluded patents on living things (except some varieties of plants—more on this in chapter 6). The world of patents and industry was foreign to biologists.

The impetus for commercialization, then, came not from Boyer or Cohen, but from Niels Reimers, the director of Stanford's Office of Technology Licensing. The job of this office was to seek out inventions by the Stanford faculty that might be patentable and to pursue commercialization with a view towards making money for the University. This program was a product of Stanford's long history of entrepreneurialism and close interaction with industry. Tipped off by an article about Boyer and Cohen's work in the *New York Times*, Reimers approached Cohen in May 1974. Cohen was taken by surprise. Only gradually was Reimers able to convince Boyer and Cohen that patenting the invention had advantages both for Stanford and for the further development of the invention. This began a difficult process that lasted six years as Reimers and Stanford's patent attorneys battled with the US Patents and Trademarks Office.

Meanwhile, others were becoming interested in recombinant DNA technologies. Cohen's friend Glaser suggested the idea to Cape and Farley, cofounders of Cetus. They were unconvinced of its value, but word of recombinant techniques got to some of Cetus' investors at the Silicon Valley venture capital company Kleiner & Perkins. In particular, a young and ambitious Kleiner & Perkins employee named Robert Swanson thought that genetic engineering was "the most important thing I have ever heard."[4] Swanson worked hard to convince Cetus to pursue the new technology—he became so committed to the cause that when Cetus decided not to pursue recombinant DNA and Kleiner & Perkins sold its shares, Swanson found himself out of a job. Now struggling to pay his rent, Swanson refused to give up on the idea of commercializing recombinant DNA techniques. Cold-calling biologists to assess the possibilities, Swanson eventually stumbled into a meeting with Boyer in January 1976. Now the stuff of Silicon Valley lore, a ten-minute meeting in Boyer's lab turned into a three-hour conversation over beers in a local pub. Boyer was open to starting a company to test the industrial applications of his invention, but he had no idea how to do so; Swanson provided that expertise.

Swanson moved quickly and by March he had a business plan and was looking for investors. Soon after, he had convinced his old employers, Kleiner & Perkins, to put up $100,000 as seed money. Genentech was incorporated on April 7, 1976. Perkins believed the chances of success were 50–50. There were two immediate priorities. First, if a patent was granted on recombinant DNA, it would be owned by Stanford. Swanson knew that in order to make Genentech viable he would have to secure a license to use Stanford's technology. Preferably, Stanford would grant Genentech an exclusive license, sealing a deal that would ensure that only Genentech could commercialize recombinant DNA. Worried about the growing public controversy over recombinant DNA, Stanford decided to wait, refusing to grant any licenses.

Second, Genentech needed to demonstrate that its technology could make money. The most obvious application of Boyer and Cohen's techniques was to use bacteria as "genetic factories" to make human proteins. However, although the recombinant DNA techniques themselves were reliable, no one knew whether a recombinant bacterium could actually express a gene spliced in from another organism. Of course, bacteria had equipment for taking its own genes and making them into active proteins, but there were good reasons

4. This quote is attributed to Swanson by Thomas Perkins. Quoted in Sally Smith Hughes, *Genentech: The Beginnings of Biotech* (Chicago: University of Chicago Press, 2011), 32.

for thinking that just putting a foreign gene inside the bacterium would not be sufficient to get the bacteria to actually make the corresponding foreign protein. In August 1976, Swanson set up an agreement through which Genentech paid Boyer's lab to conduct a "proof of concept" to show that bacteria could really produce human proteins.

For the test, Boyer and his coworkers chose somatostatin, an obscure brain hormone. Determined to produce a marketable product as quickly as possible, Swanson had wanted to go with insulin (which had more immediate medical applications in the treatment of diabetes), but the biologists insisted on a simpler protein for which the corresponding DNA could be easily synthesized in the lab. At first the experiments were a complete failure—the bacteria appeared to be producing the protein but immediately breaking it down. Swanson took the news so badly he took himself to an emergency room. By August 1977, however, a new experimental approach was yielding results— somatostatin was being synthesized in the laboratory. This was a remarkable achievement: Boyer and his collaborators had achieved a world-first in a short time frame, working on a limited budget, and in a semi-commercial context.

But somatostatin had no commercial possibilities. The real goal was getting bacteria to make insulin, for which there was a worldwide market. Two talented academic teams (one at Harvard and one at the University of California, San Francisco) were already competing towards this end. Nevertheless, with the proof-of-concept complete, Swanson was able to raise more funds through private equity. By April 1978, this money allowed Genentech to lease its own laboratory space and to begin hiring its own scientists. The space Swanson chose, a 10,000-square foot airfreight warehouse in South San Francisco, was commuting distance from downtown San Francisco and Berkeley, and close to Stanford, the high-tech companies in Silicon Valley, and the international airport. It was also just isolated enough from the political controversies over recombinant DNA that raged in the less industrial parts of the Bay Area. With the lab barely set up, and with just three full-time scientists, Genentech won the race, successfully collaborating with scientists at the Beckman Research Institute in Los Angeles to produce human insulin in August 1978. This was an astonishing result. In less than one year the fledgling company had beaten out two of the world's leading molecular biology teams.

The reward was a contract with the pharmaceutical giant Eli Lilly. Lilly had for years dominated the market for pharmaceutical insulin, obtaining its supply from pigs and cows. In the 1920s, the Canadian scientists Frederick Banting, J. J. R. McLeod, Charles Best, and James Collip had worked out a technique for extracting and purifying insulin from the pancreas of calves.

However, bacteria-produced human insulin would have several advantages. First, there were some patients who rejected or suffered allergic reactions to the animal-derived product. More importantly, the number of diabetics was expected to increase and the uncertain agricultural supply was unlikely to be able to keep up. Lilly was extremely enthusiastic about finding a way to manufacture insulin. Four days after Genentech's laboratory produced the molecule, Lilly signed a multimillion dollar contract. Lilly obtained worldwide exclusive license to manufacture human insulin using Genentech's process. Genentech got an upfront fee of half a million dollars and 8% royalties on Lilly's insulin sales. Overnight, Genentech transformed from a tiny research start-up to a new and significant player in the pharmaceutical game.

This story is important not only because it is one significant marker of the beginning of the biotechnology industry, but also because it became a model for how that industry would operate. Swanson worked hard to ensure that Genentech's contract with Lilly meant that although Lilly would obtain the rights to use Genentech's recombinant DNA technology, their use of this technology was limited to recombinant insulin. Lilly was not allowed to "steal" the technology to make other drugs based on recombinant DNA techniques. The agreement also allowed Genentech to protect itself from the high cost and high-risk process of putting the new drug through the Food and Drug Administration approval process (the new insulin was finally approved only in 1982). For its part, Lilly protected its investment by setting up a benchmark system for making payments to Genentech. Only when Genentech had met specific milestones on the road towards producing insulin in quantity and at high purity would the contract advance. This general arrangement—small start-ups making cash by licensing their new drugs and technologies to big pharma—persists to the present day.

By the end of 1980, the path to profitability for biotech was cleared by two important legal developments: the Diamond vs. Chakrabarty decision by the Supreme Court permitted the patenting of genetically engineered organisms and the Bayh-Dole Act offered new incentives for government-sponsored researchers to seek patents on their inventions (these topics are both covered in detail in chapter 6). The Chakrabarty decision, combined with Genentech's success with insulin, provided a solid basis from which the company could make an initial public offering (IPO) of its stock. On October 14, 1980, one million shares were issued at $35 each. Within twenty minutes the value of Genentech stock had reached $89 per share, closing the first day at $70. This doubling of market capitalization was remarkable for a small company that, even in 1980, had no products on the horizon for several years. The biotech boom had begun. Genentech followed up insulin with successful drugs for growth hormone deficiency (Protropin, 1984), dissolving blood clots (Activase, 1987), cystic fibrosis (Pulmozyme, 1993), non-Hodgkin lymphoma (Rituxan, 1997), metastatic breast cancer (Herceptin, 1998), asthma (Xolair, 2003), psoriasis (Raptiva, 2003), and lung cancers (Tarceva, 2004). In 1990, the Swiss pharmaceutical giant Hoffman-La Roche bought a 60% share of Genentech. In 1999, Roche bought all remaining shares in Genentech but then resold a large fraction in a series of public offerings between July 1999 and March 2000. The first of these sales (of 17% of Genentech stock) amounted to the third largest stock offering of 1999, after Goldman-Sachs and Pepsi, and rendered Roche a profit of $1.94 billion. In 2009, Hoffman completed its purchase of Genentech for a total price of $46 billion. Biotech had become big business: a handful of young scientists and entrepreneurs around Silicon Valley laid the groundwork for a remarkable new economic phenomenon.

ACADEMY AND INDUSTRY AND PAJARO DUNES

The birth of the biotechnology industry marked a significant economic event, but it also had profound consequences for the development of science and technology. Most critically, biotechnology seemed to be cutting across the boundary between pure and applied science. Cohen and Boyer's original invention had emerged from fundamental biological research on bacterial plasmids and restriction endonucleases—they had taken up these research programs expecting to publish scientific papers, not to patent. Yet, within ten years, the work had led to a commercial product worth millions of dollars. It was not just that fundamental research was traveling off campus, but also that industry labs seemed in some cases to be doing work better and faster than their academic counterparts. Genentech beat Harvard and UCSF to insulin. Likewise, one of the major breakthroughs in molecular biology in the 1980s—the invention of the polymerase chain reaction (PCR)—took place at Cetus Corporation, not in a university. To some, especially those who stood to make money, the biotech industry seemed to be ushering in a new era of productivity and innovation in science and technology.

Others were alarmed by these developments. Some scientists and university administrators worried that the money and influence flowing from industry threatened academic freedom. They believed researchers would direct their work towards ends they believed might be profitable, rather than following their curiosity or intuition. Public universities, in particular, had a mission to ensure that their research benefitted everyone, rather than enriching a few individuals or a private corporation. Patenting of research was not necessarily in the university's or the public's interest. During Boyer's work on the proof of concept for somatostatin, Genentech had paid UCSF for the overhead on the use of its labs and also paid the salaries of Boyer's postdoctoral researchers. This meant that Boyer was leading a team that was conducting research *in* a public university laboratory but *for* a private corporation in which Boyer himself had a large stake. Such tangled webs of employment, contracts, consultancies, and scientific advisory positions became increasingly common for biologists in the biotech era. Many were (understandably) worried about potential conflicts of interest. They feared that the lure of profits would skew scientific results and that biologists would increasingly keep results secret in the hopes of gaining a patent or a contract.

In 1982, with the biotechnology industry growing rapidly, the presidents of several elite universities decided to meet to discuss these issues. In the small beach town of Pajaro Dunes, on Monterey Bay, the leaders of Harvard, Stanford, the California Institute of Technology, the Massachusetts Institute of Technology, and the University of California attempted to draw up guidelines for regulating the relationship between the academy and industry. Worried that the rise of this new species of firm was "corrupting" academic biomedicine, the presidents saw biotech as threatening "pure" science, as reducing free and open inquiry, as increasing secrecy, skewing research goals, misplacing public monies, and compromising education. The outcomes, however, were rather lackluster. After several days of deliberation, the presidents could agree on only vague and general guidelines that were not binding on the institutions represented. Biotech continued to enroll and entangle academics apace.

Part of the problem was that the ideals the presidents were trying to defend were just that: ideals. Interviews with some of the first scientists to enter the biotechnology industry in the 1970s and '80s suggest that in many cases they had found *more* openness and freedom in the private sector than in university settings. In academia, the tough competition to get research grants from funding agencies like the National Institutes of Health meant that work constantly had to be justified in terms of its applications—this "grantsmanship" was far from the curiosity-driven ideal. Moreover, university budgets were low and universities burdened scientists with administrative responsibilities.

Many biologists moving into industry reported on the increased freedom that came with bigger budgets and more support staff. Within this framework, researchers in industry often did have far more scope to pursue work that interested them and it was often in the interests of their employers for them to publish it in order to establish priority for a patent claim. Perhaps most importantly for many, the opportunity to see the results of their work manifested as medicines *increased* their sense that they were contributing to the public good. Profits may have driven investment in biotech, but the industry also succeeded because it was a good place, perhaps the best place, to do science in the late twentieth century.

CONCLUSIONS

One important fact that we have learned from the history of science is that the boundary between pure and applied science (and between science and technology) has always been a highly permeable one. In the eighteenth century, natural history and botany were directed towards the discovery of economically significant plants; in the nineteenth century, geophysics was funded by the state in order to reveal mineral resources and survey territories; later chemistry was funded by manufacturers of dyes; and in the twentieth century, physics has benefitted from the enormous market for electronics. We have seen examples earlier in this book of how biology too was directed towards a range of commercial ends (agriculture, hormones, vitamins, food and drink). There are countless examples of the industrial character of much scientific work. So, we need to think carefully before we characterize the biotech industry as a complete break with the scientific past. The experiences of early biotech scientists corroborate the notion that the supposed world of "pure" science was already shot through with "applied" complications. The Pajaro Dunes model of industry corrupting pure-academic science is too simplistic. Academia and industry have never been separate spheres; we might even want to go so far as to say that a distinction between pure and applied science makes no sense—where is (or was) this so-called pure science to be found?

But it seems fair to say that *something* changed about the way biology was done in the 1970s. If it was not just a straightforward transformation from pure to applied biology, then what was it? I will suggest three possible answers here. First, the biotech industry offered a new kind of space for biological work. The emergence of biotech companies meant that biomedical research could be pursued in an economic and social context that was quite distinct from university labs or pharmaceutical companies. The influence of the freewheeling, entrepreneurial Silicon Valley culture on these new spaces meant new opportunities for thinking and working differently.

Second, as biologists began to move back and forth between these academic and industrial labs, they acquired new sets of skills. The commercial world required them to raise venture capital, meet strict deadlines, manage teams, and think about product development and marketing. As biology became an entrepreneurial discipline, biologists had to adopt a new self-image. By the 1980s, *what sort of a person* a biologist was had begun to change.

Third, the biotech industry drew biology more deeply into regimes of capital speculation. Biotech companies were not engaged in direct selling of products for a profit. Rather, biotech is involved in a high-capital game of speculation and risk—biologists began to become directly involved in these scientific futures markets. This involves more than applying biology; this was the beginning of a transformation to a new view of life, "enterprised up," as a productive force to be harnessed, technologized, and capitalized. In other words, the "biotech revolution" involved an important change in scientific and cultural attitudes towards living things: the biotechnology industry became not just about making something useful out of living stuff, but about leveraging the productivity of living material into market value.

FURTHER READING

On the emergence and history of Silicon Valley see Christophe Lécuyer, *Making Silicon Valley: Innovation and the Growth of High Tech, 1930–1970* (Cambridge, MA: MIT Press, 2008). On the founding of Varian Associates see John Edwards, "Russell and Sigurd Varian: Inventing the Klystron and Saving Civilization," *Electronic Design*, November 22, 2010). On the differences between East and West Coast styles of doing business see AnnaLee Saxanian, *Regional Advantage: Culture and Competition in Silicon Valley and Route 128* (Cambridge, MA: Harvard University Press, 1994) and Tom Wolfe, "Two Young Men Who Went West," in *Hooking Up* (New York: Farrar, Straus & Giroux, 2000), 17–65.

The best account of the discovery of recombinant DNA and the founding of Genentech is Sally Smith Hughes, *Genentech: The Beginnings of Biotech* (Chicago: University of Chicago Press, 2011). The founding and development of the early biotech company Cetus is narrated in Paul Rabinow, *Making PCR: A Story of Biotechnology* (Chicago: University of Chicago Press, 1996). An excellent overview of the early days of the biotechnology industry is Nicolas Rasmussen, *Gene Jockeys: Life Science and the Rise of the Biotech Enterprise* (Baltimore: Johns Hopkins University Press, 2014). The Regional Oral History Office of the Bancroft Library (University of California, Berkeley) has an extremely rich collection of relevant oral histories in their "Program in Bioscience and Biotechnology Studies" that can be found online: http://bancroft .berkeley.edu/ROHO/projects/biosci/.

The historical relationship between universities and industry is discussed in general terms in Jean-Paul Gaudillière and Illana Löwy, eds., *Invisible Industrialist: Manufactures and the Production of Scientific Knowledge* (New York: St. Martin's Press, 1998) and Philip Mirowsky, *Science-Mart: Privatizing American Science* (Cambridge, MA: Harvard University Press, 2011). The specific effects of the biotechnology industry on this relationship are examined in Martin Kenney, *Biotechnology: The University-Industrial Complex* (New Haven, CT: Yale University Press, 1986) and in Arnold Thackray, ed., *Private Science: Biotechnology and the Rise of the Molecular Sciences* (Philadelphia: University of Pennsylvania Press, 1998).

For a broader understanding of the economic transformations wrought by the biotech industry see Kaushik Sunder Rajan, *Biocapital: On the Constitution of Post-Genomic Life* (Durham, NC: Duke University Press, 2006); Kaushik Sunder Rajan, ed., *Lively Capital: Biotechnologies, Ethics, and Governance in Global Markets* (Durham, NC: Duke University Press, 2006); and Stefan Helmreich, "Species of Biocapital," *Science as Culture* 17, no. 4 (2008): 463. The effect of changing industrial entanglements on what it means to be a scientist are examined in a longer historical context by Steven Shapin, *The Scientific Life: A Moral History of a Late Modern Vocation* (Chicago: University of Chicago Press, 2010).

6 PATENTING LIFE

INTRODUCTION

In the 1980s, it became possible to own living things in new ways. Of course, humans have long owned horses, sheep, dogs, and even (at some times and places) other humans. But in the 1980s it became possible not just to own a particular plant or mouse or bacteria, but to own the *plans* to particular kinds of organisms—that is, to own the exclusive rights to build and reproduce whole varieties of plants, bacteria, or animals. As we saw in the last chapter, this development ensured that the biotech industry could protect its intellectual property and become a viable commercial proposition.

This was achieved through the application of patent law to organisms. In the United States, there had been some precedent for this: in 1930, Congress had enacted the Plant Patent Act to protect certain varieties of plants, such as those created by Luther Burbank (see chapter 2). But apart from this, patents were applicable only to *inventions*—products of human ingenuity. Organisms were not inventions and therefore not patentable. This chapter examines the reasons why this changed and discusses some of the social, economic, political, and ethical implications of this new understanding of life.

WHAT IS A PATENT?

The aim of patents is to encourage invention by providing an economic incentive to inventors. If you invented a new kind of mousetrap, for example, in order to make money from it, you need to sell it. But, as soon as you put it on the market, it is available for others to examine and reverse engineer. Without patents, your invention could quickly be copied and you would make very little money. Patents attempt to protect the ability of inventors to profit from their own inventions. Once a patent is granted, the inventor is given an exclusive right under the law for a limited period of time to control the invention. Importantly, a patent does not necessarily grant a right to make or use an invention (for instance, if you invent a new kind of bomb, it doesn't give you a right to blow things up with it, or even to manufacture it). But it is a right to stop others making, using, selling, or importing it without your permission.

But patents are not only supposed to bring private benefits to inventors.

They are also supposed to serve the public good. The idea is that they reduce secrecy. Without patents, one way of making money from your new mousetrap might be to keep the plans and mechanisms secret. This would make it harder for others to copy. However, it would also diminish the possibility that someone could improve on your mousetrap. Usually, to get a patent you need to make the plans to your invention public. This allows other inventors to build on your ideas. In this way patent law aims to encourage innovation built on innovation, helping the economy and bringing more new ideas to fruition. By granting rights to the inventor, everyone can benefit.

But this requires a fine balance. If too many rights are reserved for inventors, they may be able to profit unfairly (this is the reason patents have a limited lifespan—usually about twenty years, depending on the jurisdiction). If inventors are granted too few rights, they might have little incentive to invent in the first place. In practice, this means limiting the kinds of things that can be patented. First, you shouldn't be able to patent things that are too obvious. Say, for example, you took a regular mousetrap and just enlarged all the parts and tried to patent it as a "rat trap." This is a fairly obvious change and doesn't represent real innovation. Since this change doesn't contribute very much to the art and science of rodent killing, granting a patent on the rat trap would reward you for very little work. Since you would now have a monopoly on the larger traps, you could increase prices. Now everyone has to pay more for something that is not much of an improvement over what they had before. In this case, giving you a patent wouldn't serve the public interest at all.

Second, patents must be useful. If an invention doesn't have any use (or at least any known use), then it can't serve the public good. Without this rule, inventors could rush to the patent office with thousands of mousetrap designs, patenting them all in the hope that one might be of use. Patents should reward those who demonstrate specific uses, not mad scientists or speculators.

Third, usually "products of nature" should not be patentable. Naturally occurring things can perhaps be *discovered*, but not invented. It would be unfair for someone to profit from objects that they found lying around—such things are usually considered public property. In a case in the nineteenth century (*Ex parte Latimer*, 1889), an individual attempted to obtain a patent on a fiber derived from pine needles. The judge in the case decided that the fiber was a product of nature. He argued that this would be the same as granting a patent "upon the trees and forests and the plants of the earth, which of course would be unreasonable and impossible."[1]

1. *Ex Parte Latimer* Dec. Com. Pat 123 (1889).

These three rules—non-obviousness, utility, and product of nature—are designed to maintain the balance between private profit and public good.

CRISIS AND COMPETITION

To understand the reasons for the changes in patent law that took place in the 1980s, it is necessary to examine the economic, political, and technological context in which they took place. By 1980, the significant potential of biotech was already apparent. As discussed in chapter 5, the 1970s had seen not only the discovery of recombinant techniques by Boyer and Cohen, but also attempts to develop and commercialize other biotechnologies (particularly interferons). Genentech was founded in 1976 and other companies were soon to follow (Biogen 1978, Amgen 1980, Genzyme 1981, Chiron 1981). By 1980, it seemed that an important new industry was beginning to emerge, especially in and around Silicon Valley.

The 1970s was also a time of economic uncertainty in the United States and Western Europe. First in 1973 and again in 1979, the United States experienced oil crises: petroleum prices skyrocketed as countries in the Middle East slowed supply and the United States deregulated its oil industry. This caused inflation and slowed the economy ("stagflation"—a combination that economists had thought impossible). America's economy, so prosperous through the 1950s and 1960s, seemed to be losing its competitive edge, especially as more and more manufacturing moved overseas. One particular worry was Japan. The Japanese economy seemed to be booming. Its automobile manufacturers outcompeted American brands and Japan was becoming increasingly competitive in the manufacture of cutting edge technologies, such as electronics and computers.

To many in the United States, it seemed to be just a matter of time before the Japanese economy led the world. This was cause for alarm. Politicians and policymakers were desperately trying to find ways to keep America innovative and competitive amidst this tough economic climate. One solution was to make sure America capitalized on its scientific and technological discoveries. The government spent millions of dollars every year on basic research, funding university scientists through grants from the National Institutes of Health (NIH) or the National Science Foundation. But how much of this work was transformed into useful technology? Universities were quick to point out that some opportunities were missed because of the way the system worked. Under the rules of the research grants, any patents that emerged from the work belonged to the government. The taxpayer had funded the research so the government should own the results, the logic went. The problem was that the government didn't have the resources to commercialize the inventions. And the inventors themselves had little to gain by developing their inventions since they wouldn't own the patents.

In practice, the result was that many potential technologies languished. Up to 1980, the US government had accumulated 28,000 patents, of which only 5% were licensed to companies for development (in other words, only 5% stood any chance of making money). Intent on tackling this problem, two US senators—Bob Dole (Kansas) and Birch Bayh (Indiana)—sponsored a bill that would dramatically reform the regulations concerning federal grants and patents. The University and Small Business Patent Procedures Act, usually known simply as the Bayh-Dole Act, was adopted as law in December 1980.

The new Act contained a number of reforms intended to promote commercialization. It required universities receiving federal grants to actively work to commercialize inventions, to file for patents, and to share royalties with the inventor. But the main change was that the intellectual property (IP) now belonged to the university, not the government. If a faculty member working under a government grant invented something patentable, the university now stood to gain. It had a large incentive to develop the invention in order to bring money to the institution that could be used for education or research. Bayh-Dole was conceived and passed into law specifically with biotechnology in mind. The US government wanted to make sure that the economy would benefit fully from the new discoveries and inventions that were emerging from biology.

DIAMOND V. CHAKRABARTY

However, the question of whether biological stuff could count as intellectual property (IP) was still not settled. The economy could benefit from innovations in biology only if biotech companies were commercially viable. Many business analysts believed that this depended largely on whether these companies would be able to protect their IP. If, for instance, a new company engineered an organism that could produce a new drug, they needed to be able to prevent others from copying that organism and outcompeting them. In other words, the organism needed to be protected under a patent. By 1980, this legal technicality—the patentability of organisms—became critical to determining whether the biotech industry would be able to survive. As is often the case with the law, the changes to patent law depended on a landmark court case. The case, known as Diamond v. Chakrabarty, refers to Sidney A. Diamond, the Commissioner of Patents and Trademarks at the time of the case, and Anand Chakrabarty, a microbiologist who worked for General Electric (GE).

At GE's Research and Development Center in Schenectady, New York, in

1972, Chakrabarty had developed a new bacterium. This new bacterium was especially good at eating oil. Chakrabarty's idea was that it could be used for cleaning up oil spills. There were several naturally occurring bacteria (belonging to the genus *Pseudomonas*) that could be used to attack oil spills. However, since different species of bacteria specialized in breaking down the different components of oil (oil is in fact a complex mixture of hydrocarbon substances) a mixture of different bacterial species had to be used. The problem with this was that the different species competed with one another, reducing the total amount of oil degraded.

Chakrabarty's bacteria had significantly improved oil-eating capabilities (it consumed oil up to 100 times faster). Although his work did not involve recombinant techniques—that is, it did not involve moving genes between species by splicing plasmids—it did make use of plasmids themselves. Specifically, Chakrabarty took advantage of the ability of bacteria to exchange whole plasmids. Since the genes for oil-eating occurred on plasmids, he found a way of getting the genes from four different species of bacteria inside a single bacteria. This generated a new species that had superior oil-eating abilities since it now contained four different genes that could simultaneously degrade multiple components of the oil.

In 1972, Chakrabarty applied for a patent on his invention, assigning the rights to GE. The patent was denied by the US Patents and Trademarks Office (USPTO) on the grounds that living things are not patentable. When GE appealed, the Patent Office Board of Appeals agreed with this initial assessment, but the Court of Customs and Patent Appeals reversed the decision, arguing the bacterium should be patentable. Finally, the Commissioner of Patents and Trademarks appealed to the Supreme Court. The court agreed to hear the case in 1979; it was argued on March 17, 1980, and decided on June 16, 1980.

The key to the decision was whether the US Patent Act had sufficient scope to include living things. It is worth noting that Chakrabarty had also filed for patent protection both on the *process* of making his bacteria and for a method of delivering it to an oil spill (mixing it with straw). These were granted patents—it was only the patent on the actual bacterium itself that was at issue. The lower courts and the dissenting opinions in Chakrabarty argued that a bacterium should not be granted a patent because living things simply did not fall within the scope of patent law: the law didn't mention living things so they shouldn't be included.

The court's role in a case like this is to try to discern the *intentions* of the law. Not only were living things not mentioned, but some living things had been especially singled out later via the Plant Patent Act of 1930 and its updated form, the Plant Variety Protection Act of 1970. The fact that Congress had had to make a *separate* law for plants implied that living things did not fall within the scope of the original patent laws.

Those who disagreed with granting Chakrabarty's patent thought that it was inappropriate to extend patents into realms that lawmakers had not intended. The proper course of action was to wait for Congress to decide the issue and explicitly change the law to cover living things, if it saw fit. The Commissioner of Patents also tried to make the argument that the safety concerns surrounding genetic engineering were compelling reasons to reject the patent. All the judges agreed that this was *legally* irrelevant.

Those who supported the patent claim saw more latitude for extending the reach of patent protection. Chakrabarty's patent was not a "product of nature," they argued, since it exhibited characteristics significantly different from any bacteria found in nature: it was "not nature's handiwork, but his own." The Court's role, was to decide whether the bacteria fell within the language of "manufacture" or "composition of matter" specified by the law. The original authors of the patent law (most notably Thomas Jefferson) intended patents to include "anything under the sun that is made by man." Chakrabarty's invention fit within this broad criterion. On this view, the fact that Congress had not explicitly included living things in the formulation of the law was irrelevant: the whole point of patents was to cover *unforeseen* inventions. After Congress passes a law, the courts have broad power to interpret it in accordance with precedent — it was not necessary to wait for Congressional approval.

The Court's decision, five to four, was in favor of granting the patent. Diamond v. Chakrabarty cleared the way for more patents to be issued on organisms. Over one hundred patents had been held up at the Patent Office, awaiting the Chakrabarty decision. Amongst them was Herb Boyer and Stan Cohen's patent on a "Process for Producing Biologically Functional Molecular Chimeras" (that is, recombinant DNA). This patent was granted on December 2, 1980. On the strength of Chakrabarty, Genentech made a public offering of its stock on October 14, raising almost $40 million (see chapter 5). Many other companies were soon to follow. Protected by patents, biotech could now get down to business.

HIGHER ORGANISMS

The Chakrabarty decision also opened the door for more and more ambitious patent claims on living things. Chakrabarty's bacterium was, after all, just a bacterium. It seemed to be a far cry from patenting a "higher" or multicellular organism such as an ant, a fish, or a cow. However, as biotechnologies developed during the 1980s, patent law also gradually adapted to allow patenting on a wider and wider variety of life forms.

One of the first steps in this direction came from an unlikely quarter: oyster breeding. In 1984, three marine biologists working for Coast Oyster Company—including Standish K. Allen Jr.—applied for a patent on a new variety of oyster. The oyster was derived from Pacific oysters (*Crassostrea gigas*) but had been treated with chemicals that gave the oysters three chromosomes each rather than two (this is called triploidy). These oysters were found to grow faster and could be sold by oyster farmers year round. In other words, they had significant utility and commercial potential.

The patent was denied on two grounds. First, and unfortunately for Allen, he had published his method for making triploid oysters in a scientific journal in 1979. As far as the Patent Office was concerned, this made it part of the "art of oyster breeding." Anyone, they argued, could simply have looked up the technique and done it themselves. Inventions based on published techniques are usually ineligible for patents due to the criteria of nonobviousness. Second, the Patent Office said that the Chakrabarty decision did not apply to higher organisms such as oysters. Allen and his colleagues appealed the decision to the Board of Patent Appeals and Interferences. In 1987, in a case known as *Ex Parte Allen*, the Board upheld the original decision and no patent was granted. But in an addendum to their decision, they reversed the position on higher organisms: there was no barrier, they said, to patents being granted on living animals.

Ex Parte Allen set the precedent that for even more dramatic and controversial patent disputes. In 1981, Harvard Medical School recruited the molecular biologist Philip Leder (1934–). As part of his recruitment package, DuPont agreed to provide $6 million in sponsorship for Leder's lab. In return, Harvard would get to keep any patents that might arise from Leder's work, but DuPont would get an exclusive license on any subsequent technologies.

In the early 1980s, Leder's work focused on understanding breast cancer at the genetic and molecular level. This work used mice as a model organism: a better understanding of breast cancer in mice would lead to advances for human breast cancer medicine. Together with Timothy Stewart, Leder began using recombinant DNA techniques to modify mouse embryos. Their aim was to create a stock of mice with breast cancer for their studies. First, Leder and Stewart found a cancer-causing gene (called an oncogene) in a virus (some viruses are known to cause cancer). Next, they isolated a segment of DNA (called a mammary promoter) that acts to turn on genes only in breast tissue. They spliced the promoter to the cancer gene and cut and pasted this complex into a mouse embryonic cell. Finally, they implanted the modified embryo into a female mouse to make it pregnant. This process would result in offspring that would express the cancer gene in their breast tissue. In fact, the resulting baby mice did not automatically have breast cancer, but only a heightened susceptibility for breast cancer. By breeding these mice, the lab created a whole lineage of mice with these special genetic characteristics.

Leder and Stewart were not aiming to make a patentable product. But by 1983, they had realized that the mice might have commercial potential. For instance, the mice might be useful for other researchers studying breast cancer. In addition, the special susceptibility of these mice made them ideal for testing both anti-cancer compounds and drugs, as well as for testing carcinogens. In essence, the fact that these mice were more vulnerable to cancer made them useful for speeding up all kinds of cancer-related tests.

Leder and Stewart brought their work to the attention of the Harvard Medical School Office of Technology Licensing and Industry Sponsored Research, and Harvard filed for patents in 1984. Claiming a patent on the *process* of making the mice was straightforward. But Harvard's lawyers insisted that "the work's most apparent and compelling manifestation was the animal itself."[2] In other words, to fully protect the invention, the mouse (not just the process of making it) needed to be patented. In fact, the patent claims included patents not only on the mice Leder and Stewart had bred, but on all transgenic mammals (excluding humans) containing any activated oncogene *or any of their descendants*.

This was certainly a broad and aggressive claim. In the United States, the patent for Oncomouse (as the genetically engineered mouse came to be known) was granted to Harvard in April 1988 on the basis of *Ex Parte Allen*. The decision was not appealed. Oncomouse suffered more difficult journeys in other jurisdictions, especially in Canada and Europe. In Canada, the patent was eventually granted in an amended form in 2003 (the major case was Harvard College v. Canada [Commissioner of Patents], in which the claim was initially rejected). In Europe, the patent was eventually rejected after more than twenty years of legal battles between Harvard, the European Patent Office, and various opposition groups.

The European story is particularly interesting. There, the mouse patent was opposed under article 53(a) of the European Patent Convention that pre-

2. Paul Clark quoted in Daniel J. Kevles, "Diamond v. Chakrabarty and Beyond: The Political Economy of Patenting Life," in *Private Science: Biotechology and the Rise of the Molecular Sciences*, ed. Arnold Thackray (Philadelphia: University of Pennsylvania Press, 1998), 65–79. Quotation p. 75.

cludes patents on "inventions, the commercial exploitations of which would be contrary to '*ordre public*' or morality."[3] Unlike the United States—where judges could set aside any moral qualms—this meant that ethical arguments were directly relevant to the legal situation in Europe. Opponents of the patent argued against the law on economic (granting patents on plants and animals would foster monopolies, especially over food), religious (life has a sacred quality that should not be owned or considered a "composition of matter"), environmental (genes could spread from lab animals to wild varieties and cause ecological damage), and animal rights (subjecting mice to cancer was cruel) grounds. By showing that granting a patent could be socially and economically damaging to Europe, opponents mounted a compelling case against the patent.

It was also argued that the patenting of a mouse represented the beginning of a slippery slope that would ultimately lead to patents on human beings. Although this might seem alarmist, it signaled widespread discomfort with the idea that whole varieties of animals might be owned. Oncomouse, for some, has become a symbol of the excesses and dangers of biotechnology and its attempts to own, commercialize, and exploit living things.

PATENTING DNA

Perhaps those who worried about the patenting human beings weren't too far off the mark. Just a few years after Oncomouse had been granted its patent in the US, the USPTO began to receive thousands of patent applications for human DNA. Such claims had begun as early as 1977 when the University of California had filed patents covering the DNA for insulin and human growth hormone. These were granted in 1982 and 1987, respectively. The legal argument was based on precedents that allowed the patenting of products derived or purified from natural sources such as hormones or vaccines (for example, human adrenaline was granted a patent in 1906). Competing claims to these substances led to prolonged litigation between the University of California, Genentech, and Eli Lilly in the 1980s.

But in the early 1990s, accelerating work on the Human Genome Project (see chapter 12) led to a rapid increase in DNA patent claims. Craig Venter was one of the biologists at the NIH working at the cutting edge of DNA sequencing research. He developed a method of rapidly identifying human genes. When a gene is expressed, DNA is transcribed into mRNA and mRNA is translated into protein. If mRNA is collected from a living cell it provides a

3. European Patent Convention, Part II, Chapter I, Article 53. http://www.epo.org/law -practice/legal-texts/html/epc/2013/e/ar53.html.

sort of snapshot of the genes being expressed at a particular moment. Venter's method took the collected mRNA, transcribed part of it *back* into DNA (this is called complementary DNA or cDNA), and then sequenced it. These short readouts of DNA were essentially random fragments of human genes. Venter and his team called them expressed sequence tags (ESTs). Using new automatic sequencing machines, Venter's lab was able to produce 50 to 150 tags per day.

In 1991, Venter and the NIH applied for patents on thousands of ESTs. In fact, the NIH claimed patents over the entire genes based on the tag, even though they represented only a small fragment of the whole gene. Scientists, the public, and Congress all expressed dismay at the idea of patenting human genes. James Watson said that "virtually any monkey" could do what Venter's lab was doing. It was "dumb, repetitive work," another critic claimed, through which someone could lay claim to most of the human genes.[4] If patents were to be granted on genes at all, then it should be to those who do the hard work of figuring out the function of proteins that the genes encode, not for the trivial work of sequencing ESTs. Owing to the ongoing controversy, the NIH eventually decided to withdraw the applications based on ESTs.

But this was hardly the end of attempts to patent DNA sequences. Through the 1990s, many biotech and pharmaceutical companies continued to file thousands of patent applications on gene sequences. The market value of companies such as Human Genome Sciences, Incyte, and Millennium skyrocketed in anticipation of the value that this genomic property could generate as genetic targets were turned into drugs. At times overwhelmed by this deluge of patent applications, in 1999, the USPTO was moved to issue new rules clarifying the circumstances under which DNA sequence could be patented. Under the new rules, the applicant had to not only identify a previously undiscovered piece of sequence, but also specify the sequence's product (i.e., a protein), specify what the product does (i.e., its use), and demonstrate a method for actually using it for the specified purpose. Thousands of pending patent claims were denied, but these rules did not stop ongoing and aggressive attempts by researchers and companies to patent as much DNA sequence as possible.

Although the value of gene patents declined in the 2000s, estimates suggest that the number of patents related to RNA or DNA granted in the United States is around 50,000 (about 3,000–5,000 of these are on human genes,

4. The quotations can be found in Leslie Roberts, "Genome Patent Fight Erupts" *Science* 254 (11 October 1991): 184–186.

nearing 20% of all human genes). In many cases, several different patents have been issued to different individuals or companies on the same gene (for example, the DNA, the mRNA, and the protein for a single gene have all been granted *separate* patents in some cases). This leads to "stacks" or "thickets" of patents that make it extremely difficult to sort out who owns what. Although such claims on genetic territory may rarely be enforced, patent stacking might still deter biomedical scientists from conducting research on particular segments of DNA (out of fear that any discoveries might end up belonging to someone else).

The most widely reported controversy over gene patenting involved the company Myriad Genetics. In 1994, the University of Utah, the National Institute of Environmental Health Sciences, and Myriad filed for a patent on the BRCA1 breast cancer susceptibility gene. Unlike Venter's ESTs, researchers at the University of Utah had worked to characterize and identify the functions of this gene. In 1995, Myriad filed a patent on another breast cancer gene, BRCA2. Myriad also obtained patents on genetic tests based on the BRCA1 and BRCA2 genes and has attempted to enforce its patent rights. This meant that Myriad could, until its patents expired, enforce a monopoly on BRCA1 and BRCA2 testing, charging high prices for its tests. In 2001, Myriad's lawyers began sending letters to laboratories threatening legal action against those conducting BRCA1 and BRCA2 testing without a license from Myriad. Trying to force potential breast cancer victims to pay thousands of dollars for a genetic test prompted widespread outrage. Eventually, a group of patients decided to challenge Myriad's patents, resulting in a protracted legal battle (Association for Molecular Pathology v. Myriad Genetics).

After a series of lower court decisions and appeals, this case was decided by the Supreme Court on June 13, 2013, and resulted in the unanimous overturning of Myriad's patent claims. The justices affirmed that the BRCA1 and BRCA2 genes constituted naturally occurring objects and as such were not patentable material. However, the Court asserted that certain kinds of DNA molecules made in the lab and not occurring in nature, such as complementary DNA, would not fall under their ruling and hence could be patented.

Myriad lost the case and the ruling appeared to reverse the USPTO rules on gene patenting that had granted patents in cases where DNA sequences had clear utility (such as a test for a disease). However, the judgment also left significant space for further uncertainty and controversy. The court's distinctions between "naturally occurring" DNA and "man-made cDNA" (and between "informational" and "physical" entities) may not withstand close scrutiny. In particular, it raises sticky questions about what kind of work constitutes the "making" of an object—Myriad's isolation of a gene does not, according to the Supreme Court, constitute labor sufficient for the granting of a patent; but making cDNA out of mRNA in the laboratory does.

It is likely that this is not the final word about the patenting of genes. Critics of gene patenting argue that patents impede biomedical research, especially the development of diagnostic tools. In theory, an academic lab could be sued for studying or using a patented gene in its research. Moreover, patents could constrict the free sharing of scientific information and thereby discourage some avenues of scientific work. Researchers might be deterred by the possibility of being sued or undergoing long and expensive licensing negotiations. In 2007, the science fiction author Michael Crichton argued that gene patents "are now used to halt research, prevent medical testing and keep vital information from you and your doctor."[5] Patents increase costs of testing and block competition, he argued. Partly with Crichton's help and influence, this led to the introduction of a bipartisan bill into Congress in 2007 (HR977, usually known for its sponsors, Representatives Xavier Becerra and Dave Weldon). The law would have banned all future patenting of DNA sequences (no such law has yet passed Congress).

There are also many who support the ongoing use of patents to protect biotechnological inventions involving DNA sequence. They argue that the money derived from patents can be spent on further research, that patents encourage the investment of resources and money in DNA-based research, and that this protection prevents wasteful duplication of work by reducing secrecy. Supporters of patents also point to the fact that several useful therapies have been developed under the auspices of patent protection. They argue that basic and applied scientific research on genes and DNA has continued at a swift pace, despite the fears of the patent critics. No one has been sued for using a patented gene in their academic work and in practice patents don't seem to be blocking research, patent advocates contend.

OWNING LIFE

These arguments continue. Most scientists agree, however, that biomedical inventions require some legal protection: entirely eliminating the ownership of biological things would discourage innovation and limit investment in promising research. The important problem, however, is finding the right balance between work and reward, between private profit and public good. The objections to patenting ESTs, for instance, were based mostly on

5. Michael Crichton, "Patenting Life," *New York Times*, February 13, 2007.

the premise that these patents represented an unfairly large reward for relatively little effort. In that case, the balance seemed tilted too far in favor of the inventor.

The patent law was designed to achieve such a balance. However, things like genetically engineered bacteria and DNA sequences are very different sorts of objects from mousetraps or machines envisioned by the creators of patent laws. Are these objects natural or artificial? Made by nature or made by humans? Are they organisms or inventions? The special characteristics of these biological things have led to misinterpretations and mismatches that have skewed the balance of IP. Our existing patent laws don't seem quite up to the task of dealing with these hybrids. Getting a patent on a human gene seems quite different from getting a patent on a mousetrap. In the latter case, someone can always build a better mousetrap. But basically we're all stuck with the same genes—no one can build a better human gene (at least not yet).

This suggests that an alternative model for rewarding innovation in biomedicine is needed. One option advocated by some biomedical scientists is to reward significant biomedical innovations with large cash prizes, rather than patents. This is modeled on the X-Prize. Governments or nongovernment organizations could offer these prizes for particular breakthroughs (such as discovering a vaccine for HIV). The IP associated with the discovery would then be placed in the public domain and made available for everyone to use, either for treating diseases or further research. Innovation would continue to be rewarded, but far more of the benefits of a discovery or invention would flow immediately to the public.

Another possibility involves the modification of patenting rules according to the idea of "acceptable intellectual property." Acceptable IP, as proposed by the sociologist Stephen Hilgartner, would recognize that IP not only is about encouraging innovation for private profit, but also must serve the public good. It would weigh the benefits of biotechnology against the potential social costs. In other words, it would involve a more holistic approach to assessing patentability, in a way that is more similar to the kinds of considerations required under section 53(a) of the European Patent Convention. Acceptable IP would provide a way to open up debates over IP beyond the narrow technical, legal, and economic terms in which they are usually framed. This would allow the effect of patents to be considered in a broader social and political context, weighing, for instance, the effects of a patent on health care costs or the environment or social justice.

FURTHER READING

The landmark legal decisions referred to here include the following: Ex parte Latimer, 1889 Dec. Com. Pat. 123 (patenting a fiber derived from pine needles); Diamond v. Chakrabarty, 447 U.S. 303 (patenting of a genetically engineered bacteria); Ex parte Allen 2USPQ 2d (patenting a modified oyster); Harvard College v. Canada (Commissioner of Patents) [2002] 4 S.C.R. 45, 2002 SCR 76 (Oncomouse); European Patent Office Board of Appeals decision T315/03 (July 6, 2004) (Oncomouse).

There is an extensive legal, sociological, and historical literature on the patenting of organisms. Only the most useful of these are listed here. On the Chakrabarty decision see Daniel J. Kevles, "Ananda Chakrabarty Wins a Patent: Biotechnology, Law, and Society, 1972–1980," *Historical Studies in the Physical and Biological Sciences* 25, no. 1 (1994): 111–135; and Daniel J. Kevles, "Diamond v. Chakrabarty and Beyond: The Political Economy of Patenting Life," in *Private Science: Biotechnology and the Rise of the Molecular Sciences*, ed. Arnold Thackray (Philadelphia: University of Pennsylvania Press, 1998). On *Ex parte Allen* and the patenting of his oyster see Michael W. Fincham, "Publish or Patent: An Oyster Makes Legal History," *Chesapeake Quarterly* 9, no. 2 (2010), http://www.mdsg.umd.edu/CQ/V09N2/online1/. For details on the patent process for recombinant DNA see Sally Smith Hughes, "Making Dollars out of DNA: The First Major Patent in Biotechnology and the Commercialization of Molecular Biology, 1974–80," *Isis* 92 (2001): 541–575. On Oncomouse see Daniel J. Kevles, "Of Mice and Money: The Story of the World's First Animal Patent," *Daedalus* 131, no. 2 (2002): 78–88; Donna J. Haraway, *Modest_Witness@Second_Millenium. FemaleMan©_Meets_Oncomouse^TM: Feminism and Technoscience* (New York: Routledge, 1996); and Fiona Murray, "The Oncomouse That Roared: Hybrid Exchange Strategies as a Source of Distinction at the Boundary of Overlapping Institutions," *American Journal of Sociology* 116, no. 2 (2010): 341–388. For a shorter overview of patenting life forms see Daniel J. Kevles, "Patents, Protections, and Privileges: The Establishment of Intellectual Property in Animals and Plants," *Isis* 98, no. 2 (2007): 323–331.

On the early controversies over human gene patents see Leslie Roberts, "Genome Patent Fight Erupts," *Science* 254, no. 5029 (1991): 184–186; and Robert Cook-Deegan, "Gene Patents," in *From Birth to Death and Bench to Clinic: The Hastings Center Bioethics Briefing Book for Journalists, Policymakers, and Campaigns*, ed. Mary Crowley (Garrison, NY: Hastings Center, 2008), 69–72, http://www.thehastingscenter.org/publications/briefingbook/detail.aspx?id=2174. Arguments for and against gene patenting are presented in Michael A. Heller and Rebecca S. Eisenberg, "Can Patents Deter Innovation? The Anticommons in Biomedical Research," *Science* 280, no. 5364 (1998): 698–701; and Michael Crichton, "Patenting Life," *New York Times*, February 13, 2007. The Human Genome Project also produced a significant literature on the pros and cons of gene patenting. Representative works include Rebecca S. Eisenberg and Robert P. Merges, "Opinion Letter as to the Patentability of Certain Inventions Associated with the Identification of Partial cDNA Sequences," *AIPLA Quarterly Journal* 23, no. 1 (1995): 1–52; Rebecca S. Eisenberg, "Genomics in the Public Domain: Strategy and Policy," *Nature Reviews Genetics* 1 (2000): 70–74; and Misha Angrist and Robert M. Cook-Deegan, "Who Owns the Genome?" *New Atlantis* 11 (Winter 2006): 87–96. On the relationship between human gene patenting and genomics companies in the 1990s and 2000s see Paul Martin, Michael Hopkins, Paul Nightingale, and Alison Kraft, "On a Critical Path: Genomics, the Crisis of Pharmaceutical Productivity, and the Search for Sustainability," in *Handbook for Genetics and Society: Mapping the New Genomic Era*, ed. Paul Atkinson, Peter Glasner, and Margaret Lock (London: Routledge, 2009), 145–162.

For the early history of Myriad and its work on BRCA1 and BRCA2 see Shobita Parthasarathy, *Building Genetic Medicine: Breast Cancer, Technology, and the Comparative Politics of Health Care* (Cambridge, MA: MIT Press, 2007). The Myriad Genetics case can be found as Association for Molecular Pathology v. Myriad Genetics 569 U.S. 12–398 (2013). The media coverage of the case includes Jim Dwyer, "In Patent Fight, Nature 1; Company 0," *New York Times*, March 30, 2010; Andrew Pollack, "Despite Gene Patent Victory, Myriad Genetics Faces Challenges," *New York Times*, August 24, 2011; Reuters, "Court Reaffirms Right of Myriad Genetics to Patent Genes," *New York Times*, August 16, 2012; and Adam Liptak, "Justices, 9–0, Bar Patenting Human Genes," *New York Times*, June 13, 2013. Others are A. K. Rai and R. Cook-Deegan, "Moving Beyond 'Isolated' Gene Patents," *Science* 341, no. 6142 (2013): 137–138; R. Cook-Deegan, "Robert Cook-Deegan's Viewers' Guide to the Superbowl of Gene Patent Cases," *Cancer Letter* 40, no. 11 (2014): 10–14; and J. M. Conley, R. Cook-Deegan, and G. Lázaro-Muñoz, "Myriad after Myriad: The Proprietary Data Dilemma," *North Carolina Journal of Law and Technology* 15, no. 4 (2014): 597–638. Analysis of the final result from various social science perspectives can be found at http://ipbio.org/patenting -life-genes-generations.htm and http://americanscience.blogspot.sg/2013/03 /the-ontology-of-patent-law-part-i.html.

For broader reflections on notions of property in the age of information see James Boyle, *Shamans, Software, and Spleens: Law and the Construction of the Information Society* (Cambridge, MA: Harvard University Press, 2001). For the idea of prizes for biomedical innovation see James Love and Tim Hubbard, "Prizes for Innovation of New Medicines and Vaccines," *Annals of Health Law* 18, no. 2 (2009): 155–186. For the concept of "acceptable intellectual property" see Stephen Hilgartner, "Acceptable Intellectual Property," *Journal of Molecular Biology* 319 (2002): 943–946.

PART IV

GENETICALLY
MODIFIED FOODS

INTRODUCTION

In chapter 2, we saw how humans have been manipulating plants at least since the beginnings of settled agriculture. In order to make larger, faster growing, more nutritious, and tastier foods, humans have developed ever more sophisticated ways of selectively breeding fruits, tubers, and grasses. In the early twentieth century, the remarkable creations of Luther Burbank showed how powerful these methods could be in reshaping and recombining living matter. Whether or not we want to label these efforts *biotechnology*, we must at least acknowledge that these schemes had aims very similar to our more modern manipulations of food plants.

This chapter and the next discuss the history, politics, and economics of genetically modified foods (GMFs). Chapter 8 focuses especially on the economic structures surrounding and supporting GMFs. This chapter does two things. First, it explores the broader historical context within which GMFs emerged. This context includes not only the long history of human attempts to modify and control plants, but also the debates about recombinant DNA in the 1970s (discussed in chapter 4) and the Green Revolution during the Cold War. It is hard to understand the form and the vehemence of debates about GMFs without understanding this background.

Second, this chapter examines the *regulation* of biotechnologies. Much of the controversy around GMFs has centered on how they should be regulated. But these debates also raise a set of more general questions about biotech: what risks do biotechnologies pose? How should those risks be determined and by whom? What ethical frameworks or standards should we use to decide whether a technology is "safe"? And, ultimately, what rules should be put in place to make sure biotechnologies do not jeopardize human health or the environment? These are complicated questions on which scientists, policy-makers, and ethicists have had much to say. This chapter does not provide a set of definitive answers. Rather, it aims to analyze these debates and show why these problems remain extremely difficult to resolve.

GREEN REVOLUTION

Before the development of recombinant biotechnology in the 1970s, there were several large-scale attempts to apply science and technology for the improvement of agriculture. In the Soviet Union in the 1930s, for instance, Joseph Stalin attempted to increase agricultural output through collectivization, mechanization, and increased use of fertilizers. In addition, Stalin supported the theories of Trofim Lysenko (1898–1976), who claimed that treating seeds with moisture and cold could induce winter plants to grow in the springtime. His processes of "vernalization," he claimed, even allowed plants to pass on their newly acquired traits to their offspring. Soviet propaganda continued to celebrate miraculous successes for Lysenko's techniques into the 1960s, even though his methods caused widespread crop failure and starvation.

Americans too believed that the application of modern science and technology to agriculture was the key to economic development and political stability. The Soviet Union successfully tested an atomic bomb in August 1949. From this point, the Cold War tension between the Soviets and the United States could quickly escalate into nuclear war. Rather than start an all-out, mutually annihilating exchange, both sides turned their attention to the developing world. For the Soviets, they could show their strength by converting Asian, African, and Latin American nations to communism. For the United States, on the other hand, it was imperative to slow or stop the spread of revolution.

At least initially, the advantage lay with the Soviets. Asia, Africa, and Latin America contained large populations of poor farmers for whom Marxist revolution (which would likely institute land reforms) held great appeal. Communism promised to create an equal society in which the ordinary workers (including farmers) would become empowered and better off. Following the arguments of W. S. Gaud at the US Agency for International Development, many in the West believed that stopping the spread of communism meant making farmers richer. While farmers could barely grow enough to feed themselves, they would opt for communism. But if their plots of land could be made more productive, they would remain content and continue to support a capitalist system. A Green Revolution could forestall a Red Revolution.

This was the theory behind the development and deployment of agricultural technologies by the United States in the developing world. The first experiments took place in Mexico. In 1943, the Rockefeller Foundation set up the Cooperative Wheat Research Production Program. The aim was to work with the Mexican government and farmers to increase the yield of wheat. The

Rockefeller Foundation recruited a number of American plant scientists to Mexico, including the agricultural microbiologist Norman Borlaug (1914–2009). Over the next ten years, Borlaug and his colleagues tested more than six thousand different crosses between different varieties of wheat. Through careful experiments and backcrosses they produced varieties that were far more resistant to disease. Modern fertilizers caused wheat varieties to produce so much grain that they became top heavy and fell over in the field. Borlaug solved this problem by crossing Mexican strains with a Japanese dwarf wheat (called Norin-10) to create thick-stemmed plants that would not collapse.

Overall, the Rockefeller program in Mexico was a remarkable success. By 1963, the country produced six times more wheat than it had in 1944 and was able not only to provide for its own wheat needs but also to export its grain.

Emboldened by these achievements, the United States attempted to bring the Green Revolution to Asia, a crucial battleground of the Cold War. Here, the key would be rice. In 1960, the Ford Foundation and the Rockefeller Foundation funded the creation of the International Rice Research Institute (IRRI) in the Philippines. The aim was to transform the production of rice in the same way that Borlaug had transformed the production of wheat. Research focused on breeding a variety of rice that would be disease resistant and able to produce substantially higher yields (with the assistance of fertilizers). By 1962, IRRI had begun testing a cross between disease-resistant rice and a Taiwanese dwarf variety called *Dee-geo-woo-gen*. The resulting variety, IR-8, eventually proved very successful in increasing rice yields in the Philippines.

But the IRRI aimed to transform rice production beyond the Philippines too. In the mid-1960s, India experienced severe droughts leading to crop failures and impending famine. Green Revolution varieties of wheat were imported from Mexico to Punjab, and later IR-8 rice was successfully introduced into southern India. By 1975, Green Revolution varieties accounted for over a quarter of all rice and almost three quarters of all wheat planted in South Asia. By 1980, 40% of rice cultivated in East Asia, Southeast Asia, and the Pacific belonged to the new varieties.[1] As in Mexico, the Green Revolution increased yields in India sufficiently to transform it from a net importer to a net exporter of food crops. High-yield varieties—combined with new methods of irrigation, improved rural infrastructure, and increased use of

---

1. D. Gollin, M. Morris, and D. Byerlee, "Technology Adoption in Intensive Post-Green Revolution Systems," *American Journal of Agricultural Economics* 87, no. 5 (2005): 1310–1316.

fertilizers, insecticides, and pesticides—significantly reduced the severity of food shortages in India after 1970.

Despite the success of the Green Revolution in increasing crop yields in Latin America and Asia, it also had some negative side effects. Some historians and environmentalists have argued that, in the long term, the Green Revolution has done more harm than good. First, critics point out that increasing the food supply will never solve the world's food problems. Without population control, more food will lead to increased population growth, which will ultimately lead to further food shortages. On the basis of this view, agricultural technology will never be able to keep up with population growth. This view is often called the *Malthusian criticism* (after the English clergyman Thomas Malthus [1766–1834] who first pointed out the problems raised by fast-growing populations and slowly increasing food supply).

Second, in many places the Green Revolution disrupted traditional planting and harvesting cycles. Crop rotation was replaced with monocultures (just growing a single crop) since high-yield varieties could be harvested more than once a year. This combined with the effects of fertilizers, herbicides, and insecticides to substantially reduce biodiversity in areas under cultivation. The diet of some farmers was negatively affected. Crop rotation had provided food variety, now gone, and some animals (e.g., frogs) that inhabited fields and paddies and provided supplementary sources of protein were killed off by pesticides. In some cases, the result was malnutrition for farmers.

Third, the Green Revolution had significant economic costs. Farmers needed to purchase seeds, fertilizers, and pesticides, as well as machines such as tractors and combine harvesters. Those who could not afford new technologies either had to take loans or be forced off the land by increased competition. In India and elsewhere the Green Revolution has contributed to urbanization and an increasing gap between rich and poor in the countryside. Moreover, some critics have pointed out that substantial fractions of this money have flowed from the developing world to Western companies that manufacture machines and chemicals.

Finally, the Green Revolution has had a substantial environmental impact. Increased irrigation has dried up rivers and increased soil salinity. Fertilizers have increased nitrogen levels in waterways, encouraging algal growth that depletes oxygen and kills other aquatic life. The long-term use of pesticides also has negative effects on human health (including links to various forms of cancer).

The legacy of the Green Revolution will continue to be controversial. But both its successes and its failures have created the backdrop against which debates about genetically modified crops have developed. Many of the worries about GMFs are closely related to concerns about the massive transformation and industrialization of agriculture in the twentieth century.

WHAT ARE GMFS?

The creation of new plant varieties in the 1950s and '60s was a high-technology enterprise. Apart from the hybridization methods such as back-crossing and line breeding, plant scientists engaged a variety of techniques for rapidly producing beneficial genetic mutations and chromosomal changes. This included the use of chemical mutagens such as ethyl methanesulfonate and dimethyl sulfate as well as chemicals such as colchicine for doubling the number of chromosomes in a plant cell. Plants were also subjected to various forms of radiation including x-rays, ultraviolet light, gamma rays, and cosmic rays (seeds were taken into space to expose them to the highest energy radiation). Irradiation techniques were pioneered by Lewis Stadler (1896–1954) at the University of Missouri in the 1920s. After World War II, the US Atomic Energy Commission developed radiation breeding techniques, touting them as a "peaceful" use for atomic energy.

Exposing plants and seeds to radiation produced many useless mutations, some of them lethal to the plant. The radiation more or less randomly mixed up the genetic material, so the outcome could not be predicted or controlled. But the idea was that amongst all the mutations, a few plants would gain beneficial traits. These individuals could then be crossed with standard varieties to produce new and improved foods. In the early 1970s, gamma ray breeding experiments at the University of California Davis produced a semi-dwarf high-yield variety of rice. This became the variety called Calrose 76 — about half the rice currently grown in California derives from it. The Ruby Red Grapefruit also derives from radiation. The original red grapefruit was a natural mutant but biologists irradiated it to produce even stronger colors: Star Ruby in 1971 and Rio Red in 1985. These artificial mutants constitute a significant fraction of all the red grapefruits now cultivated.

These examples show that the genetic modification of food plants began well before the invention of recombinant techniques in the 1970s. However, recombinant techniques improved biologists' ability to alter plant genes in two important ways. First, they could be applied with more specificity. Rather than just blasting seeds with radiation or chemicals and hoping for beneficial mutations, plant molecular biologists could pick out specific genes for cutting and pasting from plant to plant. This is not quite as straightforward as it seems, since specific genes (for disease resistance, for example) would first have to be identified within the plant genome before recombination could take place. Discovering the function of genes, or identifying which gene (or genes) is responsible for a particular function, remains a difficult task. Never-theless, compared to mutation techniques, molecular recombination made genetic manipulation of plants a far more controllable and predictable prac-tice.

Second, recombinant techniques allowed genes to be cut and pasted *be-tween* different species of plants (or, even between plants and animals, bac-teria, or fungi). Moving genes between species is known as *transgenic* modifi-cation (as opposed to *cisgenic*, where genes are rearranged within a particular species). It is transgenic recombination that has made GMFs particularly controversial. Older mutation techniques could usually hope only to re-arrange the existing genes within a single plant species, occasionally intro-ducing a beneficial mutation. These mutated individuals could be hybridized with other plants in the same species to produce a new variety.[2] Recombi-nant techniques allowed molecular biologists, in theory at least, to take a gene from a frog and activate it in rice. Even more dramatically, it meant that mo-lecular biologists could create their own genes (in practice often by slightly modifying ones found in plants, animals, and bacteria) and insert those into a plant. In general terms, this is possible because the structure of the genetic material (DNA) is the same across all organisms.[3]

A BRIEF HISTORY OF GMFS

One of the first applications of genetic engineering to food was the development of recombinant bovine somatotropin (rBST). Bovine somato-tropin is a growth hormone found in cows. The agricultural technology com-pany Monsanto had been trying to use BST, harvested from slaughtered cows, since the 1960s. When the hormone was injected into dairy cows it produced significant increases in milk production. However, getting BST from dead cows proved too expensive for the scheme to be effective. In 1977, Genentech isolated the gene for BST and in 1979 the biotech start-up struck a deal with Monsanto to license and develop its work into a synthetic version of BST.

2. Under some special conditions, it is possible to hybridize plants from different species. But this results in a wholesale combining of two genomes to form a new species, rather than a modification of one species. An example of this is the crossing of wheat and rye to make a crop called triticale.

3. In fact, this is not quite true. There are important differences between the way DNA works in different kinds of organisms. However, the similarities are enough that molecu-lar biologists have discovered ways of making some genes work in very different kinds of organisms.

Over the next decade the chemical company spent $300 million developing and testing rBST. This provided the first major GMF battleground: Monsanto insisted that their product worked on cows exactly the same way as natural BST, while the company's opponents argued that the artificial substance could pose health and environmental dangers. After a prolonged review, the FDA approved Monsanto's application for rBST use in 1993 and the company began marketing the product the following year.

In the 1980s, companies also began to experiment with techniques for developing new varieties of crops using genetic engineering. At first some of the methods were quite crude: using "gene guns" (small particles covered with DNA and shot into plant cells) and viruses to transfer genes. Later, plant biologists brought the full power of recombinant DNA and other molecular techniques to bear on plant seeds.

The first commercially grown genetically engineered food granted a license for human consumption was a tomato. The Californian company CalGene developed the tomato—called the *Flavr Savr*—that they hoped would have a longer shelf life than conventional varieties. In tomatoes, an enzyme called polygalacturonase degrades the cells walls of the fruit and causes it to quickly soften. CalGene inserted a gene that blocked the production of polygalacturonase in *Flavr Savr* tomatoes. Most commercially produced tomatoes are picked while still green and artificially ripened using ethylene gas. CalGene's plan was that Flavr Savrs could be ripened on the vine. This would improve the color and flavor without reducing shelf life.

The Flavr Savr tomato did not live up to expectations. CalGene's application to the US Food and Drug Administration (FDA) was approved in 1992 and the first Flavr Savr tomatoes became available in supermarkets in 1994. But the first varieties of Flavr Savrs did not produce as high a yield as conventional tomatoes. And although the extra gene seemed to extend the shelf life of the tomato, it did not make the tomatoes firmer. Farmers had to invest in special equipment and expensive shipping crates in order to protect the Flavr Savrs from damage. All this made the first GMFs unpopular with growers and by 1997 the Flavr Savr was no longer being cultivated. CalGene never became profitable and it was acquired by Monsanto in 1997.

Since the mid-1990s, Monsanto has been the major player in GMF development and deployment. Between 1995 and 2000, the FDA and the US Environmental Protection Agency (EPA) gave the company permission to market genetically modified cotton, potatoes, soybeans, maize, and canola. This first generation of these GMFs utilized two kinds of genes. First, the potato, cotton, and maize were modified by the addition of a gene from the soil-dwelling bacterium *Bacillus thuringiensis* (Bt). Bt produces a toxin that kills insects. By splicing the toxin-producing gene from the bacteria into the plant, Monsanto engineered crops that were more resistant to insects and re-quired less pesticide during cultivation. In the other modification, soybeans, canola, and maize were given a gene that made them resistant to the herbi-cide *glyphosate*. This chemical was discovered by chemists at Monsanto in 1970 and is the active ingredient in its popular broad-spectrum herbicide known as Roundup. Again, Monsanto found a micro-organism resistant to glyphosate and transfected the resistance gene into the crops. This innova-tion allowed farmers to use Roundup on their fields without worrying about killing off their crops. Moreover, since this reduced the need for mechanical weeding, rows of crops could be planted closer together and the overall yield increased. Monsanto called their crops "Roundup Ready." Monsanto has also developed methods of engineering varieties with both Bt and herbicide resis-tance genes—this is known as "stacking" traits.

Since 2000, Monsanto and other companies (including Syngenta, BASF, Asgrow, DuPont, and DeKalb Genetics) have developed genetically modified squash, papaya, alfalfa, and sugarbeet (for a more complete list see box 7.1). Most of the modifications have been targeted at improving cultivation (in-sect resistance, herbicide resistance, disease resistance, drought resistance) rather than directed at improving factors that might influence consumer choice (such as flavor, color, or nutritive value). The notable exception to this is Golden Rice. This variety of rice was developed during the 1990s at the Swiss Federal Institute of Technology. The aim was to create a rice plant that synthesized beta-carotene in the grains. Beta-carotene is a precursor to vita-min A and is responsible for the yellow-orange color of fruits and vegetables such as carrots. A rice enriched with beta-carotene could have significant potential to reduce the incidence of vitamin A deficiency in developing coun-tries. Development of Golden Rice has now been taken up by the IRRI, but in 2014 it is still not available for human consumption (partly due to staunch opposition by anti-GMF activists).

The genetic modification of foods extends beyond the plant world. In addition to the use of rBST in cows, in the 1990s AquaBounty Technolo-gies produced a genetically modified Atlantic salmon. Genes from different salmon species were added to the fish genome to allow it to grow throughout the year and reach market size sooner. In 2015, the AquaAdvantage salmon was approved for use by the FDA.

In 2013, there were twenty-seven countries cultivating a total of 175 million hectares of genetically modified crops (an area about one and a half times

Box 7.1 Some Genetically Modified Foods Approved for Use in United States

- *Roundup Ready Soybean* (Monsanto, 1995): Genes from petunia, soybean, bacteria, and viruses to resist glyphosate herbicide
- *Freedom II Squash* (Seminis Vegetable Seed, 1995): Genes from bacteria and viruses to resist watermelon mosaic 2 virus and zucchini yellow mosaic virus
- *Laurical Canola* (Monsanto, 1995): Genes from California bay, turnip rape, bacteria, and viruses to produce high lauric acid oil for soap and food products
- *Knock Out Corn* (Syngenta, 1995): Corn, bacteria, and virus genes produce Bt toxin to control insect pests (European corn borer)
- *Endless summer tomato* (DNA Plant Technology, 1995): Genes from tomato, bacteria, and viruses to alter ripening
- *SeedLink Chicory* (Bejo Zaden, 1997): Genes from bacteria to generate male sterility and facilitate hybridization
- *Sunup / Rainbow Papaya* (Cornell University / University of Hawaii, 1997): Genes from bacteria and viruses to resist papaya ringspot virus
- *New Leaf Plus Potato* (Monsanto, 1998): Genes from bacteria and viruses to control Colorado potato beetle (Bt toxin) and resist potato leafroll virus
- *Sugarbeet* (Monsanto / Syngenta, 1999): Genes from bacteria and viruses to resist glyphosate herbicide
- *CDC Triffid Flax* (University of Saskatchewan, 1999): Genes from *Arabidopsis* and bacteria to resist sulfonylurea herbicide and grow in soils with herbicide residue
- *Roundup Ready Canola* (Monsanto, 1999): Genes from bacteria, viruses, and *Arabidopsis* to resist glyphosate herbicide
- *LibertyLink Canola* (Bayer, 2000): Genes from bacteria and viruses to resist glufosinate herbicide
- *Bollgard II cotton* (Mycogen/Dow, 2002): Genes from bacteria and viruses to control insects and resist glufosinate herbicide
- *LibertyLink Rice* (Bayer, 2004): Genes from bacteria and viruses to resist glufosinate herbicide
- *Roundup Ready Alfalfa* (Monsanto, 2005): Genes from bacteria and viruses to resist glyphosate herbicide
- *Mavera High Value Corn with Lysine* (Renessen, Monsanto, Cargill, 2006): Genes from bacteria to produce higher levels of lysine in corn for animal feed

Note: This is an incomplete list. For more, see the source.
Source: http://www.isaaa.org/gmapprovaldatabase/default.asp

the size of China). About 40% of this land area was in the United States, with significant use also in Brazil, Argentina, India, Canada, and China.[4] By land area, soybeans, maize, cotton, and canola make up 99% of all GMFs. In the US, the vast majority of maize (about 90%), cotton (90%), and soybeans (93%) planted each year are genetically modified varieties.[5] These figures lead to an estimate that over 70% of all processed foods in US supermarkets contain some genetically modified ingredients, although this is a highly contested and controversial figure.[6]

CRITICISM OF GMFS

From the outset, the genetic modification of food crops has provoked significant popular opposition and public fear. Criticism of GMFs has intersected with environmental activism, consumer activism, public health activism, and anti-globalization protests. As such, opposition to GMFs is a complex phenomenon that has roots in critiques of the Green Revolution, the environmental and consumer movements of the 1960s, the recombinant DNA debates of the 1970s (see chapter 4), and more recent concerns about unchecked power of multinational corporations.

The main arguments against GMFs fall into four categories. First, critics have raised concerns about the possible impact of GMFs on human health. Since GMFs could be considered *new* kinds of foods, the body might metabolize them differently from non-GMFs. This could produce toxic or carcinogenic effects. Alternatively, the introduced genes could provoke harmful allergic reactions in some individuals. Even if the foods are not actively harmful, it may be the case that genetic modification somehow makes the foods less nutritional, adversely affecting human diets (although scientific testing has not produced any evidence of such problems). These concerns are exacerbated by the fact that currently available GMFs are designed to benefit

4. Clive James, *Brief 46: Global Status of Commercialized Biotech/GM Crops* (Ithaca, NY: International Service for the Acquisition of Agri-biotech Applications, 2013), http://www.isaaa.org/resources/publications/briefs/46/.

5. This is derived from USDA, Economic Research Service 2013 data sets available at http://www.ers.usda.gov/data-products/adoption-of-genetically-engineered-crops-in-the-us.

6. The 70% figure was reported by Reuters in 2013. See Richard Schiffman, "Food Fight: Vote on GMOs Could Alter US Food System," *Reuters*, November 1, 2013, http://blogs.reuters.com/great-debate/2013/11/01/food-fight-vote-on-gmos-could-alter-u-s-food-system/.

the consumer only indirectly (or not at all): the primary benefit accrues to the farmer (providing increased yields through increased resistance to pests) and agricultural biotechnology companies (through sales of GM seeds).

Second, GMFs have provoked a wide range of arguments that focus on their impact on the environment. GM plants could be especially invasive or persistent, outcompeting other species and altering ecosystems. Bt GMF varieties designed to kill insect pests could prove harmful to other forms of wildlife. Or they could lead to the emergence of destructive "superpests" and "superweeds" resistant to insecticides and herbicides. There is also concern that genes introduced into GMFs could spread to other non-GMF plants, to weeds, to soil microbes, or to viruses (this process is known as *horizontal gene transfer*). GM crops have also begun to dramatically change farming methods. What consequences will this have for local environments? Certainly, these are not all bad: GMOs can reduce the amounts of pesticides and herbicides that farmers need to use. However, both insects and weeds may evolve in response to the use of GMOs and herbicides, producing resistant strains that once again require increasing amounts of chemicals to control.

Third, GMFs are criticized on religious and moral grounds. Using recombinant techniques to create new species is likened to playing God: it is argued that genetic modification is a great extension of human powers and it is likely to have unintended negative consequences. For some, GMFs are a technical solution to the moral problem of providing enough food for all of the world's people. Others point out that the world already has enough food and that famines result from inequitable distribution, rather than overall shortage. Given that, an ethical argument might dictate that money spent on GMF technology might better be spent in more practical ways.

Fourth, GMFs have significant implications for the economic organization of agriculture and the economics of the food supply. These economic issues, perhaps because of their complexity, are often sidelined in discussions of GMFs. However, they are crucial for understanding how and why GMFs are being developed and how they are likely to affect society. For this reason, these issues receive special treatment in chapter 8.

RISK AND REGULATION

The last four paragraphs have given only the scantest overview of the arguments against GMFs. Public concerns about GMFs have made them the subject of hundreds of tests, field trials, and studies as well as thousands of pages of reports by governments, companies, nongovernmental organizations (NGOs), and activists in attempts to assess their safety. Few of these tests have produced definitive evidence of any danger of GMFs to human health; the scientific consensus endorses the safety of GMFs. Yet, the public continues to fear genetic modification of foods.

In all these discussions, the central concept that has emerged is that of *risk*. Many of our everyday activities in modern society carry risk. When we drive a car we run the risk of having an accident that could be detrimental to our health and detrimental to our financial situation. Driving also contributes to the risk of polluting the atmosphere with greenhouse gases and exacerbating global warming. At least some of these risks (such as the risk of having an accident) can be fairly accurately measured and effectively hedged (for instance, by buying health or automobile insurance). Other risks (such as that of global warming) are harder to identify, more complex, and harder to measure. The sociologists Ulrich Beck and Anthony Giddens have argued that assessing risk (and figuring out how to appropriately distribute it) has become one of the defining features of modern societies. In particular, the management of risk has become a central role of governments.

Risk talk has certainly dominated the discussion of GMFs. What are the *risks* of planting a particular crop? What is the *risk* that GM corn will increase the incidence of cancer? What is the *risk* that genes from GM cotton will spread into weeds? Is it a 50% chance or 1%? The key assumption is that in order to decide whether or not to use GMFs, it is necessary to *know the risks* of particular actions. Regulation and governing GMFs is usually understood to depend on *identifying and measuring risk*.

In the European Union, for instance, GMFs can be grown or imported only after extensive evaluation by the European Food Safety Authority (EFSA). The EFSA has in-house scientific staff that conducts an environmental risk assessment to determine whether a permit for a particular food should be issued.[7] This process has four parts (as shown in table 7.1): risk identification (What could go wrong? How could harm occur?), consequence assessment (How serious could the harm be?), likelihood assessment (How likely is harm to occur?), and risk estimation (What is the level of risk?). The risk estimate is derived from a table that compares the consequence assessment and the likelihood assessment (table 7.2). Outcomes that are highly likely and have major consequences are high risk, while conversely outcomes that are unlikely and marginal in impact have "negligible" risk.

Although this framework is particular to Europe, it is fairly typical of how

7. In practice, however, the European Union has maintained a moratorium on approving new GMFs since 2001.

TABLE 7.1 RISK ASSESSMENT FRAMEWORK

What could go wrong?
How could harm occur?
Risk identification
(Risk scenarios)

How serious could the harm be?	How likely is the harm to occur?
Consequence assessment	Likelihood assessment

What is the level of risk?
Risk estimation

Source: European Food Safety Authority.

TABLE 7.2 RISK ESTIMATION

LIKELIHOOD ASSESSMENT		RISK ESTIMATE			
	Highly likely	*Low*	*Moderate*	*High*	*High*
	Likely	*Low*	*Low*	*Moderate*	*High*
	Unlikely	*Negligible*	*Low*	*Moderate*	*Moderate*
	Highly unlikely	*Negligible*	*Negligible*	*Low*	*Moderate*
		Marginal	Minor	Intermediate	Major
		CONSEQUENCE ASSESSMENT			

Source: European Food Safety Authority.

risk assessments for GMFs (and other technologies) are carried out.[8] But such evaluations have several problems. How does the EFSA decide which risks to measure? The idea is to consider a range of "what if" scenarios. In practice, only those scenarios considered "potentially substantive" and linked to recombinant technologies by "observable pathways" are actually evaluated. But

---

8. Amongst developed nations, the United States is the exception. Responsibility for testing and regulation is divided between the Office of Science and Technology Policy in the US Department of Agriculture (via the Animal and Plant Health Inspection Service that regulates transgenic plants), the FDA (which regulates products produced by biotechnology, including foods and drugs), and the EPA (which regulates pesticidal plants and genetically engineered microbial pesticides). These agencies have worked cooperatively with industry to promote the introduction of GMFs and to simplify regulations. One major difference with other jurisdictions is the fact that in the United States biotechnology companies are responsible for conducting *their own* testing.

how is it possible to know whether a risk is substantive without conducting an assessment first? And how is it possible to know whether a pathway exists without an investigation? In other words, the environmental risk assessment runs into the problem that it is impossible to anticipate all the possible ways in which things could go wrong.

The environmental risk assessment is based on scientific methods. Evaluations use experimental data from lab work and field studies, statistics and probability, inference from models, and extrapolation from historical data. While all these methods are important, it is possible that they might *limit* the kinds of consequences or outcomes that can be included in the risk assessment. Possibilities that are not easily measured, not easily quantified, or not easily modeled may be omitted from evaluations. Before the global financial crisis of 2008, financial institutions had developed sophisticated mathematical methods for measuring the risks that homeowners would default on their mortgage payments. When confronted with complex feedback from social, political, and economic events, some of these measures failed badly, contributing to the rapid downward spiral of the worldwide economy.

GMFs, like global finance, are complex sociotechnical systems. The summary of the arguments against GMFs in the previous section suggests some of the overlapping and interwoven levels on which they act. Of course, plants themselves are complex organisms: biologists still have much to learn about how they work. Our own bodies are also complex in the way that they metabolize food and respond to allergens. Ecosystems made up of interconnected webs of plants and animals are more complex still. On top of all this, GMFs are embedded in complex social, legal, political, and economic circumstances that may also cause them to be used in unanticipated ways. All this complexity means that many of the consequences of GMFs are highly uncertain. Risk assessment—identifying and quantifying risk—has been the preferred way to manage this uncertainty. But risk assessment cannot eliminate uncertainty—it cannot predict and account for all possible outcomes of using GMFs. The possibility of "unknown unknowns" certainly does not mean we should abandon GMFs, but it does imply that we should look for ways to supplement risk assessment frameworks with other ways of understanding what the wider consequences of GMFs may be.

BEYOND RISK

Many scientists and advocates for GMFs remain frustrated with the continuing debates in the media and ongoing public fears. They point to the fact that the thousands of studies of GMFs have found no conclusive evidence of harm to human health. And they argue that GMFs have remarkable potential for feeding growing populations, increasing food security, and reducing pesticide use. Many have become increasingly dismissive of any concerns about GMFs, arguing that the public (or the media) is merely ignorant of the facts or just "irrational." One group of US scientists wrote in the journal *Science* that the public needs to "get beyond popular biases" when it comes to GMFs.[9]

But such views take for granted the possibility that the methods of testing and regulation work effectively. However, there may be a range of political, economic, and technical reasons why no studies have yet shown any harm from GMFs—requirements may not be stringent enough, timeframes may not be long enough, and so on. And of course, more testing costs more money and it's not in the interests of corporations to prove their own products are harmful. So worries about "popular biases" should be balanced by worries about political and economic biases in the system of testing and regulation.

The limitations of risk assessment do not mean that the correct course of action is simply to abandon GMFs. Nor do they mean that science and risk assessment have nothing to tell us about the safety (or otherwise) of GMFs. Rather, we need to supplement risk assessment with more inclusive ways of making decisions about technologies. Scientific supporters of GMFs often assume that risk analysis is the only valid way of assessing technologies. The sociologist Brian Wynne has criticized the risk analysis framework on the grounds that it places too much trust in science and scientific analysis. In considering the consequences of GMFs using the methods of "predictive control" and "probabilistic quantification," risk analysis assumes that costs and benefits can be measured and denies uncertainty. In framing itself as "rational," it also portrays other approaches (for instance, those based on values or ethics) as "irrational." Risk analysis attempts to reserve the right to make decisions about technology to scientific experts.

Ironically, this actually undermines public trust in science (including public acceptance of GMFs). Scientific hubris is one of the public's main concerns about new technologies. There are plenty of examples of scientifically endorsed "safe" technologies that later turned out to be dangerous (nuclear power plants and thalidomide, for instance). Risk analysis implies that science is in control. It does not seem to acknowledge that there are some things that might be out of control or beyond prediction. Wynne argues that risk analysis "ignores the very issue raised by typical public concerns, however,

9. N. V. Federoff et al., "Radically Rethinking Agriculture for the 21st Century," *Science* 327, no. 5967 (2010): 833–834.

which is whether we can assume that science can reliably identify future consequences, or whether, on the contrary, there are going to be consequences of which current knowledge is ignorant."[10] On this view, the public is not irrational or uninformed about GMFs. Rather, public fears and concerns raise a valid question: is scientifically based risk analysis the best way of making decisions about technology?

Some opponents of GMFs, such as Vandana Shiva, have successfully exploited public fears about science and technology. They have utilized public distrust in science to promote extreme views and create paranoia about *any* attempts to use technologies to improve agriculture. Such demagoguery is unproductive, but it is made possible partly by the way in which the debate over GMFs has been framed as a *scientific* issue: if decisions about GMFs become purely scientific decisions, one of the few ways for nonscientists to enter the conversation is to oppose science itself.

One clash over GMFs has played out around eggplants in India. In 2010, following large protests, the Indian government halted the planting of Monsanto Bt brinjal (eggplant). The main concern amongst the Indian public was that the GM eggplant would outcompete local varieties of eggplants. Regions stood to lose the unique flavors and textures of their local food. Although this was framed by the government as an issue of safety and biodiversity, what was at stake for ordinary people had more to do with *cultures* of food and eating. Such things are difficult to quantify. Where do taste and culture fit in a risk analysis framework? Such issues do not involve only scientific problems and questions.

Risk analysis has serious limitations for making decisions about technology. It does not help us to decide whether or not scientific methods themselves are reliable for assessing future consequences. It ignores the potential limitations and blind spots of scientific knowledge. And, it does not take into account the full range of political, economic, ethical, and cultural issues that should be brought into view when assessing technologies. The debates over GMFs are likely to gain wider public acceptance only after governments can find more expansive and encompassing ways of considering the potential consequences of technologies. This requires finding means of bringing the views of more nonscientists to bear on decisions regarding GMFs and developing methods for including more interdisciplinary perspectives within de-

10. Brian Wynne, "Interpreting Public Concerns about GMOs—Questions of Meaning," in *Reordering Nature: Theology, Society, and the New Genetics*, ed. Robin Grove-White, Bronislaw Szerszynski, and Celia Deane-Drummond (London: Bloomsbury T&T Clark, 2003), 221–248. Quotation p. 237.

cision frameworks. Although such models are still being worked out, policy-makers in various jurisdictions have experimented with more inclusive forms of policy making, such as "consensus conferences," "citizen conferences," and other forms of participatory and deliberative governance. These conferences provide opportunities for small groups of interested citizens to learn more about particular technologies, enter into dialogue with scientific experts, and express their views. Citizen science movements have also provided important opportunities for engaging the public in debates about science and technology. Not all of these frameworks have proved successful. Nevertheless, such modes of policy making will be critical for advancing debates over biotechnologies beyond polarizing science versus anti-science positions.

FURTHER READING

There is a large body of scholarship that deals with the Green Revolution. However, this remains a controversial topic amongst historians, environmentalists, economists, and political scientists. On the relationship between the Cold War and agriculture see Nick Cullather, *The Hungry World: America's Cold War Battle against Poverty in Asia* (Cambridge, MA: Harvard University Press) and John H. Perkins, *Geopolitics and the Green Revolution: Wheat, Genes, and the Cold War* (New York: Oxford University Press, 1997). Norman Borlaug's vision for the Green Revolution is set out in his many writings, including Haldore Hanson, Norman Borlaug, and R. Glenn Anderson, *Wheat in the Third World* (Boulder, CO: Westview Press, 1982). A sense of the controversy aroused by these debates can be found in the decidedly one-sided account of the Green Revolution in India by Vandana Shiva, *The Violence of Green Revolution: Third World Agriculture, Ecology, and Politics* (London: Zed Books, 1992). For accounts of the environmental effects of the Green Revolution see David A. Sonnenfeld, "Mexico's 'Green Revolution' 1940–1980: Toward an Environmental History," *Environmental History Review* 16, no. 4 (1992): 28–52, and Lori Ann Thrupp, *Cultivating Diversity: Agrobiodiversity and Food Security* (Washington, DC: World Resources Institute, 1998).

An excellent account of Lysenko's life and work can be found in David Joravsky, *The Lysenko Affair* (Chicago: University of Chicago Press, 1970). Atomic foods have provoked significant interest amongst both historians and journalists. See John Markoff, "Useful Mutants, Bred with Radiation," *New York Times*, August 28, 2007. Work by historians includes Helen Curry, *Evolution Made to Order: Plant Breeding and Technological Innovation in Twentieth-Century America* (Chicago: University of Chicago Press, 2016), and Paige Johnson, "Safeguarding the Atom: The Nuclear Enthusiasm of Muriel Howorth," *British Journal for the History of Science* 45, no. 4 (Special

Issue: Transnational History of Science): 551–571. An interview with Paige Johnson that covers this topic can be found here: http://pruned.blogspot.com /2011/04/atomic-gardens.html.

The literature on GMFs, like that on the Green Revolution, is vast and tends towards polemical and one-sided accounts. The best place to start is with longer histories of agricultural manipulation of plants that place GMFs in context. These include Jack Ralph Kloppenberg Jr., *First the Seed: The Political Economy of Plant Biotechnology*, 2nd ed. (Madison: University of Wisconsin Press, 2005) and Noel Kingsbury, *Hybrid: The History and Science of Plant Breeding* (Chicago: University of Chicago Press, 2011). On the manipulation of animals for various human purposes see Susan Schrepfer and Philip Scranton, eds., *Industrializing Organisms: Introducing Evolutionary History* (New York: Routledge, 2003).

For a good account of the debates over GMFs see Hugh Gusterson, "Decoding the Debate on Frankenfood," in *Making Threats: Biofears and Environmental Anxieties*, ed. Betsy Hartmann, Banu Subramanian, and Charles Zerner (New York: Ronan and Littlefield, 2005), 109–133. On the development and controversy over rBST see Keith Schneider, "Betting the Farm on Biotech," *New York Times*, June 10, 1990. For the FDA's assessment of the safety of this product see US FDA, "Report on the Food and Drug Administration's Review of the Safety of Recombinant Bovine Somatotropin," http://www.fda .gov/AnimalVeterinary/SafetyHealth/ProductSafetyInformation/ucm130321 .htm. The history of the Flavr Savr tomato is detailed in Belinda Martineau, *First Fruit: The Creation of the Flavr Savr Tomato and the Birth of Biotech Food* (New York: McGraw-Hill, 2001). More anti-corporate accounts of GMFs include Dan Charles, *Lords of the Harvest: Biotech, Big Money, and the Future of Food* (New York: Basic Books, 2002) and Marie-Monique Robin, *The World According to Monsanto: Pollution, Corruption, and the Control of the World's Food Supply* (New York: New Press, 2012). The problems raised by the extremism of anti-GM activist Vandana Shiva are discussed by Michael Specter, "Seeds of Doubt," *New Yorker*, August 25, 2014, 46–57.

Discussions of risk should begin with the notion of the *risk society*, a phrase coined by the sociologist Ulrich Beck, *Risk Society: Towards a New Modernity* (New Delhi: Sage, 1992). The risk society refers to the fact that members of modern societies must live and deal with a variety of risks created by industrial society itself (e.g., nuclear waste and global warming). A similar theme has been developed by Anthony Giddens, *Consequences of Modernity* (Cambridge: Polity Press, 1990). The emergence of risks associated with science and technology has been explored by, amongst others, Sheila Jasanoff, "American Exceptionalism and the Political Acknowledgement of Risk," in *Risk*, ed.

Edward J. Burger (Ann Arbor: University of Michigan Press, 1993). The public perceptions of risks around GM crops have been extensively examined by Brian Wynne, "Risky Delusions: How GM Science Has Imagined—and Provoked—Its Publics," in *Genetically Engineered Crops: Decision-Making under Uncertainty*, ed. Iain Taylor and Katherine Barrett (Vancouver: UBC Press, 2005), and Wynne, "Risky Delusions: Misunderstanding Science and Misperforming Publics in the GE Crops Issue," in *Genetically Engineered Crops: Interim Policies, Uncertain Legislation*, ed. I. E. P. Taylor (Philadelphia: Haworth Press, 2007), and finally Wynne, "Interpreting Public Concerns about GMOs—Questions of Meaning," in *Reordering Nature: Theology, Society, and the New Genetics*, ed. Robin Grove-White and Bronislaw Szerszynski (London: Bloomsbury T&T Clark, 2003). There is a significant literature on the development of deliberative frameworks for GMFs and other risky technologies. For a review of many of these issues see Martin Bauer, ed., "Special Issue: Public Engagement in Science," *Public Understanding of Science* 23, no. 1 (2014).

8   THE ECONOMICS OF EATING

INTRODUCTION

The controversies surrounding genetically modified foods (GMFs) usually emphasize health and environmental issues. "Risky" GMFs are explicitly or implicitly contrasted with other "safer" non–genetically modified varieties. Chapter 7 explored these debates, situating GMFs within a longer history of plant modification that includes the Green Revolution. This chapter puts GMFs in a different context: the transformation and industrialization of American agriculture during the twentieth century. In the end, the kinds of problems that this chapter raises are also related to health and the environment. But it is not genetic modification itself that gives rise to these concerns. Rather, what is at issue is the ways in which GMFs are tied to the production, circulation, and ownership of our food.

Here, once again, we are treating biotechnology as part of a broader sociotechnical system. GMFs are not merely biological or technical objects but are cultural objects (foods) and economic objects (products). They are part of a technical, economic, social, and political system that allows food and food products to be produced, circulated, distributed, and owned in particular ways. These ways have significant consequences for social justice and the distribution of economic power. Understanding GMFs requires understanding this system and how it works.

Perhaps most importantly (and obviously) GMFs have consequences for what we eat. The three transformations that this chapter describes—a transformation in agriculture, a transformation of food production, and a transformation of ownership—have made many foods increasingly affordable in the developed world. Together, these transformations have contributed to a dramatic alteration of our diets, especially increasing our intake of processed foods, including foods high in sugars and fats. This is having a detrimental impact on our health—particularly by increasing the incidence of obesity and type 2 diabetes. These problems, of course, cannot be blamed only on GMFs. Other twentieth-century social transformations (such as declining time available for cooking and the reduced importance of the family meal) are also crucial parts of this story. But, GMFs are now a critical part of an economic infrastructure of factory farming and Big Agriculture and Big Food that promotes and benefits from particular ways of eating. In the long run, the role of GMFs as an *economic innovation* may outweigh their importance as a technological innovation.

TRANSFORMATION I: AGRICULTURE

Agriculture was fundamentally transformed in the twentieth century. The United States began as an agricultural economy. Until at least the middle of the nineteenth century, the mainstays of the American economy were cotton and tobacco (grown mostly in the South) and maize (grown in the Midwest). Even as late as 1900, 41% of the workforce was employed in agriculture and a similar proportion of people actually lived on farms.

Most of these farms were family run. The head of the family owned or rented a small plot in order to grow enough food to feed his or her family. Any excess food could be sold for a profit. Historical and fictional accounts often portray these family-owned farmed in idealistic and nostalgic terms as pastoral idylls. The reality was much harsher: it was hard work, difficult to make a living, and subject to the vagaries of weather, disease, and natural disasters.

Over the course of the twentieth century, the number of Americans living and working on farms decreased dramatically. From a rural population of about 60% in 1900, a hundred years later only about one-quarter of the population lived outside cities. Those actually living on farms decreased to less than 2% of all Americans (see figure 8.1). There were many forces driving this change. These include the increasing availability of work in cities (especially manufacturing jobs), better access to services (especially transportation), and the attractions of increased social mobility.

But the flight from the farm was also driven by changes in the methods of agriculture. Beginning in the 1920s, machines came to the farm. A decade or so earlier, new methods of working developed by Henry Ford (1863–1947) and Frederick Winslow Taylor (1856–1915) had begun to revolutionize the factory. Fordism and Taylorism (sometimes known outside the United States simply as "the American system") introduced scientific approaches to production. These systems emphasized large-scale and mass production through increased efficiency, mechanization, standardization, specialization, quantification, and routinization. But what could be applied to the production line could also be applied to the field. Machines such as tractors and combine harvesters could be combined with Taylor's methods to create more efficient farms. International Harvester, a manufacturer of farm equipment, coined a slogan that captured the vision of this new movement: "Every farm a factory."

Similar movements were taking hold elsewhere, too. In 1917, a Hungarian

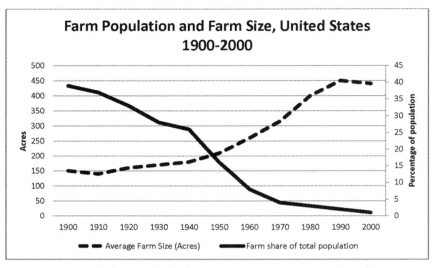

Farm Population and Farm Size, United States 1900-2000

8.1 Chart showing farm population (as a percentage of total population) and average farm size (in acres) in the United States between 1900 and 2000. Source: Illustration by author. Data from Carolyn Dimitri, Ann Effland, and Neilson Conklin, "The 20th Century Transformation of US Agriculture and Farm Policy," US Department of Agriculture, Economic Research Service, Economic Information Bulletin 3 (2005).

entrepreneur called Karl Ereky (1878–1952) published *Biotechnology of Meat, Fat, and Milk Production in an Agricultural Large-Scale Farm*.[1] Ereky wanted to make the farm into a mass-production factory to feed growing urban populations. He designed a farm for 50,000 pigs, of which 1,000 were slaughtered per day and shipped into the cities on a dedicated railway. His pigs were *biotechnologishe Arbeitsmaschine* (biotechnological work-machines), converting vegetables into meat. Likewise, in the Soviet Union, the collectivization of agriculture (beginning in the 1930s) went hand in hand with the mechanization of agriculture. Collectivized and mechanized farms would catapult the Soviet Union into the modern age, Stalin hoped.

Mechanization didn't just mean replacing bullocks with tractors, however. Purchasing this new equipment was expensive—farmers either had to have capital or be able to acquire loans. Those who could not repay their debts were

1. This was the first use of the word *biotechnology* (Ereky wrote in German: *Biotechnologie*).

forced off the land. Farmers who resisted the changes were also displaced. Machines meant that a smaller number of workers could farm much larger plots of land. This meant there was less farm work available. But increased efficiency and lower labor costs also meant that mechanized farms could sell their produce for less. Traditional farmers saw prices for their goods drop and often could no longer make a living from their land. Smaller plots were consolidated into larger ones. Around 1900 the average size of a farm in the United States was around just under 150 acres; by 2000 the average had risen to 450 acres.

Mechanization also had consequences for *what* was farmed. Farmers who lived off their land often grew a variety of crops. But on larger, mechanized farms efficiency meant growing a single crop. A monoculture meant one type of equipment, fewer types of fertilizers or pesticides, one mode of distribution, and standardized work—it afforded *economies of scale*. In 1900, the average farm grew six types of commodities; by 2000 the average was just over one type.

This transformation of agriculture has had a variety of benefits. It has meant cheaper and more plentiful food for consumers. And for those farmers able to stay on the land, it has resulted in less taxing work and a higher standard of living. But it has also caused a variety of problems. In the short term, it forced small farmers off the land and into cities, causing social dislocation. In the longer term, the industrialization of agriculture has also increased the use of fossil fuels (for both machines and fertilizers), herbicides, and insecticides on farms. It has increased the use of water for irrigation and led to concentrations of waste products (including animal waste) in the surrounding areas.

All these changes have had significant environmental impacts, especially the contamination of water supplies and fisheries. The industrialization of livestock farming (mostly sheep, cattle, pigs, chickens, and turkeys) has also led to the increased use of antibiotics and growth hormones. Many argue that these chemicals have significant effects on human health. The confined spaces used for such farming are conducive to the spread of diseases (mostly animal to animal, but occasionally to humans). For many consumers, they also raise ethical questions regarding cruelty to animals.

In the United States and most other developed nations, the industrialization of agriculture was mostly complete by the 1970s. In many parts of the developing world, however, these dramatic changes are still underway, often contributing to the impoverishment of farmers, the growth of urban slums, and degradation of the environment.

TRANSFORMATION II: PRODUCTION

If the first transformation altered the way food was grown, the second has altered the way it gets to our plates. In 2006, Michael Pollan's book *The Omnivore's Dilemma: A Natural History of Four Meals*, asked the seemingly simple question, "What should we have for dinner?" If a twenty-first-century American wanted to eat a healthy and ethically responsible meal, what should be on the table? This turns out to be a question that demands a surprisingly complex answer.

Pollan asks his readers to imagine walking around the aisles of a supermarket. Where do the various food items come from? For some items, mostly in the produce section, the answer is relatively straightforward. A potato comes out of the ground, a lemon is grown on a tree, etc. Some of these items might even have stickers on them indicating that they were grown in Idaho or California. For most meat and fish too, the items are recognizable, more or less, as particular parts of particular animals. Things get more complicated elsewhere. When we come to the breakfast cereals, or the condiments section, or the frozen dinners, or the candy, it is more difficult to discern their provenance. Are these items made of bits of plants or from animals? Which ones? And from where? As Pollan puts it, "You have to be a fairly determined ecological detective to follow the intricate and increasingly obscure lines of connection linking the Twinkie, or the nondairy creamer, to a plant growing in the earth someplace."[2]

But what if we try to do this "archeology of food" the other way around? Let's start with a farm product—like corn—and follow it to the supermarket (the result is summarized in figure 8.2). We can begin in field on a farm in Iowa. A single variety of (non–genetically modified) DuPont Pioneer corn is growing all around us. This is not the type of corn you would grill or boil for corn on the cob—it is not intended for human consumption. Rather, most of it is harvested and transported to enormous grain elevators elsewhere in the Midwest, most of them owned by large industrial food producers like Cargill or ADM (Archer Daniels Midland). In the elevators, much of it is wasted or disposed of.

Some of the corn is used for industrial cattle farming, where it is consumed by pigs or cows. This often takes place in Concentrated Animal Feeding Operations (CAFOs) in places like Kansas. These enclosures are specially designed to rapidly fatten animals on corn feed. Each steer (a young castrated male bovine) can convert 32 pounds of grain into four pounds of muscle, fat,

2. Michael Pollan, *The Omnivore's Dilemma: A Natural History of Four Meals* (New York: Penguin, 2006), 17.

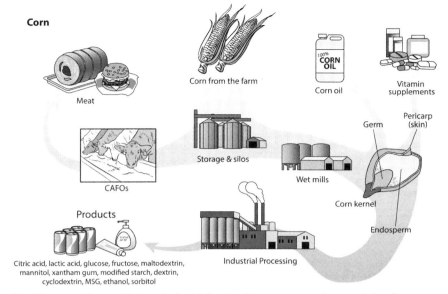

Corn

Meat

Corn from the farm

Corn oil

Vitamin supplements

CAFOs

Storage & silos

Wet mills

Germ

Pericarp (skin)

Corn kernel

Endosperm

Products

Industrial Processing

Citric acid, lactic acid, glucose, fructose, maltodextrin, mannitol, xantham gum, modified starch, dextrin, cyclodextrin, MSG, ethanol, sorbitol

8.2 The uses of corn. Modern industrial agriculture uses corn for a variety of purposes. Corn is fed to livestock in Concentrated Animal Feeding Operations (CAFOs) to produce meat. Various parts of the corn are broken down into chemical constituents and reconstructed in various food and nonfood products. Source: Illustration by Jerry Teo.

and bone per day. Like Ereky's farm, the CAFOs are factories that use animals to convert vegetable matter to meat with high efficiency. The final product of this corn is, in a sense, a steak or a hamburger.

Other corn from the elevators goes to "wet mills." Here the grains are separated into parts that are used for different purposes. The skin of the kernels can be used to make vitamins and other nutritional supplements and the germ can be pressed to make corn oil. The endosperm is the most valuable component. It is subjected to a range of mechanical and chemical processes (including drying, fermentation, centrifugation, ion exchange, and filtering) to make a range of products (including citric acid, lactic acid, glucose, fructose, maltodextrin, ethanol, sorbitol, mannitol, xanthan gum, modified starch, dextrin, cyclodextrin, and monosodium glutamate). These sorts of ingredients are widely used in processed foods. High fructose corn syrup (derived from these processes) is a ubiquitous sweetener for supermarket and fast food products.

A similar story can be told for soybeans. In fact, most of the processed foods in the supermarket are reconstructed from corn and soybean products: corn supplies the carbohydrates, soy supplies the protein, and both supply the fats and oils.[3] "Step back for a moment," Pollan reminds us, "and behold this great, intricately piped stainless steel beast: This is the supremely adapted creature that has evolved to help us eat the vast surplus of biomass coming off America's farms."[4]

Just as impressive is the reconstitution of these food parts into edible and attractive food items. Industrial food chemists have discovered ways to reconstruct these protein, carbohydrate, and fat elements into edible products using gels, emulsions, foams, and suspensions. The results are Cool Whip, Cheez Whiz, Twinkies, Coca-Cola, and so on.[5] The reasons for these remarkable feats in deconstructing and reconstructing food are largely economic. A box of corn-based cereal uses corn meal, cornstarch, corn sweetener, and some other chemical products derived from corn. This is a total of about *four cents* worth of corn. Assembled, packaged, and marketed in an attractive form, it can be sold for *four dollars* per box.

These processes deliberately disrupt the chain that links food to farm. They disconnect what we eat from things that grow in or on the ground. For the food producers, this has practical and economic benefits: if corn in one area is affected by drought or disease it can be immediately replaced by corn from elsewhere. Or, more dramatically, if soybeans become a less viable source of protein, another crop can be substituted without altering the final product — all that is necessary is finding another protein source. Reconstructing food from basic elements makes production less dependent on any particular crop, any particular region, or any particular supplier (especially in terms of price: if one crop gets too expensive, this system makes it possible to switch to another).

The decoupling of food from plants and animals has also allowed food producers to reconstruct foods in almost any way they desire. Making a profit means providing foods that are appealing to the consumer. This has been in large part responsible for the amazing diversity of products and food choices that are available to the modern consumer. In practice, however, this has

---

3. For a description of the soy protein extraction process see: http://foodprotein.tamu .edu/separations/protein.php.

4. Michael Pollan, *The Omnivore's Dilemma: A Natural History of Four Meals* (New York: Penguin, 2006), 90.

5. These soy and corn elements are also used as ingredients in nonfood products such as adhesives, asphalts, resins, cleaning materials, cosmetics, inks, textiles, paints, paper coatings, pesticides, plastics, and polyesters.

usually meant building food products that are high in sugars, fats, and salt. Changing the production of food has contributed to changes in our diets and our health.

TRANSFORMATION III: OWNERSHIP

The third transformation has affected the way in which food plants are owned. We have already seen how both the Green Revolution and the industrialization of agriculture in the West deployed all sorts of technologies: tractors, harvesters, and chemical fertilizers and pesticides. In many cases, farmers who wished to grow specific varieties of crops also had to buy their seeds from agricultural supply companies. These arrangements have made farming a more *capital intensive* activity, and they have made farmers dependent on these suppliers.

Until recently, however, suppliers could not exercise control over the seeds or the plants once the farmer had purchased them. Most farmers had the option of *saving seeds*—they could collect and save some of the seeds from their crops and use these to replant their crops the following season. This is how agriculture has been practiced through most of its history—the farmer would save some seed for the next season.

Through the twentieth century, more and more ownership rights were granted over plants. The Plant Patent Act (PPA, 1930) and the Plant Variety Protection Act (PVPA, 1970) allowed plant breeders some level of protection over their creations. This meant that they could own not just individual plants or individual seeds, but had some rights over *all* plants of a given type. A patent on a kind of mousetrap usually gives you a right not only to sell a particular mousetrap that you build, but also to prevent others from making and selling similar mousetraps without your permission. The PPA and the PVPA worked similarly for plants: they gave the breeder of a plant not only the right over a particular plant, but also some rights over all plants of that type. Another plant breeder could be prevented from simply buying your plant, allowing it to reproduce, and then selling it.

However, the scope of the PPA and the PVPA were limited. The PPA protected only asexually reproducing identical plants—plants that bred true from a single patent. Any mutation from the original was not protected. Sexually reproducing plants and tubers were offered some protection (but not a patent) under the PVPA. The PVPA allows the certification of plants that are new, distinct, uniform, and stable, protecting varieties for 20 or 25 years. However, the PVPA makes exceptions for both research and for saving seed (farmers who save the seed of a PVPA certified plant for their own use are not infringing). In the case of plants, lawmakers had seen fit to create a balance between the rights of inventors (plant breeders) and consumers (farmers) that was distinct from patenting.

With Diamond v. Chakrabarty (see chapter 6), agricultural biotechnology companies saw an opportunity to extend their proprietary rights over plants. This was first tested in 1985 in a case known as *Ex Parte Hibberd*. Kenneth A. Hibberd, Paul C. Anderson, and Pauline Hubbard were plant biologists working for Molecular Genetics, Inc. (based in Minnetonka, Minnesota). They created a genetically modified form of maize that produced high levels of the amino-acid tryptophan. Most corn-based animal feed has to be supplemented with tryptophan—this new variety would not require this step.

Hibberd and his colleagues at Molecular Genetics sought a patent on their invention. That is, they wanted to protect it under the Patent Act (1952), not under the PPA or the PVPA. A regular "utility" patent offered broader protection (that is, more rights for the owner). At first the US Patents and Trademarks Office (USPTO) demurred, arguing that the existence of the PPA and the PVPA meant that utility patents were not available for plants. But the Board of Patent Appeals and Interferences saw differently, following Chakrabarty and awarding a patent on the new variety of corn. In 2001, this view was ratified by the Supreme Court in *J .E. M. Ag Supply Inc. v. Pioneer Hi-Bred International Inc.*

Since 1985, the USPTO has routinely granted utility patents on plant varieties. Thousands of such patents have been issued, particularly to agricultural biotechnology companies. In chapter 6, we discussed the patenting of genes and DNA sequences, mostly in the context of the human genome. Plant DNA has also become subject to these regimes of ownership. In fact, this has led to an escalating patent war as companies have aggressively attempted to patent not only the plants themselves, but also specific genes, sequences, proteins, and mutations. Like the human genome, this has led to complicated patent stacking in which different companies claim ownership of related biological parts (for example, one company might claim an EST, another might claim the whole gene, another the protein expressed by that gene, and another a single nucleotide polymorphism within the gene). Since food crops are some of the most valuable plants, this competitive patenting has been especially intense for these plants.

The consequence of all this is that companies that own plant patents can now exert much greater control over farmers. In particular, they have attempted to closely control the ways in which their seeds are used. In order to purchase seeds, farmers must sign "technology use agreements." These agreements allow companies to retain rights over their seeds—farmers are permitted to use them for one crop and one crop only, they are not permitted to save seeds, and they must repurchase seeds for the next crop. The agreements also grant the seed owners the right to inspect the farmers' fields and monitor crop growth for years after an initial planting.

Monsanto and other companies have attempted to use the full power of the patent law to enforce these agreements. Farmers who have reused or saved seed have been forced to pay thousands of dollars in settlements to Monsanto under the threat of legal action. In several cases that have attracted widespread media attention, Monsanto has attempted to act against farmers who claim that they have never purchased or planted Monsanto seeds. Genetic testing of the crops in these farmers' fields has revealed the presence of proprietary genetic modifications. The farmers claim that wind or other natural means has caused cross-contamination of GM seed into non-GM fields. Monsanto has proceeded to sue these farmers for violation of Monsanto's proprietary rights. In the United States, the courts have generally strongly supported patent rights, protecting the owners over the farmers.

The case that has attracted the most attention is that of Monsanto Canada Inc. v. Schmeiser. In 2004, the Supreme Court of Canada upheld Monsanto's patent rights on their GM plants. Schmeiser continued to seek remedy against Monsanto for the original contamination of his fields. In 2008, Monsanto settled these claims out of court. This was widely hailed as a victory for farmers. More recently, however, the US Supreme court upheld the rights of the patent holder (again Monsanto) in a case where an Indiana farmer had purchased and planted genetically modified seeds from a third party. The farmer, Vernon Bowman, believed that his use of the seeds did not constitute patent infringement because of *patent exhaustion* (this is a doctrine that generally limits the applicability of a patent to the first sale—once a patented item is sold the buyer is usually allowed to use and resell the article without constraint). In its decision of March 2013, the Supreme Court disagreed and supported Monsanto's right to enforce the patent.

Regardless of the details of these cases, what is clear is that all this amounts to a significant change in the property rights regarding plants and food. Things like seeds and plant genes belonged to no one. Since the 1980s, they have become subject to strong regimes of ownership that allow companies to aggressively enforce their rights against farmers. This means that decisions about what can be grown, when it can be grown, where it can be grown, and how it can be grown are concentrated in the hands of a few corporations. In other words, it allows those owners of plants to have increasing levels of control over the food supply.

Up to this point, this chapter has not explicitly been concerned with GMFs. This is because it is suggesting that we should identify GMFs as being part of broader changes in the way our food is grown, produced, and distributed. The transformations in agriculture, in production, and in ownership all move towards the same end: the concentration of control over food in the hands of a small number of corporations. GMFs should be understood as an increasingly important tool in the construction of a large economic enterprise.

This consolidation of economic power is most obvious if we look at the seeds themselves. In terms of seed supply, just three companies control 47% of the world market (DuPont Pioneer [formerly Pioneer Hi-Bred], Monsanto, and Syngenta).[6] In the late 1990s, Monsanto spent $8 billion acquiring other seed companies, ensuring that it would dominate the GM seed market. These large companies have also entered into research and development collaborations, profit sharing arrangements, cross-licensing schemes, and truces on litigation. These sorts of actions (considered by some to be monopolistic and anti-competitive) further increase profits and centralize control.

But it is not just about seeds. Companies like Monsanto, DuPont Pioneer, and Syngenta also sell agrochemicals (fertilizers and pesticides). Technology use agreements also give the companies de facto control over farmers. They have also made attempts to gain a stake in food processing facilities. In the United States, these are already owned by a small number of companies. For instance, just two companies (Cargill and Archer Daniels Midland) own most of the grain elevators and all the wet milling facilities. This suggests a trend of *vertical integration*: the merging of companies that work at different stages of the food supply chain. The goal is control over the food supply from seed to farm to the supermarket shelf.

The ultimate aim for these companies is profit. The twentieth-century transformation of food production has made it into an immensely lucrative business. The technological innovations of tractors and fertilizers went hand in hand with economic innovations that changed the who, the how, and the what of farming in the twentieth century. GMFs should be understood in this context. The biotechnological innovation of GMFs has produced a set

6. According to a report from Action Group on Erosion, Technology, and Concentration (known as ETC Group) in 2007, Monsanto has 23%, DuPont Pioneer has 15%, and Syngenta has 9%. It also estimates that measured by worldwide area devoted to GM crops, Monsanto seeds account for 87% of the total. See http://www.gmwatch.org/gm-firms/10558-the-worlds-top-ten-seed-companies-who-owns-nature.

of legal and economic innovations (beginning with Ex Parte Hibberd) that have again reshaped the means of food production. These legal and economic technologies, enabled by genetic modification, have allowed increased control over food production and distribution. GMFs have extended the transformation of agriculture and food that began early in the twentieth century. By making it possible to gain proprietary control over plants, it has enabled greater corporate control over agriculture.

The three transformations that have been described in this chapter have ultimately contributed to a fourth transformation. This is a transformation in what we eat and how we eat. The shifting patterns of food consumption over the twentieth century have much to do with a broad range of social and cultural transformations—transformations in family life (leading to diminishing time spent cooking and eating together), transformations in transport and cities (leading to more access to restaurants via cars), and transformations in media (leading to more advertising). However, the transformations discussed here have played an enabling role in shifting diets toward processed foods, fast foods, and foods that are conveniently packaged, high in sugar and fats, high volume, and available at any time. Discussions of the risks of GMFs are often centered on the direct health and environmental consequences (possible allergens, toxicity, gene transfer, etc.—see chapter 7). But, GMFs are connected in complex ways to the whole economic and social structure of agriculture, food processing, and eating. These connections are often overlooked. Yet—through dramatic changes to our diet—this system poses an even greater risk to our health and to our environment.

If GMFs are now an integral part of the larger agricultural and food supply system, then challenging GMFs is going to mean challenging the system itself. In other words, we can only solve problems with GMFs by changing what we eat and how we eat. However, it is also worth understanding that GM technology could be embedded in a *different* system. Technologies are not "bad" or "good" all by themselves—it depends on the uses to which they are put. Might it be possible to reimagine GMFs as being put to use in a different context, outside of corporate control? And might not such a use have vastly different potential? For instance, Golden Rice (see chapter 7) suggests that there may be opportunities to develop noncorporate, socially responsible genetically engineered crops. Ironically, however, the adoption of Golden Rice has been hampered by opposition from the anti-GMF movement. While we may choose to reject the current sociotechnical system in which GMFs are embedded, we should leave open the possibility of developing these agricultural biotechnologies within a different context where new ways of using and benefiting from them might emerge.

FURTHER READING

There is a variety of books that address the transformation of agriculture in the United States. The classic work is Wendell Berry, *The Unsettling of America: Culture and Agriculture* (San Francisco: Sierra Club Books, 1977), while a more recent account of the industrialization of American agriculture is Deborah Fitzgerald, *Every Farm a Factory: The Industrial Ideal in American Agriculture* (New Haven, CT: Yale University Press, 2010). Leo Marx, *The Machine in the Garden: Technology and the Pastoral Ideal in America* (Oxford: Oxford University Press, 1964) traces a longer history of the relationship between technology and the land in the United States. For an account of the development of artificial fertilizers and their impact on food production see Vaclav Smil, *Enriching the Earth: Fritz Haber, Carl Bosch, and the Transformation of World Food Production* (Cambridge, MA: MIT Press, 2004). The best account of Karl Ereky's work is M. G. Fári and U. P. Kralovánszky, "The Founding Father of Biotechnology: Károly (Karl) Ereky," *International Journal of Horticultural Science* 12, no. 1 (2006): 9–12. For those who read German, Ereky's original account of his work is Karl Ereky, *Biotechnologie der Fleisch-, Fett-, und Milcherzeugung im landwirtschaftlich Grossbetriebe: für naturwissenschaftlich gebildete Landwirte verfasst* (Berlin: P. Parey, 1919).

In my account of the transformation of food production I have drawn extensively on Michael Pollan, *The Omnivore's Dilemma: A Natural History of Four Meals* (New York: Penguin, 2006). The issues Pollan raises are explored in more historical detail in Roger Horowitz, *Putting Meat on the American Table: Taste, Technology, and Transformation* (Baltimore: Johns Hopkins University Press, 2005) and Harvey Levenstein, *Paradox of Plenty: A Social History of Eating in Modern America* (Berkeley: University of California Press, 2003). These issues are also explored in the documentary film *Food Inc.* (Robert Kenner, director, 2008). Some of the social and cultural transformations related to eating and food are accounted in Tracie Macmillian, *The American Way of Eating: Undercover at Walmart, Applebee's, Farm Fields, and the Dinner Table* (New York: Scribner, 2012).

A account of the vertical integration of the food supply can be found in William Boyd, "Wonderful Potencies? Deep Structure and the Problem of Monopoly in Agricultural Biotechnology," in *Engineering Trouble: Genetic Engineering and Its Discontents*, ed. R. A. Schurman and D. D. T. Kelso (Berkeley: University of California Press, 2003), 24–62. On the history of patenting agricultural plants see Glenn E. Bugos and Daniel J. Kevles, "Plants as Intellectual Property: American Practice, Law and Policy in World Context," *Osiris* 7 (1992): 74–104. An extremely polemical account that looks at these issues in the context of the global South is Vandana Shiva, *Stolen Harvest: The Hijacking of the Global Food Supply* (Cambridge, MA: South End Press, 2000). The story of how opposition to GMFs has hampered the adoption of Golden Rice can be found in Amy Harmon, "Golden Rice: Lifesaver?" *New York Times*, August 24, 2013.

PART V

THE BOUNDARIES
OF BODILY LIFE

9   OWNING PART OF YOU

INTRODUCTION

Biotechnology is not just about altering genes. Breakthroughs in the ways that scientists and physicians understand biology offer the possibility of removing, storing, altering, replacing, and repairing cells, tissues, and organs. This chapter and the next discuss the implications of this traffic in animal parts beyond and between bodies. This chapter examines the intertwined ethical, legal, and social problems of ownership that have arisen as we have expanded our power to maintain and manipulate life outside the boundaries of the human body. The preservation and exchange of whole organs between humans, and between humans and nonhumans, raises a special set of social and ethical questions that will be treated in chapter 10.

The ability to move body parts around outside the body is a relatively new one. Until the early twentieth century, most biologists believed it was impossible for cells or tissues from the body to live independently. In 1907, the Johns Hopkins physiologist and anatomist Ross Harrison claimed to have grown a frog nerve cell in his lab. In 1912, Alexis Carrel, working at the Rockefeller Institute for Medical Research, developed and extended Harrison's techniques. Carrel managed to grow cells taken from an embryonic chicken heart in a glass flask, keeping the cells alive and beating for over twenty years.[1] Many biologists greeted the news of Harrison's and Carrel's experiments with skepticism. Gradually, however, tissue culture (the ability to grow bodily cells outside the body) became a well-accepted part of biomedical practice.

We now take for granted our ability to perform these kinds of extractions. It is the basis for many forms of biological and research and medical testing. Not only can we grow cells outside the body, but we can freeze and store them for later use.

Our ability to take the living body apart in this way has raised two closely related problems. First, what is the legal status of these parts? If we go to the

---

1. Biologists now accept the fact that ordinary cells cannot go on living indefinitely as Carrel's experiments seemed to show. It is likely that Carrel's cells were continually being contaminated with new cells introduced with the nutrients that were continuously fed to the cells.

hospital to get the appendix removed, we usually do not claim any right of ownership over the excised tissue. It might be disposed of by the hospital, or in some cases used for medical research. But should the same rules apply to all parts? And what if the hospital or its doctors can use those parts to make money for themselves or others? What rights should we have over these alienated pieces of ourselves?

The answer to these questions is likely to depend on what we count as a meaningfully "living" part of ourselves. Second, then, is the problem of how we should understand "life" in the biotech age. Tissue culture has already stretched and bent our notion of life to include "life outside the body." What is an appropriate definition of *life* in a world where all kinds of extractions, manipulations, distortions, and mixtures become possible?

HENRIETTA LACKS

Alexis Carrel's beating chicken heart demonstrated the possibility of tissue culture (keeping cells alive and growing outside the body). But this was a difficult business: cells had to be kept at the right temperature, and supplied with the right nutrients. Even then, human cells, it seemed, couldn't be kept alive outside the body indefinitely. Once extracted, a cell could go through a limited number of cycles of division and growth (between about 40 and 60) but then the colony would die (this is known as the Hayflick limit).

Some cells, however, are exceptional. In 1951, Henrietta Lacks was a 31-year-old working-class African-American woman from Baltimore. On February 8, Lacks was admitted to a segregated ward of the Johns Hopkins Medical School due to bleeding caused by a lesion on her cervix. A biopsy quickly diagnosed her with cervical cancer. Lacks' gynecologist sewed a pouch containing radium into her cervix in an attempt to irradiate the tumor and halt its growth. Lacks died of the cancer just eight months later on October 4.

A small piece of Henrietta Lacks lived on, however. Without the knowledge or permission of the patient or her family, some cells from Lacks' tumor were sent to the laboratory of Dr. George Gey. Gey was the director of tissue culture research at Johns Hopkins and had spent most of his career attempting to grow cells in vitro. Lacks' cells constituted a remarkable discovery. Unlike other human cells, they seemed to grow continuously and without limit in culture. Their cancerous nature, it seemed, allowed them to proliferate even in harsh conditions where other cells died.

Gey used Lacks' cervical cells to create what became the first "line" of human cells—that is, generations of cells all derived from the same original parent cells (in this case the sample from Lacks' cervix). He called it *HeLa*

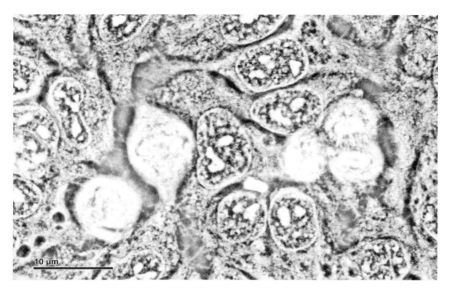

9.1 HeLa cells growing in vitro. Negative phase contrast microscopy, 4,800x magnification. Cells taken from Lacks' cervix in 1951 were transformed into the HeLa cell line. These cells had the ability to continue to grow and divide in the laboratory and have become ubiquitous in biomedical research around the world. Source: Photograph by Josef Reischig, licensed under Creative Commons Attribution-Share Alike 3.0 Unported.

(extracting the first letters from Lacks' first and last names). Gey distributed the cells for free, never attempting to patent them or limit their distribution. Nevertheless, the cells became immensely valuable. HeLa became a standard cell that could be utilized in many kinds of biomedical research—for experimenting on cell growth and division, for testing toxicity, for work with viruses, for cancer research, and for radiation studies (figure 9.1).[2] In the early 1950s, HeLa was immediately put to work in the laboratories of Jonas Salk for testing his polio vaccine. By 1954, HeLa was being marketed for sale by a biological supply company.

There are many possible tellings and re-tellings of the Henrietta Lacks story. In the 1950s, Lacks' name became associated with the polio vaccine and therefore with selfless sacrifice for the triumphant cause of medicine. In the

2. Rebecca Skloot estimates that since the 1950s there have been about 60,000 scientific articles published that use HeLa cells and that in the early twenty-first century about 300 further articles were being added to this per month.

1960s, it was discovered that HeLa cells seemed to have contaminated other human cells lines. This led to a different narrative that warned of HeLa's "vigorous" and "monstrous" qualities. Henrietta Lacks' race became an important trope for understanding the fecundity of the cell line. By the 1990s, the tale of HeLa was about property rights and economic value. Physicians and the hospitals and companies they worked for had taken cells that rightfully belonged to Lacks (or, since her death, her family) and made money. Now Lacks' race became important as a marker of exploitation. Why shouldn't Lacks' descendants be able to sue for a share of the profits?

What all these narratives have in common is the identification of the person of Henrietta Lacks with the cancerous cells scraped from her cervix. Henrietta Lacks became HeLa and, at the same time, HeLa became Henrietta Lacks. Scientific and popular reports have imagined adding up the weights of all the HeLa cells in laboratories all over the world: there is far more HeLa now, they conclude, than there ever was of Henrietta Lacks during her lifetime. Henrietta Lacks lives on in her cells, distributed across thousands of laboratories and freezers, in various growing, multiplying, and changing forms.

The stories of Henrietta Lacks raise ethical questions. Shouldn't Lacks' doctors or Gey have asked her or her family for permission to use her cells? Shouldn't Lacks' identity have been kept private? Didn't Lacks' cells constitute a form of property from which her descendants are entitled to a share of the profit? But the answer to all of these questions depends on where we draw the boundaries of bodily life. These ethical questions do not have straightforward answers because we are still in the process of working out where these boundaries lie. Biotechnologies, especially the technologies of tissue culture, are reshaping those boundaries. Henrietta Lacks' "life" extended beyond her death and beyond her body—across time and space into labs and freezers down to the present day. Behind the ethical and legal questions is the fact that what we understand as "part of ourselves" and "separate from ourselves" depends more and more on biotechnology.

JOHN MOORE

One place where such boundaries are often worked out is in the courts. If your neighbor builds a garden shed on your side of the property line, you can challenge their ownership in front of a judge. Similarly, if someone else is making money using something you believe to be yours you can take it up with a lawyer. Biotechnology makes it possible to use and profit from bodies in all sorts of new ways: banking tissues, making cell lines, and so on. In legal proceedings, we can watch in high-definition as new boundaries and limits of these uses are negotiated. As biotechnology shifts the boundaries of bodily life, the legal system acts as a means of keeping track of where the edges are.

One of the most prominent of these cases involves a leukemia patient—John Moore—and his doctor, David Golde. Since Dr. Golde worked at the University of California Los Angeles Medical Center, the case is known as *Moore v. The Regents of the University of California* (technically, the owners of the hospital). The details of the case are quite complicated, but its significance makes it worth trying to work through some of the medical and legal technicalities.

John Moore lived in Seattle where, in 1976, he was diagnosed with a rare but treatable form of leukemia known as "hairy cell." In October, Moore's doctor in Seattle sent him to Los Angeles to be treated by David Golde, a world expert on this form of leukemia. On October 5, Moore arrived at the UCLA Medical Center and Dr. Golde took blood and bone marrow samples to confirm the diagnosis. This was indeed confirmed within a few days and on October 8, Dr. Golde advised Moore that he needed to remove his spleen. This would, in all likelihood, effectively control Moore's leukemia. Moore signed a consent form for the surgery.

Over the next seven years, between 1976 and 1983, Moore repeatedly flew down from Seattle to visit Dr. Golde at UCLA. During these visits, Dr. Golde drew further samples of blood, blood serum, bone marrow, and sperm. Dr. Golde took great care to make sure that Moore's bodily fluids did not fall into anyone else's hands. When Moore asked whether he could have samples taken nearer his home in Seattle, Dr. Golde began to pay for Moore's flights and to put him up at the Beverly Wilshire hotel. Dr. Golde told Moore that these visits allowed him to monitor his continued health, and did not reveal that they were also contributing to an ongoing research project.

Some time around August 1979, Dr. Golde established a cell line based on Moore's T-lymphocyte cells, which had been extracted from Moore's bone marrow. The cell line was called *Mo* (after Moore). Dr. Golde knew that the rare nature of Moore's disease meant that Moore's cells had unique value: the cells produced a unique protein and also carried a virus that was useful for HIV research. In 1981, the University of California applied for a patent on the *Mo* cell line, naming Golde and one of his collaborating scientists as the "inventors." The University of California licensed the use of the cell lines to both the Genetics Institute (a biotech company) and Sandoz Pharmaceuticals. In return the companies provided funds for Golde's lab and the Genetics Institute compensated Golde with 75,000 shares (worth about $3 million). The patent was granted in 1984.

Moore grew suspicious only when Dr. Golde's office began hassling him to sign a consent form that referred to rights over "products" from his blood. Beginning to realize the extent of the deception, Moore sent the form to a lawyer. In 1984, Moore sued Dr. Golde, the University of California, the Genetics Institute, and Sandoz Pharmaceuticals. The plaintiffs (that is, Moore) tried to make their case in two distinct ways. First, they argued that the doctors and the hospital had breached Moore's trust by not informing him of their financial and other interests in his cells. Second, they claimed that the profiting from Moore's cells amounted to "conversion." This asserted that Moore had a property right to his cells—essentially that they still legally belonged to him.

Of the two claims, the first was less controversial. *Fiduciary duty* is a legal form of an ethical principle. It holds that people in special positions of trust and responsibility—such as doctors, lawyers, and bankers—must not breach that trust. Under the common law, a fiduciary relationship requires higher standards of behavior than an ordinary duty of care. A patient entrusts a physician with his or her care and should be able to expect that the physician has the patient's care as a priority.

In the Moore case, the Supreme Court of California found that Dr. Golde's research and financial interests breached his fiduciary duty. The fact that Dr. Golde stood to gain both money and scientific prestige by extracting Moore's bodily tissues and cells represented a conflict of interest. In other words, Dr. Golde put himself in a situation where he did not seem to be placing Moore's care as his first priority. In particular, by not properly disclosing his financial and research interests in Moore's cells to Moore himself, Golde breached his duty to provide "informed consent."

Moore's second claim proved more problematic. His attorneys argued that the sale and use of Moore's cells amounted to *conversion*. Again, this is a legal term that requires some explanation. Conversion occurs when party A makes use of something that belongs to party B, resulting in some sort of profit or gain for A. For instance, let's say B owns a field in which he grows corn. One fine day, A comes along, picks all the corn, sells it at a farmers market, and keeps the money. A has stolen B's corn—she has committed the criminal act of *theft* (and she can be punished for this by being fined or put in jail). But A has also committed the civil wrong of conversion—she has enriched herself using B's property and B has the right to sue for the money from the corn. In fact, A may not have acted dishonestly—perhaps she believed the corn was growing on her own land—but conversion still applies. If B can prove the corn belonged to him, then he has a legal right to the money.

Moore and his attorneys were arguing that Moore's cells were like the corn—they belonged to him and neither Dr. Golde, nor the University of California, nor Sandoz had any right to profit from them without Moore's permission (which he had not given). But in order to show that conversion had occurred, Moore's lawyers had to claim that Moore's cells were a kind of *property*—something that could belong to Moore just like corn, or land, or a Mercedes Benz. But did Moore's tissues—removed from his body—really still belong to him?

The majority of the judges on the Supreme Court decided that Moore's cells could *not* be his property. Moore's attorneys had tried to argue that Moore's cells were a kind of "likeness" of him. Just as Hollywood celebrities (and other famous people) have a property right over images of their person, Moore's cells were claimed to be a genetic likeness of himself. The judges were not persuaded: Moore's cells were not really unique to Moore, they argued, since the majority of genetic material is shared between all humans. Considering cells property would not be consistent with other legal notions of property.

Dr. Golde and the University of California contended that the *Mo* cell line—from which they were profiting—was distinct from the tissues taken from Moore's person. Although the cell line used raw materials from Moore, it was an invention in its own right. The *worth* of the thing lay in the work that Golde and his colleagues had done to make Moore's tissues into a useful and commercial product. Moore's cells left in the hands of Moore himself would have had no value. As far as Golde and his colleagues were concerned, Moore was in the same position as a person who had discarded a broken-down car only later to claim ownership over it once someone else had taken the time to fix it. The judges agreed that considering cells property would fail to recognize the skill and value in the "invention" of the cell line.

Finally, the judges reasoned that other tissues taken from one's body in other circumstances do not count as property. At least in California, organs and blood, once removed from your body, are not your property. Even a cadaver does not belong to the family of the deceased. The judges noted that if bodies and body parts *were* to be considered property all kinds of problems would result: biomedical research would grind to a halt as labs attempted to track down the various "owners" of fleshy parts. Treating cells as property would be bad *policy*, the judges thought.[3]

3. Not all the justices were in complete agreement. Four of them supported the arguments as I've described them here. One other judge agreed, but for different reasons. And two of the justices disagreed with significant parts of the decision. In such situations, the majority opinion prevails.

From the facts of the case, there are plenty of reasons to think that Moore's tissues and the *Mo* cell line were fundamentally similar. Dr. Golde often behaved as if they were. He knew, for example, that if another researcher had access to Moore's cells they could produce the same valuable proteins as his cell line—what was valuable about the cell line was what was valuable about Moore's cells themselves.[4] But the law had made up its mind: the *Mo* cell line was different from Moore himself and therefore not his property.

This decision supported—and further enabled—the burgeoning biotech industry of the 1980s. If individuals like Moore could claim property rights over their cells, this might disrupt the industry's ability to patent and monetize biomedical objects. The judges in the Moore case were certainly aware of this. A large part of their information about cells and cell lines came from a report by the US Office of Technology Assessment called *New Developments in Biotechnology: Ownership of Human Tissues and Cells* (1987). This report was written explicitly to spell out the *economic* implications of owning human tissues and cells. What the Moore case reveals is that the boundaries of Moore's body—what belonged to him and what did not—became not just a technical question, not just a philosophical question, and not just a legal question. Most importantly, it was an *economic* problem. How we conceptualize the limits of our bodies is dependent—in complicated ways—on how we (and others) can make money out of them.

TED SLAVIN

By the time John Moore realized that his cells were valuable, it was already too late. Someone else had patented them and was already making money. But what if Moore had realized the value of his cells before his operation in 1976? The case of Ted Slavin is a story that contrasts with Moore's and suggests how things might have been different if Moore knew more about the value of his own body.

Ted Slavin was born with hemophilia, a hereditary disease that causes even small cuts and abrasions to bleed for days or weeks. The cause of hemophilia is a defective version of the protein that usually causes blood to clot or coagulate. In the 1950s, the only treatment for hemophilia was to receive regular blood transfusions with blood from healthy persons whose blood contained the fully functioning protein. At that time, the ability to screen donated blood

---

4. Dr. Golde was especially worried about others gaining access to Moore's cells because this would invalidate his claim to a patent. Objects already available for sale or in the public domain are not patentable.

for dangerous diseases was limited. Hemophiliacs were constantly at risk of being infected by viruses carried in the donated blood.

This is exactly what happened to Slavin—through transfusions he was exposed to the hepatitis B virus over and over again. But Slavin did not succumb to the disease. Somehow, his immune system successfully battled the virus, producing vast numbers of hepatitis B antibodies that kept the virus in check. Slavin's hepatitis was diagnosed in the 1970s only when huge numbers of antibodies were discovered in his blood.

Slavin realized that his blood, full of the powerful antibodies, might be valuable. Several pharmaceutical companies were trying to find a vaccine for hepatitis and Slavin's antibodies could provide a crucial clue. Slavin began to sell his blood serum for $10 per milliliter, offering up to half a liter ($5,000 worth) to anyone who would buy it. The labs and companies that Slavin contacted quickly became his customers.

At first, Slavin just needed the money—his ability to work had been curtailed by disabilities caused by both hemophilia and hepatitis. But he also wanted someone to find a cure for hepatitis. He contacted the National Institutes of Health and found the most promising hepatitis researcher, Baruch Blumberg. Blumberg, who worked at the Fox Chase Cancer Center, had already won a Nobel Prize for his work on hepatitis B. Slavin got in touch and offered to provide him with all the antibodies he might need for his research for free. Blumberg accepted the offer. Over many years of donations, the Fox Chase lab used Slavin's blood serum to help uncover the link between hepatitis B and liver cancer and, ultimately, to develop the first hepatitis B vaccine.

But Slavin's entrepreneurialism didn't stop there. Soon, he had more customers than he could supply. Surely, he thought, there might be others out there like him with blood full of valuable antibodies. Slavin managed to track down other hepatitis B carriers and started a company—Essential Biologicals—that sold their blood. Slavin and Moore both had valuable cells. The difference was that Slavin realized their value *before* they were removed from his body. He had information, and he was able to turn that information into profit. In Moore's case, his doctor withheld information. Once his cells were removed from his body, he could no longer control them.

Slavin's case shows us that the ability to control your body and its parts depends on *knowing* about them. This is really a political issue—in the end, defining the boundaries of one's own body depends on access to knowledge. When knowledge is concentrated in the hands of doctors or biotech companies, those individuals or corporations stand in a position of economic and political power over our bodies.

CONCLUSIONS

Our new understanding of life signifies profound changes in our perceptions of the human body. In particular, parts of our bodies that were previously considered to be "waste" or "leftovers" (such as Moore's spleen or Slavin's blood) have increasingly become something that is valuable, that can be owned, and that can be subjected to patents, entrepreneurialism, and profit. Biotechnology has the potential to transform "waste" into value in unexpected ways. This unanticipated value opens up new economies that have new possibilities for exploitation, control, inequality, and injustice. Some people may increasingly gain ownership and others may be increasingly owned. It is not only genes that become subject to markets and regimes of biocapital, but cells, tissues, and bodily fluids, too.

By giving up our bodily parts to biomedical examination and biotechnology we have been able to generate all kinds of new knowledge, including tests and treatments for human diseases. Allowing people to assert ownership over "waste" body parts would put severe restrictions on this biomedical research, just as the judges reasoned in the Moore case. Waste is not valuable in and of itself; it must be collected and repurposed to be made valuable. How much of this value should accrue to the original owners of bodily parts and how much should properly belong to the repurposers? "Waste" is an ambiguous category in which the rules of ownership are not always clear. In the case of body parts, this is further complicated by the fact that tissues are marked (genetically or otherwise) by their origins. Information contained in your tissues, even once removed from your body, is information *about you*. Some of these problems can be solved by enforcing rules of informed consent—people should be told when their tissues are going to be used in research and have the right to refuse. But this does not solve the problem entirely: is it possible to give consent for some uses of your tissues and not others? How would it be possible to keep track of who has given consent for what? Biotechnological repurposing of bodily parts has created difficult practical questions about the limits of ownership, consent, and privacy.

FURTHER READING

For the full story of Henrietta Lacks and her immortal cells see Rebecca Skloot, *The Immortal Life of Henrietta Lacks* (New York: Crown, 2010). For an account that places Lacks' story in the longer history of tissue culture see Hannah Landecker, *Culturing Life: How Cells Became Technologies* (Cambridge, MA: Harvard University Press, 2007). A version that focuses more narrowly on HeLa is Hannah Landecker, "Immortality, *In Vitro*: A History of the HeLa Cell Line," in *Biotechnology and Culture: Bodies, Anxieties, Ethics*, ed. Paul Brodwin (Bloomington: Indiana University Press, 2001).

Skloot and Landecker also provide the best analyses of John Moore's and Ted Slavin's stories. On the Moore case see Hannah Landecker, "Between Beneficence and Chattel: The Human Biological in Law and Science," *Science in Context* 12, no. 1 (1999): 203–225. A highly readable version of Slavin's story appears as Rebecca Skloot, "Taking the Least of You: The Tissue-Industrial Complex," *New York Times Magazine*, April 16, 2006. For those interested in the case law, the Moore case is: *Moore vs. Regents of the University of California* (51 Cal. 3d 120; 271 Cal. Rptr. 146; 793 P.2d 479). R. Rao, "Genes and Spleens: Property, Contract, or Private Rights in the Human Body?" *Journal of Law and Medical Ethics* 35, no. 3 (2007): 371–382, provides an analysis of different legal frameworks under which ownership of human body parts might be regulated. A broader analysis of the political economy of tissues can be found in Catherine Waldby and Robert Mitchell, *Tissue Economies: Blood, Organs, and Cell Lines in Late Capitalism* (Durham: Duke University Press, 2006). For a more theoretical reflection on these transformations see Paul Rabinow, "Severing the Ties: Fragmentation and Dignity in Late Modernity," in *Essays on the Anthropology of Reason* (Princeton, NJ: Princeton University Press, 1996), 129–152.

INTRODUCTION

This chapter is concerned with hybrids and hybridity. Hybrids are mixtures, crosses, or chimeras. They often transgress or transcend well-established social and cultural boundaries. Such boundary-crossing objects are often disruptive. They can be sources of fear and anxiety. Boundaries and categories help us to make sense of the world around us — we believe we will better understand the world if we can keep it arranged in neat boxes. We hope that boundaries keep things we want inside inside and things we want outside outside. The world, however, is not always so easy to divide up and manage. Chapter 9 examined the technologies of tissue and cell culture and how these generate dilemmas over ownership and consent as they shift the boundary between the inside and outside of human bodies. This chapter confronts how we think about and value parts that participate in more extreme boundary-crossings: whole organs moving not just between human and human, but also across species.

Many "monstrous" creatures are hybrids: vampires are part human and part bat, and werewolves, the minotaur, centaurs, and many other mythical creatures straddle the divide between human and animal. These creatures are considered monstrous exactly because they disrupt our normal categories. The ability to cross or ignore boundaries can make objects seem difficult to control or contain, and hence dangerous. Many nineteenth- and twentieth-century anxieties about race were based on fears of mixing: black people and white people were considered to occupy distinct categories. Interracial mixing crossed a social and cultural divide. Overcoming racial prejudices has meant — amongst other things — changing how society classifies people.

Biotechnology presents us with many kinds of crosses. We have objects that are animal/human, animal/plant, living/nonliving. These objects are disruptive at least partly because they are hybrids: they defy or confound our usual categories. That can make us nervous. How do we treat something that is part animal and part human? Like an animal or a human? Should a human cell outside a body be valued and used more like a human body or more like a bacterial cell? Coming to terms with biotechnology is partly about understanding where hybrids come from, why they seem disturbing, and how to deal with them. Rather than trying to draw ever-sharper boundaries and categories, it is about finding ways to accommodate mixes and crosses within our worldview and work out how society should deal with them.

Some kinds of medical technologies and devices have raised similar issues concerning hybridity. Machines such as ventilators (first used extensively in the 1950s), kidney dialysis machines (first developed in the 1940s), and artificial cardiac pacemakers (externally worn versions were developed in the 1920s and '30s) are integrated into patients' bodies, taking over some of its normal functions, in order to keep them alive. Some of these devices have become computerized and miniaturized and can be implanted more or less permanently inside the body (the first implantable pacemakers became available in 1958; automatic insulin pumps are another example). Prosthetic limbs, too, have been developed into sophisticated devices that interface with the patient's body and are even able to receive, interpret, and act on signals from the brain.

These technologies challenge the boundary between body and machine, and between animate and inanimate. They too raise some tricky questions. For example, when a ventilator is used to prolong a person's life, is switching off the ventilator equivalent to killing that person? And where does the boundary lie between replacing a part of the body and enhancing the body? (This is a live issue for sporting bodies such as the International Olympic Committee, which must decide what kinds of prosthetics and devices are permitted.) But biotechnologies have the ability to cross and blur boundaries to an even greater extent than medical machines: their biological nature means that they can integrate more seamlessly, becoming living and growing parts of bodies. They act on the molecular level, and they can disperse throughout an organism (rather than, for example, being limited to a single limb or organ). This chapter will explore some of the questions that biotechnological hybrids raise and suggest some places we might look for answers.

CRYOPRESERVATION

The ability to extract and exchange bodily parts has depended on the techniques of freezing tissues. The possibility of preserving and examining life outside the body using extreme cold occurred to Robert Boyle (1627–1691) in the seventeenth century. The Italian natural philosopher Lazarro Spallanzani (1729–1799) also experimented with freezing organisms in the eighteenth century. However, this work was limited by the fact that cells that were frozen were often reduced to mush when thawed. For a long time, it was thought that freezing the water present inside any cell led to the formation of ice crystals that would puncture the delicate cellular membranes. In the mid-twentieth century, scientists hypothesized that the destruction of cell membranes occurred as a result of changes in salinity levels that resulted from temperature change. Either way, the frozen cell was fundamentally damaged and the thawed-out cell could not be restored to life.

One of the most important biomedical discoveries of the twentieth century occurred almost by accident. In 1949, Christopher Polge and Audrey Smith were working at the National Institute for Medical Research at Mill Hill on the outskirts of London. The Mill Hill complex included a farm and "farm laboratory" devoted to research on agriculture and especially livestock. The aim of Polge and Smith's work was to find a way of freezing and thawing chicken sperm in such a way that the sperm remained viable enough to fertilize an egg. Such a technique would have significant advantages for animal husbandry since it would allow breeders to exercise even greater control over reproduction.

At first, Polge and Smith's work met with little success: thawed sperm moved around well enough, but no viable embryos could be produced. A mislabeled laboratory bottle led Polge and Smith to the discovery that freezing sperm in a solution of glycerol seemed to preserve its viability. The first chicks were produced from frozen sperm in 1950.

Polge and Smith's discovery kick-started a new field of research: cryobiology, and, with it, the technique of cryopreservation. During the 1950s, freezing techniques were extended first to cattle sperm, then to livestock embryos, and soon after to human tissues. In 1951, Polge's work led to the birth of the first calf, "Frosty One," conceived with cryopreserved sperm. This work intersected with research on human fertility—freezing sperm and embryos offered the possibility of greater control over reproduction. The first human pregnancy achieved with frozen-and-thawed sperm occurred in 1954. In 1973, Polge, now working with Ian Wilmut, succeeded in bringing to term the first calf, "Frosty Two," from a cryopreserved embryo.[1] In 1984, biologists assisted in the first conception and birth of a human baby from a frozen embryo.

Cryobiologists discovered techniques for slow and controlled-rate freezing as well as more advanced cryoprotectants, such as DMSO (dimethylsulfoxide). They developed methods for freezing bone, skin, brain, cartilage, and cardiovascular tissue. Cryopreserving blood and whole organs was more difficult. During the 1960s, Arthur Rinfret developed specialized polymers and "vitrification" freezing techniques that succeeded in preserving blood. Attempts to deep-freeze whole human organs were less successful. So far,

1. Wilmut went on to perfect the techniques of somatic cell nuclear transfer that led to Dolly the sheep in 1996 (see chapter 16).

cryobiologists have not succeeding in defrosting a functioning organ. Nevertheless, in 1967, James Bedford (a retired psychology professor) was the first human being whose whole body was placed into cryonic "suspension." At present, the damage to organs resulting from the freezing and thawing process cannot be reversed. However, advocates of whole-body cryopreservation (including those selling such services) argue that eventually medical technology will reach the point where such damage can be repaired.

TISSUE BANKS

Much of this early work on cryopreservation was performed at the US Navy Medical Research Center in Silver Spring, Maryland. In 1950, this facility established a rudimentary "tissue bank" that stored skin for grafts and bone that could act as scaffolding for tissue growth. Since that beginning, the banking of tissues—for both biological research and therapeutic purposes— has become commonplace all over the world. Hospitals, research institutes, and private companies as well as the military store tissues.

The technologies of cryopreservation have now made it possible to store not only sperm and blood, but also corneas, cancerous tumor cells, fat, appendixes, ovaries, skin, sphincters, and testicles. Not all this tissue belongs to dead people: tissues can be extracted during routine medical tests (biopsies, for instance), operations, and clinical trials. Of course, not all of these are put into tissue banks (in many jurisdictions, the patient's consent is necessary in order to do so), but a report from 1999 estimated that in the United States alone there were about 307 million tissue samples extracted from about 178 million patients and that this number was growing by roughly 20 million samples per year.[2] More recent and more definite numbers are hard to come by, but this suggests that a significant fraction of people who live in the developed world will have their tissues preserved somewhere at some time in their life.

The dilemmas arising around tissue ownership don't apply to just Ted Slavins and John Moores (see chapter 9)—many of us have a stake. Who owns all this disembodied flesh and blood? Mostly, it belongs to the collectors. Usually we give up ownership of our parts when they are physically separated from us. But this doesn't mean tissue banks can do anything they want with their collections. The acquisition, appropriate storage, and use of tissues raises serious ethical questions. For instance, under what circumstances can a sample be collected? Must the donor be informed? Must the donor give (informed)

---

2. Elisa Eiseman and Susanne B. Haga, *Handbook of Human Tissue Sources: A National Resource for Human Tissue Samples* (Santa Monica, CA: RAND Corporation, 1999).

consent? Who should have access to the samples? What if the samples contain or are linked to information that can be used to identify the donor? Can a sample collected for one use be reused for a different purpose?

The answer to these questions depends on our attitudes towards our body and its parts, as well as our attitudes towards property, sharing, and community. A bank is a place where we deposit our money in order to keep it safe and in order to earn interest on it. But banks also serve a social purpose—the pooling of money allows banks to make loans, generating investments that benefit the economy as a whole (and, at least in theory, society as a whole). The notion of a tissue bank is a metaphor: we deposit our tissues not with the hope of retrieving them in the future, but rather to provide society with a resource that it might distribute for the collective good.

The donation of blood and organs has traditionally been understood as an act of civic gift-giving: rolling up one's sleeve not necessarily to help a relative or friend, but to help the community at large. A large supply of blood suggests not only healthy individual bodies but also a healthy body politic. In the days after the September 11, 2001, attacks on New York, for instance, thousands of people lined up to donate blood—donating bodily parts was a means of showing solidarity, support, and resilience in the face of terrorism. The giving and exchange of blood—and sometimes other body parts too—is a symbol of nation, citizenship, and community.

But what happens when bodily parts become less "gifts" and more "commodity?" When biotechnology companies, hospitals, doctors, and drug companies begin to treat bodily parts as property, other people too begin to see their bodies as their own personal capital. Recently, for instance, there has been a trend towards *private* banking of blood and tissues. Rather than donating their organs to public banks, some individuals have opted to set up their own tissue account, saving parts of their body for future regenerative or therapeutic use (this is called autologous donation). One increasingly common practice is the banking of the umbilical cord blood of newborns (which contains stem cells) for possible future medical use.

Treating the bodies as commodities often leads to an erosion of altruism and community. Individuals may be more likely to try to keep their tissues for themselves or for their immediate family. Even more worrying, creating a private market in tissues seems increase the flow of tissues from poor to rich (and from the global South to the global North)—the wealthy can afford to buy their health at the expense of those who must sell their body parts to survive. At present, tissues are treated both as gifts and as commodities, depending on the tissue and the context. In the United States, reproductive material (sperm and eggs), for instance, can be donated or legally sold. Whole organs, on the other hand, can only be legally gifted, but thousands are purchased anyway through a worldwide black market.

Biotechnologies have increased the possibilities and incentives for human parts to circulate and be exchanged outside the body. As a side effect, they have also increased the extent to which these parts may be treated as commodities. But this is not just a problem of who owns what. It is also a problem about how individuals in a society relate to one another: do we want to live in a society that encourages altruism or one where every exchange is reducible to money and markets? Finding a way to justly regulate the exchange of tissues will depend on finding a way to manage circulation without completely succumbing to the imperatives of commodification. It will mean taking into account not just the value of blood and tissues in markets, but also their symbolic and sentimental value as gifts and sacrifices.

XENOTRANSPLANTATION

The ways in which organs circulate from human to human suggests much about how people relate to each other. Likewise, the transplanting of organs from animals to humans (xenotransplantation) can tell us much about how we relate to other species around us. This too has a history that predates modern biotechnology.

Alexis Carrel's pioneering attempt to grow chicken heart cells outside the body (discussed in chapter 9) was just the beginning of an even more ambitious project. For Carrel this was just a first step towards being able to maintain whole organs outside the body, or even to replace organs with mechanical substitutes. In the 1930s, Carrel worked with the pilot Charles Lindbergh (the first person to fly nonstop across the Atlantic) to create a "perfusion pump" designed to keep a whole organ alive outside the body by "perfusing" them with blood. The pump worked, but it couldn't prolong the life of organs indefinitely. And, it was still a far cry from developing a fully mechanical organ substitute.

Carrel had long known that a more likely solution to the problem of failing human organs was to be found in animals. "The ideal method would be to transplant on man organs of animals easy to secure and operate on, such as hogs, for instance," Carrel wrote in 1907. But he also realized the difficulties this might entail: "It would in all probability be necessary to immunize organs of the hog against the immune serum."[3] The problems of organ rejection were already apparent. Carrel knew of some of the early European at-

---

3. Alexis Carrel, "The Surgery of Blood Vessels, etc.," *Johns Hopkins Hospital Bulletin* 18 (1907): 18–28.

tempts to transplant organs between humans and animals. Between 1902 and 1910, Ernst Unger (in Germany) and Mathieu Jabouley (in France), amongst others, attempted various transplantations between rabbits, pigs, goats, monkeys, and humans. None of the patients survived more than a few days. Almost no more such surgical experiments were attempted for forty years.

The main problem with these attempts at xenotransplantation was, as Carrel suspected, the rejection of the organ by the immune system. The human immune system is designed to attack any foreign cells that it discovers within the body. By recognizing special molecules on the outside of cells, it is good at distinguishing familiar from foreign. An organ introduced from an animal (or indeed from another human) immediately triggers this immune response, causing the body to attack the transplanted organ. Usually this results in organ failure and a quick death for the patient.

By the 1960s, significant advances had been made in techniques for transplanting organs from human to human. Much of this success resulted from the discovery of drugs that suppressed the immune system, allowing the organ to establish itself in the body. The success of these operations led to an increased demand for organ transplants and led to an acute shortage of human organs available for transplant. One obvious solution was to once again turn to animals.

In 1963, Keith Reemtsma at Tulane University in New Orleans transplanted kidneys from first a baboon, then a rhesus monkey, and finally a chimpanzee into patients with kidney failure. The chimpanzee recipient survived nine months, demonstrating that xenotransplantation was feasible. In 1964, Thomas Starzl in Denver, Colorado, transplanted six more baboon kidneys. The patients all died within 100 days. Such experiments continued through the 1960s and 1970s, almost always involving patients for whom no human donor could be found. Further advances in immunosuppression, particularly the discovery of ciclosporin in 1972 (and later tacrolimus in 1992), continued to make surgeons hopeful that xenotransplantation would succeed. Nevertheless, to date there has been no long-term survival of a patient receiving an animal-to-human organ transplant.

But implanting whole organs is not the only way to use animals for medicine. Tissues and cells can also be transplanted to treat burns (skin grafts), diabetes (transplant of islet cells), neurological disorders such as Parkinson disease (transplant of neural cells), and even AIDS (transplant of immune system cells). Biologists now distinguish between *in vivo* xenotransplants (including implants of live cells, tissues, or organs) and *ex vivo* uses, in which transplanted or implanted materials have contact with (nonhuman) animal cells outside the body. For instance, patients with liver or kidney failure can be restored to health by circulating their blood through a living animal (usually a baboon). The animal's liver and kidney do the work usually performed by the human organs. Human skin cells, insulin, and dopamine have also been produced outside the body using animal cells as "feeders."

More recently, engineers and biologists have worked together to develop "bioartificial" devices that use animal cells to perform organ functions outside the body. In these devices, the patient's blood is directed to flow through a chamber containing, for instance, cultured pig liver cells. The cells filter the blood, performing the usual function of the liver. These techniques have enjoyed success, especially in prolonging the life of a patient before a human transplant organ becomes available.

The major driver for xenotransplantation is the critical shortage of human organs. In the United States, 50% of potential organ recipients die on the waiting list. Animal organs could alleviate this problem, as well as providing new treatments for diseases such as diabetes and Parkinson disease. But there remains a profound discomfort with introducing animal parts into human bodies. Medical anthropologists who study people who have undergone (human to human) transplants report that the patients often sense the foreignness of their new organ, feeling it is inside of them but not part of them. What will be the consequences of transplanting animal organs and tissues?

Even if we don't end up feeling part animal, the very idea of this kind of hybridity challenges our everyday social and cultural categories. If the heart of a pig or the kidney of a monkey can exist inside our bodies, sustaining our life, then this puts us into new sorts of relationship with those animals. They become closer to us. Xenotransplantation has the potential to change ideas about kinship and relatedness—our similarities and differences to other creatures must be reassessed.

The boundaries of the human become more fluid and more open too. If an animal organ can be put inside our bodies, then it must share some fundamental similarities with our own biology. At least at the level of tissues, differences in terms of genes or appearance may become less important than biocompatibility. This may even provoke some rethinking of ideas about difference amongst humans.

All this is speculation because at present xenotransplantation is in its infancy; its lack of success has meant that society has not had to explicitly deal with its consequences. But this is likely to change. Xenotransplantation raises some important ethical questions: What guidelines should exist to ensure the safety of xeno procedures? How should we balance the therapeutic benefit to humans against the use of animals as medical and experimental subjects? How will it be possible to ensure fair access to xenotransplantation? Satisfactory answers to these ethical questions will require coming to terms with new relationships and boundaries between animals and humans. The meanings of safety/danger, same/different, and nonhuman/human are being redefined by these technologies as we go along. In other words, resolving the problems posed by xenotransplantation will mean becoming more comfortable with hybrids.

GROWN TO ORDER

The ethical questions surrounding xenotransplantation are also complicated by the changing relationships between biology and business. Increasingly, the transfer and exchange of tissues and organs is a business. In this business, human and animal parts are the goods. In most nations, including the United States, the buying and selling of human organs is illegal. Nevertheless, in 2012, the human rights group Organs Watch estimated that between 15,000 and 20,000 human kidneys were sold illegally around the world. Human organs have become, especially for the world's poor, goods to be traded, often for cash.

One clear trend in the history of xenotransplantation is the transition from the use of monkeys and apes (mostly baboons and chimpanzees) to the use of pigs. Monkeys and apes are closer evolutionary relations (and therefore closer genetic relations) than pigs. This means that the difficulties posed by immune rejection are fewer for primates. But pigs have other advantages: monkeys and apes are more expensive to raise and care for, they are in shorter supply, and their very closeness to humans makes them targets for animal rights activists. Pigs, on the other hand, are readily available. They are already farmed for food; they breed rapidly, grow rapidly, are cheap to maintain; they are anatomically similar to humans; and their physiology has been extensively studied. This transition to pigs is a further symptom of the commodification of body parts. The use of a mass-produced animal as organ donor shows how tissues are increasingly becoming involved in the capitalist logic of markets. The demand for organs can be met by increasing the supply from animals.

Pigs are also considered ideal candidates for genetic modification. Some experiments have already been done on producing transgenic pigs that are more suitable for xenotransplantation. PPL Therapeutics, the company that funded the research that led to Dolly the sheep, has used cloning techniques to create piglets that lack some of the genes associated with provoking an immune response. Some biotechnologists look forward to a future in which it will be possible to "grow organs to order." Pigs, they foresee, could be genetically engineered to be immunological matches for specific patients, reducing or eliminating the problems associated with rejection. Need a new heart? Just have a transgenic pig grown up to meet your specifications. Even better, have a whole line of pigs created now in case you need a heart, liver, or kidney transplant in the future.

The technologies of cloning and tissue culture have also been put to use in a slightly different arena. In the last decade, a few companies have explored the possibilities of growing edible animal meat. Called *in vitro meat* or *cultured meat* or *shmeat*, the aim is to grow layers or sheets of animal muscle cells to create slabs of tissue approximating a steak of pork, chicken, or beef. At present, producing edible meat with these methods remains prohibitively expensive (about $1 million for a half-pound [250g] beef steak). But, in the long term, in vitro meat may offer ethical (no animals need be killed), environmental (it reduces the impact of livestock), health (the contents of the meat [e.g., fat] can be precisely controlled), and economic (increasing world food supply) advantages.

In vitro meat provides another example of how biotechnology has led to the commodification of body parts. Tissues, human and animal, are increasingly objects that can be bought, sold, and traded. This is a new economy of flesh. In the future, making our food and making our organs may become increasingly part of the same industry, dependent on the same technology and the same global capital and infrastructure.

CONCLUSIONS

This chapter and the last have considered many kinds of ways in which tissues can cross the boundaries of our bodies: sperm and blood can be frozen, tissues can be banked, organs can be transplanted and perhaps some time soon be replaced with those of animals. Biotechnologies have contributed to our abilities to mix and match in these diverse ways, keeping cells and organs alive (or suspended) in test tubes, petri dishes, freezers, and other bodies.

What are the consequences of these affronts to our bodily integrity? For one thing, they pose a set of ethical challenges. We need to find ways to manage the extraction, storage, and transfer of human and animal parts that are humane, are just, and do not infringe the privacy of those involved. This depends, in part, on questioning the economic value that is increasingly imposed on the body by biotechnology and medicine. As we saw in chapter 9, parts of the body are increasingly implicated in regimes of ownership. Ethical problems concerning human body parts have been deeply entangled in legal problems relating to property and economic problems relating to markets. But economic value is not the only form of value that is attached to organs and tissues: bodily parts also have sentimental value, scientific value, dignity value, and value to community. We need to consider these different forms of value alongside the economic and property value of tissues.

But how do we do this? Assigning value to life outside the body is especially difficult. Our usual categories suggest that there is a clear boundary between animal/human, living/dead, and inside/outside. These categories allow us to make clear distinctions of value: humans are (usually) more valuable than animals, living is more valuable than dead, inside (the body) is more important than outside. The blurring of these categories by biotechnologies has made these value distinctions problematic. What is the worth—in terms of dignity, community, sacrifice, respect, and so on—of a frozen egg, a blood donation, or a transgenic pig heart? We need new hybrid categories and new ways of describing the value of these chimeric entities. Working out ethical solutions to the problems of biotechnology means working out how to value tissues without just "putting a value" on them.

FURTHER READING
The literature on medical devices is dominated by discussion of the ethical and legal frameworks. However, a useful book that examines issues of enhancement, regulation, privacy, security, and bodily integrity is Mark N. Gasson, Eleni Kosta, Diana M. Bowman, eds., *Human ICT Implants: Technical, Legal, and Ethical Considerations* (The Hague, Netherlands: Asser Press, 2012). Polge's discovery of cryopreservation is narrated in R. H. F. Hunter, "Ernest John Christopher Polge: 16 August 1926–17 August 2006," *Biographical Memoirs of Fellows of the Royal Society* 54 (2008): 275–296. Historical accounts of the development of cryopreservation include Bronwyn Parry, "The Brain on Ice: Technologies of Immortality," *Studies in the History and Philosophy of the Biological and Biomedical Sciences* 35 (2004): 391–413; and Joanna Radin, "Latent Life: Concepts and Practices of Human Tissue Preservation in the International Biological Program," *Social Studies of Science* 43, no. 4 (2013): 483–508. On the impact of new cellular technologies see also Hannah Landecker, "Living Differently in Biological Time: Plasticity, Temporality, and Cellular Biotechnologies," *Culture Machine* 7 (2005). For a collection of anthropological reflections on new meanings of life and death see Sarah Franklin and Margaret Lock, eds., *Remaking Life and Death: Toward an Anthropology of the Biosciences* (Santa Fe, NM: School of American Research Press, 2003).

The history of tissue banking and the politics and "economies" of tissues have been analyzed first by R. Titmuss, *The Gift Relationship: From Human Blood to Social Policy* (London: London School of Economics Press, 1971), and more recently by Bronwyn Parry and Cathy Gere, "The Flesh Made Word: Banking the Body in the Age of Information," *Biosocieties* 1 (2006): 41–54; Catherine Waldby and Robert Mitchell, *Tissues Economies: Blood, Organs, and Cell Lines in Late Capitalism* (Durham, NC: Duke University Press, 2006); Kara Swanson, *Banking on the Body: The Market in Blood, Milk, and Sperm in Modern America* (Cambridge, MA: Harvard University Press, 2014); R. Almeling, *Sex Cells: The Medical Market for Sperm and Eggs* (Berkeley: University of California Press, 2011); and Susan E. Lederer, *Flesh and Blood: Organ Transplantation and Blood Transfusion in 20th Century America* (New York: Oxford University Press, 2008).

There is a large body of literature dealing with the social and cultural meanings of dead bodies to organ transplantation (especially from the discipline of medical anthropology). Some important works include Lesley A. Sharp, *Bodies, Commodities, and Biotechnologies: Death, Mourning, and Scientific Desire in the Realm of Human Organ Transfer* (New York: Columbia University Press, 2007); Margaret Lock, *Twice Dead: Organ Transplants and the Reinvention of Death* (Berkeley: University of California Press, 2002); Lesley A. Sharp, *Strange Harvest: Organ Transplants, Denatured Bodies, and the Transformed Self* (Berkeley: University of California Press, 2006); Nancy Scheper-Hughes and Loic Wacquant, eds., *Commodifying Bodies* (London: Sage, 2002); and Nancy Scheper-Hughes, "The Ends of the Body: Commodity Fetishism and the Global Traffic in Human Organs," *SAIS Review* 22, no. 1 (2002): 61–80. An issue of the *American Journal of Bioethics* was devoted to the regulation of markets for organ transplantation (*American Journal of Bioethics* 14, no. 10 [2014]).

Important works that deal with the value of bodies in contemporary society include Melinda Cooper, *Life as Surplus: Biotechnology and Capitalism in the Neoliberal Era* (Seattle: University of Washington Press, 2008); Margaret Lock and Judith Farquhar, *Beyond the Body Proper: Reading the Anthropology of Material Life* (Durham, NC: Duke University Press, 2007); and Ruha Benjamin, *People's Science: Bodies and Rights on the Stem Cell Frontier* (Stanford, CA: Stanford University Press, 2013).

An account of early twentieth-century attempts at transplantation and xenotransplantation by Carrel and Lindbergh is David Friedman, *The Immortalists: Charles Lindbergh, Alexis Carrel, and Their Daring Quest to Live Forever* (New York: Ecco, 2007). The literature on xenotransplantation is sparser. A good overview from inside the field is J.-Y. Deschamps, F. A. Roux, P. Sai, and E. Gouin, "History of Xenotransplantation," *Xenotransplantation* 12, no. 2 (2005): 91–109.

PART VI

MAPPING GENES, MAKING SOCIETY

The next several chapters focus on the implications of biotechnologies based on genetics. Before we get to these, however, we must discuss one of the most significant uses of genetic science in the twentieth century. Historical examples can provide useful lessons for understanding some of the social and political effects of biotechnologies. In particular, the ways in which biological ideas have been mobilized into social policy in the past may provide clues as to what is likely to happen in the future. History can supply cautionary tales about what kinds of science and what kinds of policies we might want to avoid. This is certainly true of eugenics. The history of eugenics provides context for the discussion of modern medical genetics in chapters 12 and 13 and suggests that that we need to exercise extreme caution in applying *any* biological ideas to make social policy.

Eugenics was a biological, social, and political movement that enjoyed widespread popularity between roughly 1910 and 1945. Beginning in England, it reached its strongest forms in the United States and Germany. Eugenicists argued that both socially desirable traits (attractiveness, intelligence, industriousness) and socially undesirable traits (mental illness, poverty, sexual promiscuousness, alcoholism) were caused by heredity (that is, passed on from parents to offspring). For the most part eugenicists believed there was no escaping one's biological destiny. The eugenics movement attempted to enact social policy on the basis of these biological "facts." Society could be improved, they argued, only if the socially "fit" (those with desirable traits) produced more offspring than the socially "unfit" (those with undesirable traits). Eugenic policies therefore encouraged reproduction of the "fit" and discouraged reproduction of the "unfit."

One response to this history is to argue that eugenics was simply *wrong*: that it was based on *bad science*. Biologists now consider many eugenic ideas far too simplistic. For one thing, it was based on the naïve assumptions that traits such as intelligence were based on single genes and that individuals possessing "bad" genes could be readily identified. Eugenicists often mixed scientific ideas with wild speculation in order to bolster their arguments for social policy.

But just saying eugenics was wrong misses the point. Eugenic ideas were espoused by many of the leading geneticists of the early twentieth century and aspects of its program were at least tacitly accepted by many other scientists. It was given broad institutional and financial support. Money came from the Carnegie Institution of Washington and from the Rockefeller family; members of the Protestant clergy supported the movement and hundreds of high schools, colleges, and universities taught courses on eugenics. It was also a diverse movement that had supporters from across the political spectrum. Arguing that eugenics was merely "bad science" suggests that "good science" would have produced good social policy. But history shows that science that seemed sound at one time is invariably seen to be wrong or incomplete later on. This should lead us to be extremely cautious when we try to make decisions based on current scientific thinking. Moreover, just labeling eugenics "bad science" does not explain why it had such a wide appeal to so many scientists and nonscientists. Understanding how eugenics did damage to so many people's lives requires understanding its "successes" as well as its failures.

FROM DARWIN TO GALTON

Although eugenics was developed and implemented in many countries, the roots of eugenic ideas are usually ascribed to the work of the nineteenth-century naturalist Charles Darwin (1809–1882). In his *On the Origin of Species* (1859), Darwin proposed the theory of natural selection: organisms evolve through competition with one another—those best adapted to their environment survive while some species become extinct (sometimes called "survival of the fittest"). Darwin's ideas had certainly been influenced by his observations of Victorian society. He had seen or read about England's poor dying in workhouses and slums, and he had watched some of his own children perish due to illness. But, at first at least, Darwin did not dare extend his theory to take account of human societies—its suggestion that humans were distantly related to apes was controversial enough as it was.

But others were not so reticent. Darwin's work was a revelation to many, explaining so much of the complexity of nature with one simple principle. For some, its implications for society were immediate and obvious. The individual who took up these ideas most enthusiastically was Francis Galton (1822–1911), Darwin's own cousin. Beginning in 1865, Galton began to publish his ideas about applying Darwinian theory to humans. Galton believed that by selecting "better" or "fitter" human beings for marriage (and reproduction), society could gradually be improved. After all, farmers and breeders selected the best cows or the best dogs for breeding in order to produce the most desirable offspring. Couldn't the same principle be applied to humans? Galton believed it could: it is "quite practicable," he wrote, "to produce a highly gifted race of men by judicious marriages during several consecutive generations."[1]

For this scheme to make sense, however, Galton had to show that socially valuable human traits (such as intelligence) were heritable (passed on from parents to offspring). If traits weren't heritable, then selection would have no effect.[2] Intelligence, however, was difficult to measure. Sensibly, Galton began by attempting to show that other, simpler, human traits were passed down from generation to generation. Galton set up a laboratory in London to measure the heights, weights, arm spans, and breathing capacity of thousands of individuals. The Anthropometric Laboratory, established in 1885, was especially interested in comparing measurements across generations: between grandfather, father, and son, for instance. If Galton could show that taller grandfathers produced taller fathers and taller sons, then it would strongly suggest the heritability of height, for instance.

Galton's work in this field had a lasting legacy: the development of modern statistics. Galton realized that not all tall fathers have tall sons—there was not an *absolute* or *deterministic* relationship between generations. Rather, it was perhaps only that tall fathers had a greater *likelihood* of producing sons taller than *average*. To express these relationships, Galton (and his followers) developed statistical techniques such as *correlation* and *regression*. Galton showed that traits such as height were distributed according to a *normal curve* and that selecting shorter or taller parents shifted the shape of the curve for the subsequent generation.

In the end, Galton also tried to show that the same sorts of relationships applied to less quantifiable traits. In the 1860s, Galton began compiling the family histories of "eminent men" in England (such as Fellows of the Royal Society). Galton's hypothesis was that eminence was passed through families: eminent ancestors produced eminent descendants. Galton published the results of his study in what he called "historiometry" in his 1869 book *Hereditary Genius*.

Galton's work was continued by his protégé, Karl Pearson (1857–1936). Pearson was a skilled mathematician and further developed Galton's statis-

---

1. Francis Galton, *Hereditary Genius: An Inquiry into Its Laws and Consequences* (London: Macmillan, 1869), 1.

2. If intelligence was heritable, then we would expect intelligent parents to produce intelligent children. If intelligence was not heritable, then we would expect intelligence to be randomly distributed through the population—in other words, anyone could be intelligent, regardless of parentage.

tical ideas, gaining increasing scientific credibility for the new field of eugenics. Pearson inaugurated the journal *Biometrika* in 1902 to publish eugenic ideas. In 1911, he took up the specially created post of Galton Eugenics Professor at the new Department of Applied Statistics at University College London. By the first decades of the twentieth century, eugenics comprised a well-respected and widely supported set of scientific ideas.

The plausibility and prestige of eugenics was enhanced by the ways in which it resonated with European (and especially English) notions of race and national competition. The British Empire, in particular, was justified on the grounds of racial superiority: the English ruled over Indians and other peoples because of their superior racial characteristics, Victorians believed. "History shows me only one way," Pearson espoused, "in which a high state of civilization has been produced, namely the struggle of race with race, and the survival of the physically and mentally fitter race."[3] But if international politics was determined by racial characteristics, then this meant that preserving those characteristics was absolutely essential for maintaining power.

If inferior individuals, especially the poor and the lower classes, reproduced more frequently than the better types, the English race would quickly degenerate, eugenicists believed. The result would be catastrophic: Britain would lose its special status in the world order. This "degeneration" is exactly what many eugenicists saw around them. Galton and Pearson noted with dismay that the poor *did* seem to be having more offspring than the upper classes. Many eugenic policies were aimed at averting this national disaster by discouraging the reproduction of the poor and encouraging the reproduction of the rich (Pearson suggested giving economic incentives to well-matched couples to have babies).

Galton and Pearson's work attracted much attention in Britain, especially amongst the elite. But eugenic policies were never implemented by the British state. Other countries—including Argentina, France, Germany, the Soviet Union, and the United States, and Scandinavian nations—not only produced their own eugenic ideas but also implemented them to various degrees before World War II. Many of these nations experienced rapid population growth (especially through migration) in the last decades of the nineteenth century. This expansion was accompanied by an increasing role for the state in providing health care. These two conditions increased the perceived legitimacy of state intervention to safeguard the health of the population.

In France, for instance, a French Eugenics Society was founded in 1912 and

3. Karl Pearson, *National Life from the Standpoint of Science* (London: Adam & Charles Black, 1901), 19–20.

a National Social Hygiene Office was founded in 1924 (funded by the Rocke-feller Foundation). The latter focused on measures such as treating tubercu-losis, preventing alcoholism and prostitution, and controlling sexually trans-mitted infections. In other words, eugenics was limited to attempts to combat disease and improve living conditions. In Sweden, on the other hand, eugenic ideas led to the implementation, in 1934, of a compulsory sterilization law. The law sterilized persons deemed "unfit" to have children due to perceived mental, physical, or social deficiencies. In the United States and Germany too, eugenics moved from "positive" (attempts to encourage procreation amongst "healthy" individuals) to "negative" (attempts to stop or outlaw procreation of "unhealthy" individuals).

EUGENICS IN THE UNITED STATES

It was in the United States that eugenics found its way most deeply into social policy. America's history of slavery and immigration made ques-tions and problems of race (and especially race mixing) especially pertinent. The second half of the nineteenth century saw not only the emancipation of African Americans, but also increased levels of immigration of Irish, Chi-nese, Italians, and Jews. As in Britain, this population explosion caused in-creasing anxiety about the potential dilution of racial strength.

The most influential American convert to eugenics was Charles Davenport (1866–1944). Davenport was inspired by the rediscovery of Mendel's laws in 1900. In the 1860s, an obscure Austrian monk—Gregor Mendel—had con-ducted a series of experiments on peas in his vegetable garden. His results showed that specific traits of the peas (such as color) were passed on from generation to generation according to precise numerical ratios. This theory, widely circulated only long after Mendel's death, seemed to support eugenic notions that specific traits were passed on from generation to generation.

In 1904, Davenport persuaded the Carnegie Institute of Washington to provide funds to establish an experimental station at Cold Spring Harbor on Long Island, New York. The Station for Experimental Evolution was dedi-cated to searching for Mendelian traits in humans. Davenport used the gen-erous funding to begin collecting extended family pedigrees from around the United States. These recorded the incidence of mental diseases, "imbe-cility," epilepsy, criminality, "sexual immorality," alcoholism, and other un-desirable traits. Like Galton, Davenport aimed to show that these characteris-tics were passed down through families. This was painstaking work, resulting in a widely circulated and acclaimed book, *Heredity in Relation to Eugenics*, published in 1911.

Such work was expected to have significant social implications. In par-

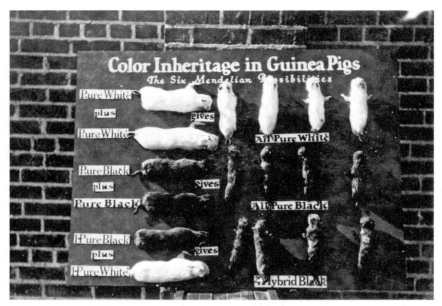

11.1 Color inheritance in guinea pigs (ca. 1925). Exhibit designed to educate the public about Mendelian inheritance and eugenics. Reproduction between two white guinea pigs produces healthy offspring. Reproduction between two black guinea pigs produces black guinea pigs. Reproduction between black and white guinea pigs, however, produces smaller and seemingly deformed offspring. Such displays were supposed to demonstrate the eugenic danger of racial mixing. Source: American Philosophical Society. Used by permission.

ticular, it showed, according to Davenport, that the mixing of races (especially between white and black) was producing inferior offspring and gradually weakening American society (figure 11.1). "The great influx of blood from Southeastern Europe," Davenport worried, would make the American population, "darker in pigmentation, smaller in stature, more mercurial, . . . more given to crimes of larceny, kidnapping, assault, murder, rape, and sex-immorality."[4] In the first place, Davenport recommended stricter rules for immigrants, checking their family history to ensure that the "nation's protoplasm" was not compromised. These eugenic ideas found expression in the Immigration Act of 1924. The new law created quotas based on national ori-

4. Charles Davenport, *Heredity in Relation to Eugenics* (New York: Henry Holt, 1911), 219.

gin. It severely limited immigration from Southern and Eastern Europe (including Jews), and prohibited immigration from the Middle East, East Asia, and India.

As eugenics grew in scientific reputation and popularity between 1910 and 1920, other policies began to be implemented too. "Race hygiene," as it was known in some contexts, began to become a worldwide movement with prominent scientist and politicians offering their support. In the United States, eugenic laws were largely adopted on a state-by-state basis. Initially, these efforts focused on encouraging appropriate marriages between "fit" individuals. For instance, some states provided prizes for "fitter families" or "better babies"—those certified as fulfilling eugenic criteria (figure 11.2). Competitions were held at state fairs and ads were run in newspapers. Education was a key component of these campaigns: eugenicists attempted to inform the public about how to spot a eugenically fit partner and warn people about the potential consequences of "bad matches."

Some of this now seems rather farfetched. We must not forget, however, that biological fitness was understood as the cause of crime and disease and

11.2 Better Babies Contest award certificate. Louisiana State Fair, Shreveport, 1913. Certificate from a contest designed to encourage matches and reproduction between eugenically fit individuals. Source: Mrs. Frank deGarno Papers, MS#1879, University of Tennessee Libraries, Knoxville, Special Collections. Used by permission.

therefore a critical social problem. Moreover, prominent scientists and scientific institutions of the time backed it up. However racist and misguided they now appear, many eugenicists sincerely believed that they were acting in society's best interest, protecting it from degeneration and ruin.

By the 1920s, eugenic measures in some jurisdictions began to go even further. Instead of just encouraging beneficial matches they began to actively intervene to prevent "bad" matches. The first step was marriage laws. As many as thirty states adopted laws that restricted marriage if either party was considered eugenically "unfit" (often using such criteria as "feeble-mindedness" or evidence of sexually transmitted disease).

For some eugenicists, even this did not go far enough. For one thing, unmarried people could have babies anyway, and thereby still affect the overall gene pool. One solution was to remove this possibility completely. The first eugenic sterilization law was passed in Indiana in 1907. But by the end of the 1920s, thirty-one states had enacted laws that forced some types of "unfit" men and women to undergo compulsory (involuntary) sterilization. In a 1927 case known as *Buck v. Bell*, the US Supreme Court declared such laws constitutional. Although men were sterilized to reduce aggression and control criminal behavior, many of the laws were directed at women as child-bearers, holding them more accountable for reproduction (about 60% of all sterilizations were performed on women).

Those subject to sterilization varied from state to state. Mostly commonly the legislation specified "imbeciles" or others deemed mentally deficient. In some cases, this extended to the mentally ill, the deaf, the blind, epileptics, and individuals with physical deformities. The racialized aspects of eugenic thinking became all too clear: African Americans and Native Americans became particular targets. Many individuals were sterilized not only against their will, but also sometimes without their knowledge (one practice was to perform sterilizations when a patient presented at a hospital for another condition or operation, including childbirth).

Between 1907 and 1963, about 64,000 forcible sterilizations were performed in the United States. Many of the sterilization laws remained in place until the 1960s and '70s, although the number of sterilizations declined from the 1940s onwards.

EUGENICS IN GERMANY

We usually associate the policies of Nazi Germany with unmitigated racism. Indeed, Adolf Hitler's racism targeted the Jews for complete genetic extermination. However, the Nazi regime also sought to expel or kill homosexuals, the mentally ill, the physically handicapped, and the otherwise sick and infirm. These Nazi racial and social policies were informed by eugenic ideas.

However, eugenics was not imported into Germany by the Nazis—its theories had long and deep roots amongst German thinkers. Eugenics in Germany was developed through the work of Alfred Ploetz (1860–1940) and Wilhelm Schallmayer (1857–1919). In his 1895 book *The Efficiency of Our Race and the Protection of the Weak*, Ploetz described an ideal society run according to eugenic principles: marriage and reproduction would be limited according to moral and intellectual capacities and disabled, sick, and weak children would be killed. Schallmayer likewise believed that eugenic policies were the key to solving social problems (such as overcrowding, epidemics, poverty, and crime) and achieving national power for the newly unified German state.

German defeat in the First World War fueled fears of the decline of the "Nordic race" and boosted eugenic thinking. In the 1920s, however, Germans looked more and more to the United States for examples of how to deal with racial and eugenic problems. For the most part, the US was happy to help. In 1927, the Rockefeller Foundation provided funds to establish the Kaiser Wilhelm Institute for Anthropology, Human Heredity, and Eugenics in Berlin. The work of its director, Eugen Fischer (1874–1967), influenced Hitler's racial thinking.

Once the Nazis came to power in 1933, they immediately began to implement their eugenic policies. In July 1933, the Reichstag passed the "Law for the Prevention of Hereditarily Diseased Offspring," requiring physicians to register known cases of hereditary disease. By 1934, the Nazis had implemented laws permitting involuntary sterilization for feeble-mindedness, mental illness, epilepsy, and alcoholism. Nazi propaganda emphasized the great cost to the state and to society of caring for physically and mentally ill people (figure 11.3). German eugenic laws were closely modeled on US examples and the Nazis closely followed eugenic developments in America. The Nazis propaganda also emphasized that Switzerland, Britain, Japan, Scandinavian countries, and the United States all had similar eugenic laws: "We do not stand alone," Nazi eugenic posters proclaimed.

In 1935, the infamous Nuremberg Laws prohibited marriages between Aryans and "unfit" persons. Jews, of course, were the main targets, but other eugenically unfit persons also fell under the scope of the law. By the outbreak of war in 1939, some 400,000 individuals had been sterilized. As the war created a growing demand for hospital beds and medical resources, sterilization was not enough: a "final solution" was required for those who were deemed to be a burden on German society. For the Jews, this meant a policy of total

11.3 Nazis and eugenics. Comparison of daily living costs for an individual with a hereditary disease and for a healthy family (from the series "Blood and Soil," ca. 1935). The left-hand texts reads: "An invalid costs the state RM5.50 per day"; the right-hand reads: "For RM5.50 a hereditarily healthy family can live for one day!" This poster was part of the National Socialist effort to improve the German people through "racial hygiene." Source: Bildagenter Preussische Kulturbesitz. Used by permission.

extermination. Other groups too—Roma, homosexuals, communists, and Slavs—were targeted for eugenic, racial, and political reasons. Between 1939 and 1941, doctors murdered 70,000 mental patients at psychiatric hospitals, gassing them with carbon monoxide. In 1941, many of these doctors were reassigned to concentration camps to assist with the mass murder of other "undesirables."

AFTER WORLD WAR II

World War II marked a dramatic turning point with respect to attitudes towards race and eugenics. The Nuremberg Trials of suspected Nazi war criminals revealed the full extent of the horrific crimes of the regime. In particular, Nazism showed how racism and eugenics had been used to justify grotesque medical experimentation, forced sterilizations, and the mass extermination of millions of people. Of course, eugenics did not disappear overnight, but eugenic thinking became less and less scientifically respectable. Some jurisdictions continued to have sterilization and other eugenic laws on their books, but they were used less and less frequently.

Racism became far less socially and politically acceptable after World War II. One of the priorities of the new United Nations (formed in 1945) was to enact a Universal Declaration of Human Rights. This document—adopted in 1948—set down a core of basic rights (including liberty, security of person, privacy, education, and freedom of thought and religion) that should be upheld "without distinction of any kind, race as race, color, sex, language, religion, political or other opinion, national or social origin, property, birth or other status."[5] Again, this did not mean that racism or other kinds of discrimination were eliminated. But it did begin to shift social and political conversations and attitudes—*equality* began to become an ideal towards which a society should aspire.

Scientific studies of human biological differences changed direction too. In the 1940s and '50s, some biologists turned to the study of heredity, difference, and evolution on a molecular level. Molecules, at first at least, seemed agnostic on questions of human difference and social behavior. Ironically, it is the success of molecular biology that is now forcing us to reconsider problems of discrimination based on biology. This will be considered in chapter 13. Also hoping to avoid controversy, other scientists turned to examining the relationship between biology and social behavior in other animal species (e.g., ants and bees).

However, some individuals in fields such as psychology, psychiatry, sociology, anthropology, and physiology still sought to connect human genes to social traits. After the war, these individuals inaugurated the field of "behavior genetics," which, at first at least, carefully avoided any discussions of race. From the 1970s onwards, work on human difference was taken up in the fields of "sociobiology" and "evolutionary psychology." Scientists in these disciplines argued that many modern human behaviors and traits (sexual behavior, gender roles, aggression, and many others) could be understood as evolutionary adaptations to our hunter-gatherer lifestyles. This left room for accounting for observed differences in social behavior (e.g., intelligence) along racial lines. After eugenics, race-based thinking in biology was transformed, but it did not disappear entirely (see chapter 21).

5. Article 2, Universal Declaration of Human Rights. See: http://www.un.org/en /documents/udhr/.

The rise and demise of eugenics provide a very good example of how science and society are linked. Eugenics emerged from nineteenth-century social attitudes towards human difference. However, in the end, it also served to legitimate those same attitudes. In other words, the science and the social attitudes supported one another. After World War II, both scientific and social ideas about race shifted together.

CONCLUSIONS: EUGENICS AS BIOTECHNOLOGY

In chapter 13, we will consider whether the history of eugenics provides any lessons for the application of modern genetics (to medical and social problems). Before addressing this question, however, it is worth reflecting on some of the similarities and differences between eugenics and the other biotechnologies discussed in this book.

First, eugenics was based on and developed from some of the best science of its time (even if it ultimately oversimplified its implications). It drew on the theory of evolution and cutting edge work in the study of genetics, including Mendel's theories of inheritance. Applying these ideas to humans required the development of new kinds of mathematics, and eugenics generated a whole field of inquiry that eventually dropped its interest in human heredity and became known as statistics. Eugenicists did let their convictions get the better of their scientific judgment, selecting data that suited their purposes and ignoring complexities (such as the polygenic origin of traits). Nevertheless, many eugenicists were respected scientists in their own fields and in wider society and they were funded and supported accordingly.

Second, like many of the biotechnologies we have encountered in this book, eugenics was aimed at control over biology and human improvement. Eugenicists sought to exert greater power over human reproduction, selecting desirable heredity traits and, ultimately, shaping a smarter, stronger, and more able society. It sprang from a political vision of a scientifically planned and technologically enhanced society. Its supporters imagined a future where biological control drove human intellectual and cultural progress.

Eugenics was not a molecular science. Darwin and Galton had little notion of how heredity worked. By the earliest years of the twentieth century, biologists knew that heredity seemed to be carried in units they called "genes" (the term was coined by Wilhelm Johannsen in 1909). But eugenicists did not know what genes were or how they worked (let alone that they were made of a molecule called deoxyribonucleic acid). Eugenicists' level of control was therefore fairly crude: they could intervene only on the level of the whole organism (that is, by picking out specific individuals for reproduction).

This also meant that eugenics employed an especially crude version of biological determinism: it suggested that all the traits of parents would be passed on wholesale to the offspring (subject to some statistical variation). "Biology is destiny"—you could not escape the lot passed down to you by heredity, your life and character were fully determined by your genes. Because eugenics could intervene only by controlling reproduction, this also meant that any "improvements" it offered were at the level of the collective or the "race." There was nothing it could do to help "unfit" individuals (except, perhaps, stop them being born). It was only society as a whole that could benefit.

These two features of eugenics—its crude biological determinism and its emphasis on the collective over the individual—have often been invoked to distinguish eugenics from more recent biomedical interventions. The next chapters assesses whether such distinctions hold up to scrutiny.

FURTHER READING

For a summary of the work of Francis Galton see Michael Bulmer, *Francis Galton: Pioneer of Heredity and Biometry* (Baltimore: Johns Hopkins University Press, 2003). Galton's own work on the inheritance of genius is Francis Galton, *Hereditary Genius: An Inquiry into Its Laws and Consequences* (Honolulu: University Press of the Pacific, 2001 [1869]). The work of Karl Pearson is described in a biography by Theodore M. Porter, *Karl Pearson: The Scientific Life in a Statistical Age* (Princeton, NJ: Princeton University Press, 2004) and the larger context of his statistical work is covered in Theodore M. Porter, *The Rise of Statistical Thinking, 1820–1900* (Princeton, NJ: Princeton University Press, 1986).

On the wider eugenics movement beyond Britain, a good starting point is the broader scientific context of the sciences of heredity as described by Staffan Müller-Wille and Hans-Jörg Rheinberger, *A Cultural History of Heredity* (Chicago: University of Chicago Press, 2012), especially chapter 5. On the movement in France see William Schneider, "Toward the Improvement of the Human Race: The History of Eugenics in France," *Journal of Modern History* 54 (1982): 268–291.

There is an extensive literature on eugenics in the United States. The most comprehensive is Daniel J. Kevles, *In the Name of Eugenics: Genetics and the Uses of Human Heredity* (Berkeley: University of California Press, 1985). Extremely readable are Diane Paul, *Controlling Human Heredity: 1865 to the Present* (Atlantic Highlands, NJ: Humanities Press, 1995); Diane Paul, *The Politics of Heredity: Essays on Eugenics, Biomedicine, and the Nature-Nurture Debate* (Albany: State University of New York Press, 1998); and Edwin Black, *The War against the Weak: Eugenics and America's Campaign to Create a Master Race* (New York: Four Walls Eight Windows, 2003). For a sense of Davenport's views it is worth going directly to the source: Charles Davenport, *Heredity in Relation to Eugenics* (New York: Henry Holt, 1911) and Charles Davenport, "The Effects of Race Intermingling," *Proceedings of the American Philosophical Society* 56, no. 4 (1917): 364–368. The Eugenics Archive has a very extensive collection of online images and resources related to the history of eugenics in America: http://www.eugenicsarchive.org. For perspectives that put eugenics within the broader contexts of early twentieth-century genetics see Elof Axel Carlson, *Mendel's Legacy: The Origins of Classical Genetics* (Stony Brook, NY: Cold Spring Harbor Laboratory Press, 2004); and Elof Axel Carlson, *The Unfit: A History of A Bad Idea* (Stony Brook, NY: Cold Spring Harbor Laboratory Press, 2001).

The German and Nazi versions of eugenics have also attracted significant attention. On pre-Nazi eugenics in Germany see Paul Weindling, *Health, Race, and German Politics between National Unification and Nazism, 1870–1945* (Cambridge: Cambridge University Press, 1989) and Sheila Faith Weiss, *Race, Hygiene, and National Efficiency: The Eugenics of Wilhelm Schallmayer* (Los Angeles: University of California Press, 1987). On the Nazi policies the best source is Robert N. Proctor, *Racial Hygiene: Medicine under the Nazis* (Cambridge, MA: Harvard University Press, 1988).

The literature and debates about sociobiology and evolutionary psychology are extensive. On behavior genetics see Aaron Panofsky, *Misbehaving Science: Controversy and the Development of Behavior Genetics* (Chicago: University of Chicago Press, 2014). The classic work of sociobiology is Edward O. Wilson, *Sociobiology: The New Synthesis* (Cambridge, MA: Belknap Press, 1975). And the lessons for humans are clearly spelled out in Edward O. Wilson, *In Search of Nature* (Washington, DC: Island Press, 1997). For criticism of sociobiology see Richard Lewontin, Leon Kamin, and Steven Rose, *Not in Our Genes: Biology, Ideology, and Human Nature* (New York: Pantheon, 1985) and Philip Kitcher, *Vaulting Ambitions: Sociobiology and the Quest for Human Nature* (Cambridge: Cambridge University Press, 1985). For a historical perspective on the rise of sociobiology see Carl N. Degler, *In Search of Human Nature: The Decline and Revival of Darwinism in American Social Thought* (New York: Oxford University Press, 1991). For prominent examples of evolutionary psychology see Donald Symons, *The Evolution of Human Sexuality* (New York: Oxford University Press, 1979); and Jerome Barkow, Leda Cosmides, and John Tooby, eds., *The Adapted Mind: Evolutionary Psychology and the Generation of Culture.* (New York: Oxford University Press, 1992); as well as more popular accounts by Steven Pinker, *The Blank Slate: The Modern De-*

nial of Human Nature (New York: Viking) and David Buss, *The Evolution of Desire: Strategies of Human Mating* (New York: Basic Books, 1995). For a critique of evolutionary psychology see David J. Buller, *Adapting Minds: Evolutionary Psychology and the Persistent Quest for Human Nature* (Cambridge, MA: MIT Press, 2006).

12   THE HUMAN GENOME PROJECT

INTRODUCTION

Biotechnology has promised and continues to promise dramatic improvements to medical diagnosis and treatment. We have already seen examples of how recombinant DNA led to new pharmaceutical products (chapter 5). Biological research too has been driven by the hope that greater understanding of the human body will further enable the development of new medical interventions. Since the rise of molecular biology and the "cracking" of the genetic code in the 1960s, many biologists have paid special attention to genes. These biologists see genes (and therefore the DNA that carries them) as the key to unlocking the most important knowledge about organisms. One analogy that molecular biologists use often is that of a computer program: DNA is like a piece of software or code. Just as the software gives instructions telling a computer what to do, DNA provides a complete set of instructions telling the organism what to do. Just as an error (or bug) in software might cause the computer to operate incorrectly, an error in DNA could cause the body to malfunction (that is, to become diseased).

This view of biology reached its apex in the 1980s and 1990s. Since DNA was considered to be so important, biologists argued that large amounts of money should be spent to understand it in full. Mapping and characterizing all the human genes would provide the basis not only for future biological research but also for biotechnologies and pharmaceutical innovations. In other words, there would be a significant payoff in terms of improved human health, they claimed. The Human Genome Project (HGP) emerged from this view of human biology that placed DNA at the center.

This chapter describes the origins of the HGP, its justifications, its methods, its results, and some of its consequences. The HGP might be considered just part of biology rather than biotechnology. But the project laid the foundation for a wide range of biotech enterprises that we will discuss in later chapters. Moreover, the HGP provides a backdrop to many of the debates about the social and political consequences of biotechnology. To some outside the project (including philosophers, sociologists, anthropologists, and some biologists), the link between DNA and bodies posited by the HGP seemed too simple. They worried that the HGP could lead to a reprise of eugenics (chapter 11), which also drew straightforward links between heredity and bodies. We will consider the implications of the HGP for discrimination in chapter 13.

Another reason to include the HGP in this account of biotechnology is because it provides a neat example of a sociotechnical system. The HGP was not just a scientific project: it required public funding (and therefore political support), involved international collaboration (and therefore global institution-building), generated spin-off technologies (that could transform the economy), and carried potential consequences for medicine, racial and ethnic discrimination, legal rights, and philosophies of human nature. Understanding the HGP means seeing it not just as a technical system that produced the human genetic code, but as a sociotechnical system that generated a particular set of social, political, and economic outcomes.

WHAT IS A GENOME?

The term "genome" was first used in 1920 by the botanist Hans Winkler (1877–1945). In its original context, it referred to the complete set of genes for an organism. Geneticists knew that hereditary information seemed to be carried on the thread-like structures called chromosomes that were contained within the nucleus of cells. These structures seemed somehow to be organized into distinct units of heritability known as genes. All the units (genes) on all the chromosomes made up a genome.

Over the course of the twentieth century, molecular biologists discovered, first, that chromosomes were actually very long molecules made out of DNA. And, that the role of some parts of this DNA is to carry instructions for the making of proteins in the body (see box 3.1). Those "protein-coding" parts of the chromosome came to be identified with genes. Genes became just those parts of chromosomes that were instructions for proteins. But molecular biologists also discovered that the chromosomes contained lots of other parts: parts that never made protein, parts that appeared to control or regulate the protein-making of other parts, parts that just seemed to repeat over and over, and other seemingly inactive or useless parts.

The word genome is now used to describe all of the DNA of a particular organism. This does not just include the genes. In other words, the genome is the entirety of the DNA contained in an organism—the complete and ordered set of nucleotide letters (As, Gs, Ts, and Cs).

Genomes can vary greatly in size. Some viruses have genomes of only a few thousand letters (or base pairs or nucleotides). Some forms of bacteria have a million base pairs (bp), other bacteria almost 10 million bp, fungi up to 100 million bp, insects between 100 million bp and 6 billion bp, most mammals around 5 billion bp (humans come in at 3 billion bp), and some flowering plants up to 100 billion bp. In other words, there is a million-fold difference in the length of the genomes of different kinds of organisms. Larger genomes do not necessarily carry more genes.

WHAT IS SEQUENCING?

Sequencing is the process of actually determining the ordering of the nucleotide letters in a stretch of DNA. For the sake of simplicity, let's consider just a single strand of the DNA (even though DNA is usually double-stranded). This strand is a long molecule. It has a "backbone" along which are strung a series of molecules. This molecule can be one of four types: adenine (usually abbreviated A), guanine (G), thymine (T), cytosine (C). These molecules (called bases or nucleotides once attached to the backbone) can occur in any order: there might be five adenines in a row, or adenine then guanine then cytosine, or guanine followed by sixty thymines (for more details see the primer on DNA in box 3.1 and figure 3.1).

A whole chromosome is very long, so a stretch of DNA may contain hundreds or thousands of bases, one after another. It is the ordering of the bases that makes up the "code" of DNA. For genes — the parts of the DNA that make protein — it is the ordering of the bases that specifies or defines how to make the protein (see the genetic code in table 3.1).

Sequencing means determining this ordering of bases on a long DNA molecule. It is like reading out the DNA, discovering the order of the nucleotides along the backbone and recording this as As, Gs, Ts, and Cs on a piece of paper or a computer screen. This is not an easy task. Molecules are far too small to be directly observed under a microscope — you cannot just look and see the As, Gs, Ts, and Cs. Instead, complicated biochemical procedures were developed to reveal the patterns of nucleotides.

In 1976, Allan Maxam and Walter Gilbert developed one method of sequencing using specially constructed radioactive A, G, T, and C molecules. These molecules were incorporated into DNA and the sequence could be determined by detecting the levels of radiation. In 1977, Frederick Sanger (1918–2013) developed an alternative method that used reduced levels of radiation and fewer toxic chemicals. The reliability and relative simplicity of the chain-termination method, as it is sometimes known, made it the dominant sequencing method (until next-generation sequencing methods began to emerge in the mid-2000s). The process of Sanger sequencing is explained in box 12.1 and figure 12.1.

Although an improvement, Sanger sequencing still required four separate sequencing reactions (one for each type of nucleotide) and the use of radi-

Box 12.1 Sanger Sequencing

In Sanger sequencing, specially constructed A, G, T, and C molecules are used. These special molecules are "chain-terminators"—once they are added to a piece of DNA, no further bases can be added on (the chain ends). Imagine starting with many identical copies of a single stranded piece of DNA (this is the DNA you wish to sequence). Mixed with these copies are many normal A, G, T, and C molecules plus some of the special G-chain-terminator variety (we'll leave out the A, C, and T chain-terminators for now). Now, if you add a special enzyme called DNA polymerase to the mix, the single DNA strands will start to try to build themselves into double-stranded molecules using the free-floating As, Gs, Ts, and Cs. According to the rules of base-pairing, As will pair up with Ts and Cs with Gs.

Each single-strand will keep on adding normal As, Gs, Ts, and Cs until by chance it incorporates one of the special G-chain-terminator molecules: at this point the strand-building will stop. When the strand incorporates a G-chain-terminator is completely governed by chance: some might get a chain-terminator on the very first time it tries to incorporate a G and stop almost immediately. Others might go on for hundreds of base pairs before stopping (remember, there are normal Gs in the mixture too, so it may incorporate lots of these before it gets a special G).

We now add a chemical to the mixture that causes all the double-stranded DNA to separate into single strands again. The result is many strands of different lengths, all ending with a G: some might be just a few base pairs long (those that chain-terminated right away), others might be much longer (those that, by chance, didn't pick up a G-chain-terminator for quite some time).

Now, we do this whole procedure again, but now instead of using the G-chain-terminator, we use the C-chain-terminator. This results in lots of strands of different lengths too, now all ended in C. And then we do the same for T and A.

The last trick is to sort all these strands by size. This can be done by a process known as gel electrophoresis. In electrophoresis you make the strands slightly electrically charged and put them in a thick gel. If you put a voltage across the gel so that one side is positive and one side negative, the charged strands will be attracted to one side (positive attracts negative). But it's hard for long strands to travel through the thick gel—in fact, the longer the strand, the harder it is to move, so the result is that longer strands end up left behind and the whole collection ends up in size order. Once this is done, you have the sequence! If the very shortest strand ended in a G, then that must be the first letter in the sequence; if the next shortest ended in an A, then A comes next; and so on. You can read off the letters up to the longest strand.

Sanger Sequencing

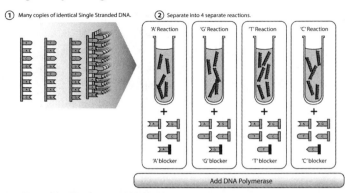

① Many copies of identical Single Stranded DNA.

② Separate into 4 separate reactions.

'A' Reaction   'G' Reaction   'T' Reaction   'C' Reaction

+   +   +   +

'A' blocker   'G' blocker   'T' blocker   'C' blocker

Add DNA Polymerase

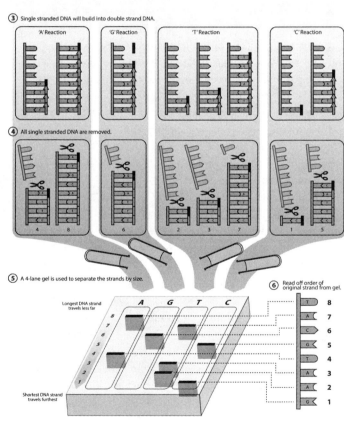

③ Single stranded DNA will build into double strand DNA.

'A' Reaction   'G' Reaction   'T' Reaction   'C' Reaction

④ All single stranded DNA are removed.

4   8   6   2   3   7   1   5

⑤ A 4-lane gel is used to separate the strands by size.

⑥ Read off order of original strand from gel.

Longest DNA strand travels less far

A   G   T   C

8   7   6   5   4   3   2   1

Shortest DNA strand travels furthest

T 8
A 7
C 6
G 5
T 4
A 3
A 2
G 1

12.1 Sanger sequencing. Refer to box 12.1. Sanger sequencing involves four separate sequencing reactions, one for each of the four bases (2). In each case, a single stranded piece of DNA is reassembled into a double strand until it incorporates one of the specially designed chain-terminator nucleotides (3). The strands can be sorted by size using gel electrophoresis. Longer strands are less mobile and will not travel as far in the gel (5). If the strands are radioactively or fluorescently labeled, they can be made to show up inside the gel. The order of the bases can then be read backwards off the gel (the strands that have traveled the least far indicate the last base, while the strands that have traveled the farthest indicate the very first bases) (6). Source: Illustration by Jerry Teo.

ation and x-ray films for visualizing the DNA strands on a gel. In the early 1980s, Leroy Hood and Lloyd Smith (from the California Institute of Technology) adapted Sanger's method by replacing radiation with four colored fluorescent dyes and gels with capillary tubes. These modifications meant DNA sequencing could be performed in a single reaction and could become highly automated. By 1987, Applied Biosystems was producing the first automated DNA sequencing machines (called the ABI 370).

BEFORE GENOMICS: MODEL ORGANISMS AND MAPPING

The HGP is sometimes described as *Big Science*. This term emerged in the 1950s to characterize the massive particle physics laboratories that were built at Brookhaven, New York; Argonne, Illinois; and elsewhere. These laboratories required substantial funding, big buildings, expensive machines, and large interdisciplinary teams to run them. The HGP imported some of these features into biology.

However, biology before the HGP was not always a small-scale activity. In particular, individual biology labs were often organized into networks of researchers working towards the solution of a particular set of problems. Often these teams were devoted to the study of certain organisms. In the early twentieth century, for example, many geneticists studied the fruit fly called *Drosophila melanogaster*. Later, many molecular biologists turned their attention to even simpler organisms, such as the gut bacteria *Escherichia coli* (*E. coli*) and a virus that infects it, *phage Lambda*. Communities of scientists working on the same organisms might communicate informally, exchange samples, and share data or techniques (even while officially remaining in competition with each other for funds and discoveries). Such activities potentially allowed for greater productivity compared to labs that were working alone.

Organisms being studied in this way are usually called "model organisms." This is because their relative simplicity allows biologists to use them as stand-ins (or models) to understand the behavior of more complex organisms (such as humans). If we can understand the way transcription (or translation, or cellular metabolism, or cell death, etc.) work in a bacteria, the logic goes, we will be a step closer to understanding how it works in humans. Moreover, the logistical, financial, and ethical considerations that would make some types of human experimentation impossible do not necessarily apply to viruses, bacteria, worms, or even mice.

In 1963, Sydney Brenner (then at Cambridge's Laboratory for Molecular Biology) introduced a new model organism: *Ceanorhabditis elegans*. Using this simple worm, Brenner hoped to shed light on the genetic regulation of the development of an organism—that is, how did genes instruct various cells to take on their various roles within an animal body and to assemble themselves in the correct positions? Brenner's students and coworkers continued to expand work on *C. elegans*, spreading to many laboratories around the world and making the worm into one of the most important model organisms for molecular biology.

One of the main activities of this worm community was to determine which genes acted in which cells in the *C. elegans* body. In the 1960s, and for most of the 1970s, direct sequencing of the genes was not yet feasible. Instead, biologists sought to determine the location of genes, assigning them to the different chromosomes and then, as more information was gathered, placing them in order along a chromosome (that is, determining the relative ordering of the genes). This kind of work dated back to early genetic experiments on fruit flies conducted in the early twentieth century by Thomas Hunt Morgan (1866–1945). By carefully breeding different mutants of flies (or worms) biologists could observe which mutations were most often inherited together—close association between two traits meant that they shared nearby locations on a physical chromosome.

By the mid-1980s, Brenner and his followers sought to make a *map* of all the worm genes. This would be a complete index of worm genes and their relative locations. At the Laboratory for Molecular Biology, this work was led by John Sulston and Alan Coulson. They were joined by Robert H. Waterston, an American from Washington University in St. Louis who had visited Brenner's lab in the 1970s. The worm project established informal rules and infrastructure (including an important database called ACeDB, later WormBase) for sharing information and credit amongst laboratories.

The *C. elegans* mapping project demonstrated the feasibility of a large-scale, international biological collaboration and set the stage for the HGP. Sulston, in particular, argued in favor of the HGP on the basis of his experiences and success with worms. In 1989, Sulston and Waterston were awarded one of the first grants (from the National Institutes of Health and the UK's Medical Research Council) to sequence part of the human genome. Over the next nine years, Sulston and Waterston not only successfully sequenced the worm genome (published in December 1998), but also contributed extensively to human genome sequencing.

Work on model organisms and especially the genetic mapping of the *C. elegans* worm form important precursors to the HGP. They laid the basis for the kinds of collaboration, sharing, and large-scale project work that the HGP entailed. This pre-history also suggests that we should think of the HGP not as a single project but as a related set of *genome projects* that did not begin or end with the human (other genome projects included the bacterium *Haemophilus influenzae*, the fruit fly, the mouse, and the chimpanzee).

BUILDING A PROJECT

In 1978, Sanger had used his sequencing method to sequence the genome of the bacteria virus (or bacteriophage) phi X 174. It contained 5,386 base pairs. But improving technology meant that more and more ambitious sequencing efforts were becoming possible. In 1985, a handful of biologists met to discuss the possibility of sequencing the whole human genome. Robert Sinsheimer, a biologist and the chancellor of the University of California at Santa Cruz, thought that this was the kind of large-scale project that would bring worldwide attention to his university. The biologists that Sinsheimer assembled to plan the project thought it was incredibly ambitious. Certainly it might be possible to build a physical map of the genome, but a complete sequence was probably not feasible without huge technological leaps.

Sinsheimer's efforts were frustrated. But he had germinated an idea that was soon taken up by others. A large-scale project for biology had some appeal for political reasons. In the early 1980s, the biotech industry was just getting off the ground. The United States was leading the world in this field, but many Americans feared that other nations might quickly catch up. This was certainly what appeared to be happening in other high-tech fields such as electronics. Japan, in particular, appeared to be manufacturing higher quality products as well as taking the lead in producing innovations. In 1981, Japan's Science and Technology Agency had begun to support a project to automate DNA sequencing that involved companies including Fuji and Seiko. A big investment into biology might give US industry, and thus the US economy, the competitive edge it required to stay ahead in biotechnology and counter the Japanese technological threat.

In 1985, Charles De Lisi (1941–) was appointed as the director of the Office of Health and Environmental Research at the Department of Energy (DOE). The DOE, formerly the Atomic Energy Commission, had long had an interest in biology and especially those parts of it related to genes and heredity. This concern grew out of research in the 1940s and '50s that was concerned with the long-term effects of atomic radiation on the human body. De Lisi believed that sequencing the human genome project would allow biologists to firmly answer questions about the genetic effects of radiation exposure. Attempting to build support for such a project within DOE, De Lisi convened a scientific workshop in Santa Fe, New Mexico (near the Los Alamos National Laboratory), in March 1986. Gradually, the project gained support, both from scientists and the DOE bureaucracy.

It was recognized that the human genome would be a project of immense scale. In 1987, Leroy Hood estimated that it could be done at a cost of between $200 million and $300 million per year for fifteen years. This would include technology development, physical mapping, sequencing of model organisms, and the institutional resources to pull all this together. Securing such a large amount of money required the approval of Congress. During 1988 and 1989, De Lisi worked to gather support on Capitol Hill. A crucial supporter was Senator Pete Domenici from New Mexico. Domenici worried what would happen to the economy of his state if the Cold War ended: what would happen to the national laboratories like Los Alamos and the other defense-related research and development industries? They needed a new mission.

One of the main obstacles faced by De Lisi was the National Institutes of Health (NIH). As the preeminent biomedical research body in the United States, they thought they should take the lead in the HGP. The NIH, however, was more cautious in its approach, sensitive to doubts amongst some biologists about the benefits of the project. Nevertheless, throughout the late 1980s, the NIH was funding human genome research on its own. By 1989, the DOE and the NIH were battling in front of Congress for leadership of the HGP. In 1990, Congress approved the creation of the National Center for Human Genome Research within the NIH, effectively handing leadership of the project to that agency. For its director, the NIH appointed James Watson (1928–), world famous for his discovery (with Francis Crick) of the structure of DNA. With Watson at the helm, the HGP had a renowned scientist and forceful public personality to drive it forward.

WHY SEQUENCE?

At the same time as the political groundwork for the HGP was being carried out, the relationship between genes and disease seemed to be getting clearer. Research showed how particular mutations in specific genes caused diseases.

One of the best examples is cystic fibrosis (CF). CF is a hereditary disease that causes thick mucous to build up in the body, affecting especially the lungs, liver, pancreas, and intestines. Although the prognosis for patients with CF has dramatically improved over the last thirty years, many still die from infections of the lungs caused by mucous build-up.

In 1988, a team at the National Institutes of Health led by Francis Collins (1950–) made a breakthrough in CF research. They identified a gene—called CFTR—that seemed to contain a mutation in many CF patients. The normal version of this gene was responsible for building a protein 1,480 amino acids long. But CF sufferers seemed to have three nucleotides in their genetic sequence altered, causing a protein that was missing one amino acid at position 508. Moreover, this protein was involved in transporting chloride ions in and out of cells in the body. The mutation seemed to be causing the protein to function incorrectly, making cells unable to transport chloride. It seemed plausible that this chemical imbalance was exactly what was causing the thickening of the mucous.

Although not all the details were worked out, it seemed that Collins and his coworkers had discovered a direct link between a molecular change and the symptom of the disease. A change in the sequence could be linked to—in the case of CF—life or death for the patient. Such a find provided a compelling case for sequencing more genes: it would allow biologists to identify the causes of other genetic diseases.

The HGP's supporters pointed to these promised medical benefits when asked to justify the project in scientific meetings, in Congress, and in public forums. But many biologists also believed that the HGP would be a great boon to biology more generally. For one thing, such a large-scale project would raise the profile of the field and bring unprecedented levels of funding for biology as a whole. Some compared it to the Manhattan Project (to build the atomic bomb) or the Apollo program—an endeavor that would galvanize the whole discipline, enhance national prestige, and boost the economy. Advances in the fundamental understanding of biology would also result, many biologists thought. A complete sequence of the human genome would provide the basis for studying not only pathology, but also the functioning of the normal body. Moreover, the advances in sequencing, computing, and informatics that the project would require would not only benefit biology, but also potentially spin off into socially useful technologies.

However, the HGP also had many opponents, both inside and outside the biological community. While some supported the broad idea behind the project, they disagreed with the way it was being implemented and especially the speed at which it was being undertaken. Some biologists believed that the HGP would take money away from other kinds of research. Traditionally, the NIH awarded grants to individual biologists to fund work undertaken at their own laboratories (usually within universities). Such awards were based on peer review of scientists' proposal. Would the money devoted to the HGP mean that less was available for this kind of funding? Science, many believed, was based on the creativity and innovation of the individual investigator. The genome project seemed likely to redistribute resources away from the small scientists and towards big, centralized laboratories.

This worried some biologists even more because the HGP did not appear to involve the kind of creative scientific work that they were used to. Sequencing could be made almost automatic—it was, as Walter Gilbert put it, "Not science, but production." Biologists pointed to the risk that students trained in sequencing would not have gained the skills to think for themselves. At both an institutional and an individual level, the HGP ran the risk of stifling scientific creativity, critics argued.

Other biologists believed that sequencing the whole human genome was a waste of time and money. The multibillion-dollar price tag would not buy the cure to diseases. For one thing, a huge amount of work would still need to be done to understand all that sequence. One letter to *Nature* argued that "sequencing the genome would be about as useful as translating the complete works of Shakespeare into cuneiform, but not quite as feasible or easy to interpret."[1] Perhaps the money could be better spent in other ways. Some suggested that mapping was the more important activity: it would be more beneficial (and cheaper) to build a complete map of all the human genes (that is, working out where they are all located on the chromosomes) before launching into full-scale sequencing. To bolster this argument, biologists pointed to the fact that only 2% of the total human genome sequence seemed to code for proteins. The rest—labeled "junk DNA"—just seemed to be noise. What was the point of spending billions of dollars sequencing 98% "junk"?[2]

All of these problems and questions were vigorously debated within the biological community from the mid-1980s onwards. In the end, the NIH and the DOE managed to gather enough political support to outmaneuver the critics and drive the HGP forward.

Ultimately, the HGP did cause significant changes in the way biology was organized. Large amounts of money were spent on technology and infrastructure, including automatic sequencing machines, supercomputers, and data infrastructure. The sequencing efforts were concentrated in a few large centers that did the vast majority of the sequencing work (the so-called G5: the Sanger Center in Cambridgeshire, United Kingdom; the Whitehead Institute in Cambridge, Massachusetts; the Genome Institute at the University of Washington, St. Louis, Missouri; the Baylor College of Medicine in Hous-

1. Quoted in Robert Cook-Deegan, *The Gene Wars: Science, Politics, and the Human Genome* (New York: W. W. Norton, 1994), 114.
2. Both the concept and the term "junk DNA" remain controversial. Estimates of the amount of nonfunctional DNA in the human genome have changed as biologists have learned more about how the genome works. The 2% figure was current at the time the debates about the HGP were taking place. More recent findings suggest that a significant portion of so-called junk DNA does have some function.

ton, Texas; and the DOE Joint Genome Institute in Walnut Creek, California). These centers depended on large grants from the federal government. They required managers to organize the staff, technicians to attend to the machines, and powerful computer systems to process and store the data. In other words, the HGP required a massive scaling up of biological work: it required new kinds of organization, new kinds of people, and new kinds of work.

The project also involved significant international collaboration beyond the United States and United Kingdom: laboratories in France, Japan, Germany, and China contributed DNA sequence. To make this work, the Human Genome Organization also needed to develop novel means of communication, cooperation, and information sharing. In 1996, the HGP-participating labs agreed on a set of rules for sharing their data. These "Bermuda Principles" (so named because of where the agreements were reached) have inspired more widespread data-sharing and "open science" practices beyond biology.

PUBLIC VERSUS PRIVATE

Even once the genome project was underway, disagreements persisted. Within the project itself, biologists had different ideas about the best, fastest, and cheapest ways of sequencing. One of the main problems presented by a full genome sequencing project is that of scale. A complete (haploid) human genome contains three billion pairs spread across 23 chromosomes. This means that each chromosome contains roughly a few hundred million base pairs of DNA. Even with the dramatic improvements in Sanger sequencing and automation that the HGP achieved, the method could reliably sequence only about 500 base pairs at once. To get around this limitation it was necessary to chop a chromosome up into thousands of fragments that were each less than 500 base pairs long and sequence each of these in turn.

This is in fact what the HGP did. The problem, however, is that it is not easy to get the pieces back into the correct order. This is just like an enormous jigsaw puzzle with millions of overlapping pieces that have to be assembled. The solution adopted by the DOE-NIH was to first build "maps" of chromosomes prior to chopping them up. A map of a chromosome identifies the ordering of specific features or sites in the DNA. Next, the chromosomes were broken into large pieces (of about 150,000 base pairs). By comparing a bit of sequence from the large chunks to the maps, it was possible to determine the order of the chunks along the chromosome. Finally, the large chunks were broken into much smaller pieces (around 500 bp, suitable for Sanger sequencing). Once the sequencing was completed, the order of the fragments within the large chunk could be determined by sophisticated computer software that searched for overlapping or matching segments and aligned them. In this way, it was possible to determine the sequence of the whole chromosome.

In the midst of the HGP, one biologist thought he could vastly improve on this effort. Craig Venter, who worked at the NIH, had already created a stir in 1991 by discovering a means of rapidly identifying genes within the genome (see chapter 6). This method (called expressed sequence tags or ESTs) posed a challenge to the HGP because it further called into question the necessity of sequencing *all* the DNA: if it was possible to rapidly identify the genes, why bother with all the other sequence?

But by the mid-1990s, Venter was thinking even more ambitiously. He thought he could sequence the human genome faster and cheaper with his own method. This method was simply to cut out many of the intermediate steps involved in the DOE-NIH plan. Using large chunks and mapping took a lot of extra time. Venter thought he could use very powerful computers to eliminate these steps: just break a whole chromosome into lots of small, random fragments, sequence them all, and then let a computer program match them all up into the correct order. Before the mid-1990s, no one had thought this was possible—computers and their software simply weren't powerful enough to organize so much data.

But Venter was confident it would work. He obtained funding from the health technology company Perkin-Elmer (in 1993 they had acquired Applied Biosystems, which produced the sequencing machines for the HGP) and set up Celera Genomics in 1998. Venter claimed publicly that Celera would beat the public project, producing the human genome faster and cheaper. Aside from the prospect of "losing the race," biologists involved in the public project worried that Celera would attempt to keep parts of the genome sequence secret, or to apply for patents, in order to profit from their work. Many biologists saw the human genome as something that belonged in the public domain, a benefit to all as a resource for future medical research. The attempt to commercialize it appalled many scientists and made Venter a controversial figure.

In response to Celera, the publicly funded HGP changed their strategy and sped up their efforts. In the end, it was called a draw. In April 2003, the public and private efforts officially concluded with simultaneous publications of the "complete" sequence in *Science* and *Nature* magazines.[3] The public effort had

3. The "end" of the HGP may be marked in several different ways. Celera and the public project announced a "working draft" of the human genome jointly in June 2000. An

taken roughly 13 years and cost $3 billion dollars. Celera had spent only $300 million (although they had the advantage that they could use the data created by the public project, freely available in online databases). The company, succumbing to pressure from the public project, ultimately changed their policy and made much of their data available for noncommercial use.

POSTGENOMICS?

The HGP was hailed as a success for international science and a boon for global health. The public project finished within its budget and two years ahead of schedule. Francis Collins, who had taken over the leadership of the public project from Watson, said of the HGP in 2000, "It is probably the most important scientific effort mankind has ever mounted."[4] Bill Clinton and Tony Blair spoke in similarly hyperbolic tones of discovering "the language in which God created life," and a "revolution in medical science."[5]

But behind the politicians and project leaders, many rank-and-file biologists were pointing to some strange surprises. In particular, the genome sequence raised interesting questions about human uniqueness. It appeared, for one thing, that approximately 97% of our DNA was identical to chimpanzees. What did this similarity mean? Where were those genes that made us humans special? Even worse, we seemed to be worryingly short of genes altogether. At the start of the project, most biologists had predicted that humans would have something over 100,000 genes. As more and more of the genome was revealed, estimates were revised downwards towards a final figure of just over 20,000. The simple flatworm *C. elegans* has approximately the same number. This suggested that genes alone might not be as important as the HGP proponents had suggested. Many biologists began to suspect that something more complex (and interesting) might be going on. Genes didn't appear to be the whole story.

In addition to this, the HGP told us very little about differences *between* humans. The HGP had worked on the premise that all human genomes were fundamentally similar—there was really *one* human genome and it would

---

"initial working draft" was actually published in February 2001. The project was declared finished in April 2003. Work on various chromosomes continued until 2006.

4. Collins' remark was quoted in dozens of news outlets—for example, in Tim Radford, "Scientists Finish First Draft of DNA Blueprint," *Guardian*, June 26, 2000.

5. The first quote belongs to William J. Clinton and the second to Tony Blair. These quotations were widely reported. A full transcript of the announcement can be found here: http://transcripts.cnn.com/TRANSCRIPTS/0006/26/bn.01.html.

not matter *whose particular* genome was sequenced. The public project had begun by collecting a large number of anonymous samples (blood from females, semen from males). A few of these were selected for sequencing such that no one would know whose DNA was actually being used. In the end, a large fraction of the HGP sequenced was produced from a single anonymous male donor from the United States. Celera's project also collected DNA from twenty-one anonymous donors from which it selected five for sequencing.[6]

After the completion of the HGP, a number of biologists argued that we needed to understand more about the genetic *variation* amongst people. This would contribute to further understanding disease (if a particular population has a lower occurrence of a particular disease, that could point the way to a genetic cause). The most notable of these projects was the International HapMap Project, begun in 2002, which has measured patterns in variation in DNA sequence taken from individuals in different parts of the world (initially, the United States, Tokyo, Beijing, and Ibadan, Nigeria) (see chapter 22).

This turn towards variation was partly motivated by another surprise: the HGP had turned up very few genes that seemed to be strongly and directly associated with particular traits or diseases. The "gene for obesity" or the "gay gene" or the "gene for autism" or intelligence or diabetes or heart disease had just not emerged. Even the gene for cystic fibrosis turned out to have a more complicated story. Since 1989, almost 2,000 different mutations in the CFTR gene have been discovered and other genes seem to have marked effects on the severity of the disease too.

Around 2000, some biologists began to realize that finding the human genome sequence might be the beginning of their work, rather than the end. The genes, by themselves, couldn't account for or explain many diseases, traits, or behaviors. Biologists also needed to know how genes are spliced by the cellular machinery, how epigenetic modification works, how the folding of DNA can silence genes, why some mRNA transcripts get degraded by cells, and so on. Moreover, perhaps it was the case that genes worked together in complex combinations or in networks that involved genetic and nongenetic components. The sequence by itself provided very few clues to how the parts actually functioned in a living organism.

The end of the HGP gave rise to a host of new subdisciplines within biology: systems biology, proteomics, interactomics, predictive biology, integrative biology, metabolomics, metagenomics, comparative genomics, and

6. In 2003, Craig Venter revealed that his DNA had been amongst the original twenty-one samples.

others. The HGP had not delivered on its promise of finding the causes of human disease. Many of these new disciplines attempted to find new approaches in order to fill in the gap between the genome and the fully functioning organism.

CONCLUSIONS: THE CENTURY OF THE GENE?

The feminist philosopher of science Evelyn Fox Keller has argued that the twentieth century was "The Century of the Gene." The notion that genes could, almost by themselves, build, program, and control organisms held a tight grip on biologists and on the public imagination. Films and novels such as Michael Crichton's *Jurassic Park* (1993) traded on the notion that DNA was all it took to reproduce a fully functioning organism (in this case, dinosaurs). The HGP was largely a product of this gene-centered thinking. It was based on the hope that the genetic sequence would reveal most of what there was to know about human biology. This view of biology has led to the kinds of determinist thinking associated (in an extreme form) with eugenics: if we are really just determined by our DNA, then there is very little we can do to overcome our genetic fate. Parents, education, and social welfare programs cannot change people's genes, so a genetic determinist view can justify reducing social and personal responsibility. In the "century of the gene" it was the gene that was to blame.

But the shortcomings of the HGP have forced both biologists and non-biologists to rethink this simple determinist view. It now seems that genes by themselves can achieve little and explain very little of our behavior. Causation now seems to be complex, multiple, and combinatoric. DNA is less a "master molecule" or a program, and better viewed as part of a complex reactive network of molecules within the cell. Although popular and newspaper accounts often seem still to place genes at the center, biologists are finding new ways to talk about biology that do not privilege DNA. The successes and failures of the HGP provide a crucial background for understanding the problems posed by genetic testing and genetic discrimination that are to be discussed in the next chapter.

FURTHER READING

The most comprehensive account of the 1980s lead-up to the HGP is Robert Cook-Deegan, *The Gene Wars: Science, Politics, and the Human Genome* (New York: W. W. Norton, 1994). This book was completed well before the HGP itself was finished—the story is brought up to date in Victor McElheny, *Drawing the Map of Life: Inside the Human Genome Project* (New

York: Basic Books, 2010). These volumes both cover the political ins and out in some detail. For more individual perspectives see John Sulston and Georgina Ferry, *The Common Thread: A Story of Science, Politics, Ethics, and the Human Genome* (Washington, DC: Joseph Henry Press, 2002) and J. Craig Venter, *A Life Decoded: My Genome, My Life* (New York: Penguin, 2008). On Venter's work also see James Shreeve, *The Genome War: How Craig Venter Tried to Capture the Code of Life and Save the World* (New York: Random House, 2005). For some interesting discussion of the perceived ethical and social issues associated with the genome project whilst it was getting underway see Daniel J. Kevles and Leroy Hood, eds., *Code of Codes: Scientific and Social Issues in the Human Genome Project* (Cambridge, MA: Harvard University Press, 1992).

There is also scholarship that covers specific aspects of the project in more detail. On the history of protein, RNA, and DNA sequencing see Miguel Garcia-Sancho, *Biology, Computing, and the History of Molecular Sequencing: From Proteins to DNA, 1945–2000* (New York: Palgrave-Macmillan, 2012). On the history of bioinformatics and its relationship to the genome project see Hallam Stevens, *Life Out of Sequence: A Data-Driven History of Bioinformatics* (Chicago: University of Chicago Press, 2013). For a brief but useful account of the "speeding up" of scientific work in the genome projects see Michael Fortun, "Practicing Speed Genomics," in *The Practices of Human Genetics*, ed. Michael Fortun and Everett Mendelsohn (Dordrecht: Kluwer, 1999), 25–48. There is an extensive literature on model organisms including Robert E. Kohler, *Lords of the Fly: Drosophila Genetics and the Experimental Life* (Chicago: University of Chicago Press, 1994); Karen Rader, *Making Mice: Standardizing Animals for American Biomedical Research, 1900–1955* (Princeton, NJ: Princeton University Press, 2004); Angela N. H. Creager, *The Life of a Virus: Tobacco Mosaic Virus as an Experimental Model, 1930–1965* (Chicago: University of Chicago Press, 2001); Sabina Leonelli, "Growing Weed, Producing Knowledge: An Epistemic History of Arabidopsis thaliana," *History and Philosophy of the Life Sciences* 29, no. 2 (2007): 55–87; and Sabina Leonelli and Rachel A. Ankeny, "What Is So Special about Model Organisms?" *Studies in the History and Philosophy of Science: Part A* 42, no. 2 (2011): 313–323.

On what has happened since the completion of the HGP see the Nature News Special Issue, "The Human Genome at Ten," *Nature* 470 (2011). For social science reflections on postgenomics see Sarah S. Richardson and Hallam Stevens, eds., *Postgenomics: Perspectives on Biology after the Genome* (Durham, NC: Duke University Press, 2015), Barry Barnes and John Dupré, *Genomes and What to Make of Them* (Chicago: University of Chicago Press; 2008), and Jennifer Reardon, *The Postgenomic Condition: Ethics, Justice,*

Knowledge after the Genome (Chicago: University of Chicago Press, forthcoming).

For critiques of genetic determinism see Evelyn Fox Keller, *Century of the Gene* (Cambridge, MA: Harvard University Press, 2000); and Dorothy Nelkin and M. Susan Lindee, *The DNA Mystique: The Gene as a Cultural Icon* (Ann Arbor: University of Michigan Press, 2004).

PART VII

GENETIC TESTING, DISCRIMINATION, AND BIOETHICS

13 GENETIC TESTING, DISABILITY, AND DISCRIMINATION

INTRODUCTION

Biotechnologies have contributed to our rapidly growing understanding of the genetic causes of disease. The Human Genome Project and subsequent genome sequencing have provided biologists with a wealth of data. Thousands of studies have attempted to link symptoms and medical histories with genetic information. This has increased our knowledge about how diseases are passed from parents to offspring. In most cases, this information has not yet allowed us to *treat* those diseases—there is no pill which we can use to alter a person's genes in order to stop them getting Parkinson disease, for instance. Nor does it usually allow us to predict with certainty which diseases a person may get. Nevertheless, this knowledge of genetics is allowing us to diagnose some diseases far more readily and to make some predictions about an individual's future health based on his or her genes.

This chapter is concerned mostly with the information gained from such genetic tests. In some cases, the tests facilitate earlier diagnosis, preventative measures, and better care. However, where genetic tests are performed prenatally (before birth), the discovery of a potentially serious genetic disorder often leads to advice from physicians or genetic counselors to terminate a pregnancy. This is usually justified on the grounds that the baby—had it been born—would have lived a life of pain and suffering.

However, such decisions raise a host of ethical questions. How much pain and suffering makes a life not worth living? (After all, there are many people in the world who *do* live with pain and suffering—are their lives not worth living?) Who gets to decide on such questions? How can we know for sure what someone's life will be like? Does eliminating some kinds of people from living constitute a form of discrimination?

These are all questions that fall into the field of *bioethics*. We will discuss the origins and overall aims of bioethics in more detail in chapter 14. For now, we are going to draw on the arguments of bioethicists to help us better understand the potential social impact of the biotechnologies associated with genetic testing.

A BRIEF HISTORY OF GENETIC TESTING

The possibility of testing for specific hereditary traits was imagined as part of the eugenics movement (see chapter 11). However, eugenicists in the first half of the twentieth century did not have the technological means to perform such tests. Instead, they relied on recognizing external traits or behaviors that they believed to be inherited. They also relied on outwardly visible traits (such as skin color) that they believed were associated or correlated with particular heredity conditions.

Nevertheless, the ability to test for genetic diseases predates the eras of genomes and modern biotechnology. Some genetic diseases leave chemical or physical traces in the body that can be discovered without a genetic test. The genetic disease phenylketonuria (PKU), for instance, is caused by a genetic mutation that causes a protein to fold incorrectly. This misfolded protein means that the body cannot break down the amino acid phenylalanine. Instead, a toxic substance builds up in the body causing brain damage and seizures. The toxic substance itself can be tested for in blood or urine, leading to diagnosis. One of the first tests for an inherited disorder was implemented in Massachusetts in 1963, when the state mandated the testing of all newborns for phenylketonuria. The aim of this *neonatal* testing was that babies diagnosed with the disease could immediately be given special care: in this case a diet that restricted the intake of phenylalanine and neutralizes the effects of the disease.

For some genetic diseases (called autosomal recessive) it is possible to test for "carrier" status. Each of us possesses two copies of each of our chromosomes—one inherited from our father, and one from our mother.[1] This means that most genes in our bodies come in two copies. In autosomal recessive disorders, the disease requires *both* copies of the gene to be mutated or corrupted (in most cases, this is fairly rare). But this means that it is possible that some individuals in the population have one "good" version of the gene and one mutated version. These individuals do not have the disease—their one good gene is enough. But they do carry around the defective copy of the gene and hence they are called "carriers."

What if a carrier reproduces with a person who has two normal copies of the gene? In that case, each offspring will get one version of the gene from the carrier and one from the other parent. In all cases, the offspring will have at least one normal version of the gene. On average, half the offspring will also

---

1. This is only strictly true for females. Males carry two copies of chromosomes 1 through 22, and then one copy each of the X and Y chromosomes. Females have two Xs instead.

be carriers, but none will get the disease. But what if two carriers have children? In this case, there *is* a chance that a child will get one defective gene from one parent and one defective gene from the other and end up with two defective genes—this means they will get the disease. In fact, on average, one out of every four of the carrier-carrier children will get the disease.

In the 1970s, several programs were established to test for carriers of autosomal recessive diseases such as sickle-cell anemia and Tay-Sachs disease. The idea behind these programs was that carriers who knew they were carriers could avoid having children with other carriers. In this way the incidence of the diseases could be reduced. Since the aim of carrier testing is to prevent reproduction between specific individuals, it might be seen to be similar to eugenic programs of the early twentieth century (see chapter 11). However, although many such programs were organized or sponsored by the state, carrier-testing programs usually do not intervene directly in reproductive choices (that is, they do not try to ban marriages or stop people from reproducing). Rather, they aim to provide individuals or couples with information from which they might make their own, informed reproductive decisions.

By the early 1980s, the notion of a "genetic disease" was becoming more widespread in both medical discourse and popular culture. People had known for a long time that diseases run in families. However, the advent of molecular genetics and the HGP led to an increase in the numbers and types of diseases that were attributed to genetic causes. Cancer, for instance, increasingly became understood as a genetic disease. Certainly, all sorts of environmental exposures also seemed to increase the risks of certain cancers. But some types of cancer were known to run in families, suggesting an important genetic component. In this case and others, biologists, physicians, and the public began to attribute more importance to genes as causal agents. This was partly due to increased knowledge of genes and their function, but was also part of a broader popular and medical acceptance of the power and importance of genes in all aspects of life (for instance, in determining identity, personality, sexuality, intelligence, and so on).

The hunt for genetic causes of diseases led to the development of genetic tests for many conditions. For instance, discoveries in the 1990s made it possible to test for the susceptibility to breast cancer by testing for particular mutations in the BRCA1 and BRCA2 genes.[2] Such tests can be conducted on adults, but many are now conducted prenatally via amniocentesis: a needle

2. It is important to stress that testing positive for specific mutations in these genes does not mean that one will certainly get breast cancer. Rather, particular mutations are statistically associated with a greater risk of getting the disease.

is used to extract fluid from the amniotic sac that surrounds the fetus at or around the sixteenth week of pregnancy. In some cases, other procedures are performed, including chorionic villus sampling (taking blood from the placenta) or cordocentesis (testing blood from the umbilical cord). Although some of these tests may allow for various types of medical or surgical interventions before birth, the main justification for the test is to give the parents a chance to abort a fetus that is diagnosed with a serious genetic disorder.[3]

In the developed world, tests for a large number of diseases and traits are routinely conducted on fetuses in utero. The tests that are actually conducted in any particular case may depend on the wishes of the parents, pertinent risk factors (including the age of the mother), and medical insurance. In 2015, nearly four thousand tests for different traits are available and the number is increasing all the time (see the partial list in box 13.1).[4] Some of these tests provide straightforward and reliable information—the child either has sickle-cell anemia or it does not.[5] Other tests, such as those for conditions such as autism spectrum disorders or hypertension, can only provide a measure of risk—a positive test does not mean that the child will definitely develop autism or hypertension, only that it has an increased risk of doing so.

There is one other form of genetic testing that should be added to the list of the three types of genetic testing mentioned here so far (neonatal testing, carrier testing, and prenatal testing). Preimplantation genetic diagnosis (PGD) occurs during an in-vitro fertilization (IVF) procedure: embryos are subjected to genetic testing before being implanted in the mother's uterus. Any embryo that is shown to possess mutations associated with disease would not even progress to the stage of pregnancy. Since this a far more powerful tool for genetic selection, PGD will be discussed separately in chapter 18.

Biotechnology companies—with partners in the pharmaceutical industry—have played a large role in developing genetic tests and packaging them into marketable forms. The testing itself has also expanded the role of physicians and hospitals in managing and overseeing pregnancies. Parents may be advised by physicians, hospitals, or insurance companies to seek the advice of genetic counselors to help navigate the vast amounts of information generated by the tests. In short, genetic testing is an industry in its own right. The

---

3. It is also argued that the tests allow parents to prepare psychologically, socially, and financially for the birth of a child who may need special care.

4. For more information see: https://www.genetests.org/.

5. It's not quite that simple—no test is perfect and genetic tests can show both false-positives and false-negatives.

Box 13.1 Some of the Available Prenatal Genetic Tests (2015)

Cystic Fibrosis
Sickle-Cell Anemia
Tay-Sachs Disease
Hereditary Haemochromatosis
Neurofibromatosis Type 1
Achondroplasia
Hemophilia A
Duchenne Muscular Dystrophy
Fragile X Syndrome
Spina Bifida
Gaucher Disease
Marfan Syndrome
Polycystic Kidney Disease
Timothy Syndrome
Autism Spectrum Disorders
Huntington Disease
Alzheimer Disease
Hypertension
Alpha-Thalassemia
Beta-Thalassemia
Congenital Adrenal Hyperplasia

For more information see the website of the National Newborn
Screening & Global Resource Center: http://genes-r-us.uthscsa.edu/
For a breakdown of what tests are conducted where see:
http://genes-r-us.uthscsa.edu/sites/genes-r-us/files/nbsdisorders.pdf

act of having children has become inextricably linked to the complex of bio-tech, Big Pharma, and bioethics.

FREEDOM AND CHOICE

The field of bioethics has generated a vast literature that considers the moral implications of genetic testing (in all its forms). Rather than attempt to summarize all of these discussions, the rest of this chapter will focus on some key examples and uncertainties that bring the arguments on both sides into focus.

Arguments in favor of genetic testing are usually framed in terms of freedom of choice. Doctrines of reproductive rights suggest that parents and especially women should have full control over when, where, how, and under what circumstances they choose to reproduce. Arguments from the perspectives of women's rights reinforce such a view. The "rights of the child" (enshrined in a 1989 UNICEF convention) might also be construed to be consistent with genetic testing, since they provide a means of maximizing the chances that children will be born healthy. Moreover, most conceptions of human rights extend to the right to access to health care and to information, especially information that pertains to one's own body (or genes).

From all these rights perspectives, then, genetic testing seems to maximize the choices available for parents and children. However, some philosophers have responded by questioning whether such "choices" are actually available in practice. Genetic testing, as recommended by the American College of Obstetricians and Gynecologists, is performed routinely in the developed world. Expecting parents often don't have the knowledge or confidence to question the recommendations and authority of their physicians.

More subtly, parents may feel social or economic pressure to undergo testing. Giving birth to a child with Down syndrome or an otherwise impaired child (while knowing that a test was available) might place families under significant financial burdens or cause social stigmatization. Diane Paul has argued that "reproductive decisions will often be driven by the conjoined interests of powerful non-state entities, such as physicians, lawyers, insurers, and biotechnology firms."[6] As with other choices in our life, we all act under certain social, economic, and political constraints. Despite the rhetoric of "free choice" in reproductive decisions, the possibilities that are practically available to us may strongly favor some outcomes over others.

HOW TO VALUE A LIFE

Another argument that seems to favor genetic testing is that it has the potential to drastically reduce human suffering. Some philosophers argue that bringing unnecessary suffering into the world is a moral wrong and that we therefore have a moral obligation to select the best possible children. By selecting (as far as possible) healthy children, it eliminates (usually by abortion) those who would have had to endure various forms of disease, disability, and hardship.

There are two ways in which to understand this argument. First, we might

6. Diane Paul, "Is Human Genetics Disguised Eugenics?" in *Genes and Human Self-Knowledge: Historical and Philosophical Reflections on Modern Genetics*, ed R. Weir et al. (Iowa City: University of Iowa Press, 1994), 67–83. Quotation p. 78.

interpret it as claiming that some lives (those of extreme disability or suffering) are not worth living. That is, it is better not to have existed than to have lived and suffered. One might object to this view on the grounds that it is *always* better to have existed than not existed (from an individual point of view, some life, no matter how bad, seems preferable to no life at all). Nevertheless, this argument justifies genetic testing on the grounds that it allows parents to pick out and eliminate those lives not worth living.

Second, we might interpret the argument as valuing health over sickness (or in the case of selecting for nondisease traits, this amounts to valuing higher capacities—such as intelligence or athletic ability—over lower capacities). Genetic testing is not turning sick babies into healthy ones; rather, it is eliminating (via abortion) the sick in favor of the healthy (this is presuming that parents will go on to have other healthy children). This means that it is tacitly placing a higher value on healthy lives and a lower value on sick lives (or for nondisease traits, a higher value on greater capacity). This argument justifies genetic testing on the grounds that is it allowing for the creation of more-valuable lives.

In both these cases we are left with a problem: "Is it really possible to place a value on different kinds of human lives?" And, if it is possible, then how are we to assign such a value? (And who gets to do so?) Such questions of value are extremely problematic: what counts as valuable varies between different people and over time (depending on things such as culture and religion). The philosopher Philip Kitcher asks us to consider the following scenario, from an imagined future in the year 2069:

> Once . . . many babies were doomed to die in infancy, there were special institutions for "defective" children, and the more enlightened nations diverted large sums from other health and education projects to provide special care for children with gender disabilities. But the progress of the reproductive responsibility movement has been heroic: Tay-Sachs is a thing of the past, Down Syndrome is virtually eliminated, congenital forms of heart disease are now extremely rare, there are fewer people with mutant tumor-suppressor genes, far fewer fat people, far fewer homosexuals, far fewer short people.[7]

At first, this scenario seems pleasant enough: all kinds of diseases have been eliminated via genetic testing and governments are saving money that can now be spent on health and education. But as Kitcher completes his list of

7. Philip Kitcher, *The Lives to Come: The Genetic Revolution and Human Possibilities* (New York: Free Press, 1997), 249.

"defects," we are perhaps a little less comfortable. Is being short a defect? Or fat? Or homosexual? Certainly the argument can be made that such people suffer more—short people might get teased in school and picked last for the basketball team, fat people may die younger, and homosexuals may face difficult social and psychological adjustments within a society where heterosexuality is the norm. But is this a justification for eliminating them from the population? Kitcher is suggesting that genetic testing could be a slippery slope: we start by eliminating those with Tay-Sachs and Down syndrome, but end up counting more and more traits as "defective."

Perhaps there is a sensible place to draw the line. Most people might agree that eliminating Tay-Sachs is good, but eliminating left-handed people is going too far. Again, we end up with the problem of measuring the value of a life. In debates over genetic testing, disabled-advocacy groups have argued that the lives of disabled individuals are no less valuable than anyone else's.[8] Disabled people also have the ability to live fulfilling lives. The world-famous astrophysicist Stephen Hawking is often cited as an example: he has lived an immensely difficult life, yet he has made remarkable contributions to scientific understanding. Would a genetic test have revealed his potential for amyotrophic lateral sclerosis and led to termination?[9] More generally, should we consider the lives of less talented people to have less value than the lives of more talented people?

Perhaps we could draw a line between lives that we consider "normal" versus those that are "pathological." The history of medicine suggests that we should be very careful in deciding what falls into these categories. As recently as the 1960s, homosexuality was considered a disease by Western medicine: "sufferers" were offered "therapy" in the form of hormones. Ideas about "normality" are constantly shifting. Indeed, if genetic testing were widely used, this would itself have significant potential to *change* what counts are normal. If, for instance, society began to eliminate individuals with very low IQs, this would, over time, shift the public perception of what sort of IQ score counted as "normal." Again, we could find ourselves on a slippery slope where the range of normality becomes narrower and narrower.

Some bioethicists and philosophers have tried to offer a rational means of assigning value to human lives. Kitcher, for example, has a scheme it which is possible to value lives by assessing an individual's ability to make and fulfill

8. These groups have also pointed out that genetic testing is actually *devaluing* their lives since it is reinforcing the view that disabled lives are less worthwhile.

9. Only 5% of ALS cases are hereditary. Nevertheless, the point still holds.

life plans, and weighing pleasant versus painful experiences. Such schemes may be helpful, but it is not clear that they can solve all the dilemmas created by the problem of weighing the value of vastly different kinds of human experiences. All in all, then, these arguments should make us extremely wary about applying genetic testing to all but the most grievous hereditary diseases.

UNCERTAINTY

Social scientists have also pointed out that genetic testing is a product of a gene-centric worldview. This worldview, arising from molecular biology, genetics, and genomics, has tended to overstate the importance of genes, especially their role in causing diseases and determining our physical and psychological traits. As noted at the end of chapter 12, the HGP contributed to a view that genes *determine* our lives. The science fiction film *Gattaca* (1997) depicts a world in which the idea of genetic determinism has run wild: those individuals who are genetically perfect dominate, while the genetically "inferior" are condemned to an existence on the margins of society.

Critics of genetic determinism have continued to point out that development (from an egg cell to a embryo to a fetus) and environment have significant roles to play in the majority of human disorders and traits. To take a straightforward example, in the case of phenylketonuria (discussed above), the patient can be fed a special diet that completely counteracts the effects of the disease. The "environment" (here as food) trumps the genes.

Even more than this, recent work in biology has suggested that our genes and our environment are in constant interaction. Our genome is constantly receiving cues from our environment that are causing it to change and respond, turning genes on and off. Our genome is tagged, folded, and twisted in ways that react to intracellular and extracellular signals. Such epigenetic markers may even be passed on from parents to offspring. Such findings suggest that ultimately we may not be able to untangle the effects of genes from the effects of the environment.

Again, such uncertainly should make us wary about genetic testing. If the environment or gene-environment interactions play such a significant role in many disorder and traits, how can we make reliable decisions on the basis of genes alone?

CONCLUSIONS: LEARNING FROM EUGENICS

Is genetic testing a form of eugenics? Those who practice it aim to influence the *kinds* of humans who are born. And they certainly do so on the basis of prevailing biological and medical knowledge about genes and heredity. As genetic testing is applied to more and more traits (particularly those where it can identify only a *risk*, rather than a certainty, of disease) the results begin to look more and more similar to eugenics.

But there are at least two important differences between the kinds of eugenics discussed in chapter 11 and modern applications of genetic testing. First, we might point out that eugenics became highly coercive, forcing people to undergo sterilizations or actively preventing them from marrying. Genetic testing does not partake in such extremes—individuals still have a choice as to whether to take the genetic tests, and a choice as to whether to act on them. It is the individual, not the state, that decides what kinds of humans are born.

Second, we can also try to distinguish genetic testing on the grounds that it is justified in terms of individual, and not collective, welfare. The reason for terminating a pregnancy on the basis of genetic testing is to prevent the suffering of a particular person. Eugenics, on the other hand, was more concerned with the overall "fitness" of the population or race or nation. The welfare of individuals was less important than the welfare of the community.

These are both important distinctions and it is certainly crucial that genetic testing continues to be centered on individual choice and individual welfare. However, this chapter has suggested that things may not be quite so simple. In the case of individual choice, we have seen that choices may often in practice be highly constrained—we need to remain vigilant that individuals (especially economically and socially vulnerable individuals) are really free to choose. And in the case of making choices about individual welfare, we have seen that in practice such decisions are very hard to make—how do we reasonably weigh the value of different kinds of lives?

The field of personal genomics is now further extending the range and the availability of genetic testing (we will examine the consequences of personal genomics in chapter 20). This direct-to-consumer genetic testing has the potential to further narrow the gap between biological knowledge and its social implications and interpretations. But the most significant lesson of eugenics is that we should be cautious in our interpretations of biology: we should be especially wary of using biological findings to inform our social policies. We should be cautious in thinking that biology can offer us satisfactory explanations for school shootings or sporting prowess.[10] If we remain

10. In the wake of the 2012 Sandy Hook Elementary School shooting in Connecticut, researchers began examining the DNA of the shooter. See Gina Kolata, "Seeking Answers in Genome of Gunman," *New York Times*, December 24, 2012. On the significance of genes in sport see David Epstein, *The Sports Gene: Inside the Science of Extraordinary Athletic Performance* (New York: Penguin, 2013).

skeptical about biology, we may have a better chance of ensuring that genetic testing does not transform into another eugenics.

FURTHER READING

On the politics and meanings of prenatal diagnosis and abortion see Rayna Rapp, *Testing Women, Testing the Fetus: The Social Impact of Amniocentesis in America* (New York: Routledge, 2000) and Barbara Katz Rothman, *The Tentative Pregnancy: Prenatal Diagnosis and the Future of Motherhood* (New York: W. W. Norton, 1986). On the history of sickle-cell anemia see Keith Wailoo, *Dying in the City of the Blues: Sickle-Cell Anemia and the Politics of Race and Health* (Chapel Hill: University of North Carolina Press, 2001). On testing for PKU and its history see Diane Paul, "Contesting Consent: The Challenge to Compulsory Neonatal Screening for PKU," *Perspectives in Biology and Medicine* 42 (1999): 207–219; and Diane Paul, "The History of Newborn Phenylketonuria Screening in the U.S.," in *Promoting Safe and Effective Genetic Testing in the United States: Final Report of the Task Force on Genetic Testing*, ed. Neil A. Holtzman and Michael S. Watson (Baltimore: Johns Hopkins University Press, 1998). On the history of Tay-Sachs disease see Michael Kaback and Robert Desnick, eds., *Tay-Sachs Disease* (New York: Academic Press, 2001).

In considering the arguments for and against genetic testing, one needs to begin with the literature concerning biological and genetic determinism. One of the classics here is Richard Lewontin, *It Ain't Necessarily So: The Dream of the Human Genome and Other Illusions* (New York: New York Review of Books Press, 2000). In addition see Richard Lewontin, Steven Rose, and Leon Kamin, *Not in Our Genes: Biology, Ideology, and Human Nature* (New York: Pantheon, 1984). An alternative perspective is provided in Barton Childs, *Genetic Medicine: A Logic of Disease* (Baltimore: Johns Hopkins University Press, 1999).

There is a vast bioethics literature that concerns itself with the rights and wrongs of genetic testing. Here, I suggest a handful of titles that provide an overview of the main issues and lay out their main arguments in particularly convincing forms. Kitcher's scheme for assessing the value of human lives can be found in Philip Kitcher, *The Lives to Come: The Genetic Revolution and Human Possibilities* (New York: Free Press, 1997). See also R. M. Dworkin, *Life's Dominion: An Argument about Abortion, Euthanasia, and Individual Freedom* (New York: Knopf, 1993). A strong case for testing and selection is made in Julian Savulescu, "Procreative Beneficence: Why We Should Select the Best Children," *Bioethics* 15, no. 5/6 (2001): 413–426. Persuasive counterarguments are offered in R. Bennett and J. Harris, "Are There Lives Not

Worth Living? When Is It Morally Wrong to Reproduce?" in *Ethical Issues in Maternal-Fetal Medicine*, ed. D. Dickenson (Cambridge: Cambridge University Press, 2002), 321–334; and J. Harris, "Rights and Reproductive Choice," in *The Future of Reproduction*, ed. J. Harris and S. Holm (Oxford: Clarendon Press, 1998). The perspective of disability rights activists is captured in Eric Parens and Adrienne Asch, eds., *Prenatal Testing and Disability Rights* (Washington, DC: Georgetown University Press, 2000) and in David Wasserman, Jerome Bickenbach, and Robert Wachbroit, eds., *Quality of Life and Human Difference: Genetic Testing, Health Care, and Disability* (Cambridge: Cambridge University Press, 2005).

There are also several works that explicitly connect the history and lessons of eugenics to modern debates on genetic testing: Troy Duster, *Backdoor to Eugenics* (New York: Routledge, 1990), Edward Yoxen, *Unnatural Selection: Coming to Terms with the New Genetics* (London: Heinemann, 1986) and Ruth Schwartz Cohen, *Heredity and Hope: The Case for Genetic Screening* (Cambridge, MA: Harvard University Press, 2008).

14 BIOETHICS AND MEDICINE

INTRODUCTION

The previous chapter examined the difficult decisions and tensions raised by the technologies of genetic testing. It also introduced a type of reasoning that belongs to the specialized discipline of bioethics. Bioethics now appears in various places all around us and increasingly dominates thinking and discussion about biotechnologies. Universities engage bioethicists to act on institutional review boards and supervise medical research; hospitals consult bioethicists about the appropriate use of particular technologies or techniques; governments employ bioethicists to advise them about appropriate policies and laws concerning issues such as stem cells, cloning, pain management, xenotransplantation, euthanasia, nanomedicine, infertility treatment, animal rights, organ donation, surrogacy, cryonics, and informed consent. Bioethicists deal with a range of biotechnologies—almost all the topics covered in this book have generated an extensive bioethics literature.

Bioethics emerged as a discipline and discourse in the early 1970s. However, as the importance of biotechnologies grew in the late 1980s and early 1990s, bioethics began to play a larger role in shaping public discourse and perceptions of biotech. The HGP's commitment to funding research on the "ethical, legal, and social implications" of the project brought bioethics into the center of this work. The importance of bioethics in setting the agenda for biotechnologies became even more apparent with the advances in cloning (1996) and stem cells (1997) that will be discussed in chapters 16 and 17.

Before we address these controversial reproductive technologies, this chapter will review the history of bioethics and suggest some limitations and drawbacks to the approaches that it offers. Although the chapter is critical of bioethics, the aim is not to suggest that bioethics is pointless or fundamentally flawed. Indeed, bioethics remains appropriate and necessary in many situations and can contribute to better decision-making about biotechnologies. Rather, the aim of the chapter is to suggest that the bioethics approach, *by itself*, is often not enough. The discourse of ethics needs to be supplemented by other kinds of perspectives and questions that go beyond asking "should we or shouldn't we?" In many cases, this means realizing that, before we can give a good answer to this question, we need to explore the historical, political, economic, and social trajectories of technologies in much greater detail.

A BRIEF HISTORY OF BIOETHICS

Medical practitioners, in traditions from Confucian to Buddhist to Islam, have long been subject to some kinds of special ethical principles. In the West, the most famous physician of ancient Greece, Hippocrates (ca. 460 BC–ca. 370 BC), is known for establishing an oath to guide medical practitioners. The Hippocratic Oath recognizes the special power of doctors over patients and the consequent need for a special moral code. Updated versions of this oath are still used by present-day medical schools.[1]

During medieval and early modern periods, the behavior of physicians was governed by Christian moral codes. The medical schools of Europe were religious institutions and expected their students to learn and practice medicine according to the teachings of the Bible and the Church. More formal ethical and professional codes did not emerge until the nineteenth century (for instance, the American Medical Association adopted a professional code of ethics in 1847). But even where such codes did exist, there were usually no institutional structures strong enough to enforce standards or sanction breaches. The local, personal character of most medical practice made such strictures largely irrelevant.

In the twentieth century, however, there were dramatic changes in both the character of medical practice and in the technologies required to carry it out. Most histories of bioethics locate its origins in the immediate post–World War II period. The revelations of Nazi medical experiments, especially as detailed in the Nuremburg trials, showed the extremes to which "science" and "research" could go. The 1950s and 1960s also saw the emergence of a host of new and disruptive medical and surgical procedures and technologies, including kidney dialysis, organ transplantation, respirators, and intensive care units. These advances created situations in which physicians (or family members, or hospitals) were forced to make difficult decisions. If a patient was unconscious, unable to breathe by themselves, and unlikely to ever recover, under what circumstances should an artificial respirator be switched off? Did such patients have a "right to die"? Was it permissible to remove a kidney from a healthy patient (exposing the donor to the inevitable risks of surgery) in order to save another life? When the demand for heart transplants

1. Many now use the "Declaration of Geneva," first adopted by the General Assembly of the World Medical Association in Geneva in 1948 and revised on many occasions since.

far exceeded the available supply, who should get first priority? Many of these innovations were also extremely expensive, raising questions about equality of access to health care.

Furthermore, in the United States, the public became aware of several examples of serious mistreatment of human subjects in medical research. In 1961, Stanley Milgram at Yale University ordered volunteers to administer what they thought were harmful electric shocks to Milgram's assistants (in fact, the electric shocks were fake and the assistants were pretending to be in pain). In 1971, Philip Zimbardo led a group at Stanford University in constructing a "prison experiment": some volunteers were assigned to be prisoners and some to be the prison guards. The experiment had to be halted after six days due to psychological and physical abuse inflicted by the "guards" on the "prisoners."

Even worse, in 1972 it was revealed that the US Public Health Service had conducted an ongoing set of experiments on African American men with syphilis. Researchers at the Tuskegee Institute in Alabama did not inform their subjects that they had the disease and did not provide them with treatment, even after it was known that penicillin was effective against the disease. Revelation of these abuses as well as concerns over new medical treatments fueled a growing public attention to the ethical issues surrounding biomedicine. Chapter 4 described the effect of the counterculture on biologists in the 1960s. The counterculture gave rise to a consumer rights movement that began mobilizing and organizing consumers to fight for safe, high quality, and environmentally friendly products. Patients, as consumers of health care, also began to organize and lobby for safe and transparent medical practices, targeting doctors, hospitals, and pharmaceutical companies.

In the zeitgeist of the 1960s physicians, as well as scientists, came under pressure to show that their work was socially useful and doing good rather than harm. Federal government spending on science was declining and Congress was threatening to exert increased control over scientific work (in 1968, Senator Walter Mondale attempted to create an "advisory commission" for overseeing scientific work). Many physicians and scientists believed that they needed to respond to the public's desire for a more responsible science and to bow to pressure for some sort of outside oversight of their work.

It was in this context that medical researchers turned to philosophers and theologians for help. Several groups outside biomedicine were already working towards creating applied versions of philosophy that could deployed in public policy. In 1969, a group of philosophers raised money to establish the Hastings Center, a nonprofit research institute based in Garrison, New York. Since 1971, the Hastings Center has published the *Hastings Center Report*, effectively the first journal devoted to bioethics. In that year too, Georgetown University (Washington, DC) established the Kennedy Institute of Ethics to focus on issues of health care and bioethics. These institutions were staffed largely by academic philosophers and theologians, many of who were looking to apply their training to more practical social problems.

The crystallization of bioethics as a discipline also depended on the work of Van Rensselaer Potter II (1911–2001). Potter, a biochemist and cancer researcher, coined the term bioethics in 1970 and the following year published *Bioethics: Bridge to the Future*. This book offered a clear definition of bioethics and a program of action for the new field. In particular, Potter and the Hastings Center worked to make sure that bioethics would above all be practical: it would "directly serve those physicians and biologists whose position demands that they make the practical decisions."[2] In other words, bioethics would not tolerate head-in-the-clouds philosophy—it was designed to help doctors, hospitals, and drug companies get their work done.

INSTITUTIONALIZATION AND EXPANSION

The 1980s and 1990s saw a rapid expansion of bioethics. Doctors, hospitals, universities, and governments increasingly turned to philosophers to provide them with advice and recommendations on a wide range of medical and biomedical issues. In 1974, largely in response to the Tuskegee revelations, the US Congress established the National Committee for the Protection of Human Subjects of Biomedical and Behavioral Research. In 1979, this committee produced the Belmont Report, outlining the principles on which medical research should be undertaken (respect for persons, beneficence, and justice) and laying down specific procedures for the conduct of such research (informed consent, weighing risks against benefits, and equitable selection of subjects). Informed consent, in particular, became a cornerstone of biomedical research and is an important contribution of bioethics.

Beginning in 1966, the NIH required that any institution receiving a grant set up an institutional review board (IRB) to oversee and manage the risks of any experiments involving humans. In the 1970s and '80s, IRBs and ethics committees became ubiquitous in universities and hospitals, bringing bioethical reasoning to bear in the day-to-day operation of biomedical prac-

---

2. The quotation is from a 1973 essay by Daniel Callahan, another of the field's founders. It is reprinted as Daniel Callahan, "Bioethics as a Discipline," in *Bioethics: An Introduction to the History, Methods, and Practice*, ed. N. Jecker, A. Jonsen, and R. Pearlman (Sudbury, MA: Jones and Bartlett, 1997), 87–92. Quotation p. 91.

tice. Through such institutions, thousands of physicians and researchers were educated in the basic tenets of bioethics.

In the United Kingdom, the birth of the first baby conceived via in-vitro fertilization in 1978 (see chapter 15) brought the ethics of assisted reproduction into the public spotlight. In 1982, the UK government convened a panel chaired by the philosopher Mary Warnock to examine issues raised by fertility treatment. The resulting Warnock Report laid the groundwork for the formation of the Human Fertilization and Embryology Authority in 1991. This statutory body was given the power to regulate all matters relating to IVF, artificial insemination, and the storage of human eggs, sperm, and embryos. The role of the Human Fertilization and Embryology Authority in setting policy with regard to assisted reproduction (and later cloning) meant that it contributed significantly to the institutionalization of bioethics in the United Kingdom.

As noted earlier, the Human Genome Project (HGP) also marked a turning point for bioethics. Realizing that the project raised new and complex ethical issues (especially regarding genetic testing and discrimination: see chapter 13), its leaders at the Department of Energy and the National Institutes of Health promised to set aside between 3% and 5% of the project budget for exploring "ethical, legal, and social issues" (ELSI). Projects funded under ELSI included those that investigated fairness in the use of genetic information, the privacy and confidentiality of genetic information, the possible psychological impacts of genetic information, the needs for educating doctors about genetics, the role of uncertainties in genetics research, the conceptual and philosophical implications (free will, for instance), and health and environmental issues. The HGP's ELSI became a model for other biomedical projects—it provided a neat way of taking care of the ethical and social concerns within the ambit of the project itself.

In 1996, President Bill Clinton further widened the power and profile of bioethics in the United States by creating the National Bioethics Advisory Commission. This group, consisting of doctors, lawyers, and philosophers, issued reports on the ethics of research involving human participants, on stem cell research, on cloning, on research involving people with mental disorders, on clinical trials in developing countries, on research involving biological materials, and on the ethics of biobanks. These reports were intended to guide the design and implementation of government policy around these issues.

In 2001, when the National Bioethics Advisory Commission mandate expired, President George W. Bush created the President's Council on Bioethics.

The members of the council and its chairperson were directly appointed by the President with a mandate to "advise the President on bioethical issues that may emerge as a consequence of advances in biomedical science and technology."[3] In 2004, the President's Council on Bioethics issued its report, "Monitoring Stem Cell Science." This report supported President Bush's ban on federal funding for stem cell research proclaimed in August 2001 (see chapter 17). President Obama disbanded the President's Council on Bioethics in November 2009, replacing it with the Presidential Commission for the Study of Bioethical Issues.

The history of these commissions suggests the increasingly powerful and political nature of bioethics: since the 1980s it has become increasingly well funded, increasingly institutionalized, and more deeply involved in contentious social and political issues.

FOUR CRITIQUES

Because of these increasingly powerful connections, it is important to critically evaluate some of the foundations on which bioethics is built. What sort of perspectives does it offer on modern medicine and biotechnology? What are its possible shortcomings or blind spots? What might be left out of bioethical accounts? This section describes four arguments that suggest some of the limitations of bioethics.

BIOETHICS AND MEDICINE

First, bioethics appears to have very close connections to the medical profession itself—to doctors, hospitals, and medical research institutions. The story of the birth of bioethics in the 1970s suggests that it evolved partly to *forestall* external critiques of medicine. By keeping criticism contained within the domain of medicine itself, bioethics aimed to disable more powerful or dangerous critiques that might have come from outsiders (such as consumer and patients' rights groups). The historian of medicine Charles Rosenberg argues that bioethics shares a "value-orientation" with medicine:

> As a condition of its acceptance, bioethics has taken up residence in the belly of the medical whale; although thinking of itself as autonomous, the bioethical enterprise has developed a complex and symbiotic relationship with this host organism. Bioethics is no longer (if it ever was) a free-

---

3. George W. Bush, "Creation of the President's Council on Bioethics" Executive Order 13237, November 28, 2001 (66 FR 59851).

floating, oppositional, and socially critical reform movement. . . . By invoking and representing medicine's humane and benevolent, even sacred, cultural identity, bioethics serves ironically to moderate and thus manage and perpetuate a system often in conflict with that idealized identity. In this sense, principled criticism of the health-care system serves the purpose of system maintenance.[4]

The shared value-orientation means that bioethics never offers the kind of critiques that might really threaten or forcefully challenge medical practices. For instance, a more radical critique of medicine might include questioning the very definitions and categories on which modern biomedicine is based. The history of medicine has demonstrated that such definitions and categories of disease have changed dramatically over time (for example, homosexuality was defined as a disease up to the 1960s and "hysteria" was a legitimate diagnosis until the end of the nineteenth century). In some circumstance, making "good" or "right" decisions might involve fundamentally questioning the medical categories of a particular time and place. For instance, we now see attempts to "cure" homosexuality with hormone therapy as morally wrong; but this assessment has less to do with the problems with the therapy itself and more to do with a fundamental rethinking of how we should classify homosexuality. Bioethics does not seem to allow for the kinds of critiques that would question or destabilize medical categories.

The close relationship between bioethics and the medical profession has meant that—by and large—it accepts the definitions and categories of that discipline. Just as historians have shown that such definitions and categories vary over time, anthropologists have demonstrated that different cultures have different categories and definitions of disease too: people in different times and places experience illness and suffering differently.

By accepting professional medical categories of disease, bioethics makes no allowance for such variation. The diseases and categories of the pathologist, the oncologist, and the neurologist set the terms for bioethical work. The medical anthropologist Arthur Kleinman argues that "the experience of illness is made over . . into a professionally-centered construct that is . . divorced from the patient's suffering."[5] In particular, bioethical decision-

---

4. Charles Rosenberg, "Meanings, Policies, and Medicine: On the Bioethical Enterprise and History," *Daedalus* (Special Issue) 128, no. 4 (1999): 27–46. Quotation p. 38.

5. Arthur Kleinman, *Writing at the Margin: Discourse between Medicine and Anthropology* (Berkeley: University of California Press, 1995), 49.

making emphasizes the ordered, clinical, high-spaces of modern medicine. Patients may experience their diseases in vastly different ways, in different settings, and on different terms.

In other words, bioethics may be relevant for doctors making decisions in modern hospitals: Should this new kind of operation be performed on this patient? Should we turn off the respirator? But patients (and their relatives and friends) have to make other kinds of decisions in other settings: Can I keep working with my illness? What should I tell my children? How is my family going to pay for the medical bills? These kinds of problems are part of the lived experience of disease and suffering too. Bioethics may offer us little guidance in these circumstances. Moreover, the dominance of bioethics can often shut out these kinds of problems and questions, suggesting that the professional-medical questions are the only ones that are legitimate or important. Since bioethics frames its questions and problems in terms of professional-medical categories, patient's own experience of disease can be sidelined or devalued.

The harshest critics of bioethics might argue that the field allows the medical profession to represent itself as ethical and responsible and that it has become a way of demonstrating that biomedicine is dutifully attending to its ethical responsibilities, while protecting itself from lawsuits and carrying out its business as usual. Such critics perceive bioethics as a sophisticated form of marketing and an integral part of maintaining and supporting the medical system itself. Although this view is probably too cynical, the close intellectual and institutional ties between bioethics and medical practice deserve continued scrutiny.

PRINCIPLISM

The second argument relates to the philosophical roots of bioethics. Early versions of bioethics were heavily influenced by religious thinking and traditions. Many of those in the first generation of bioethicists were trained in moral philosophy and theology, and especially in the tradition of Catholic humanism that draws on the work of St. Augustine and Thomas Aquinas. Since then, bioethics has broadened its intellectual roots and become increasingly secularized. Nevertheless, most of its practitioners are still trained in the traditions of Western philosophy. As such, many of the arguments in bioethics derive from that tradition: Socrates, Aristotle, Immanuel Kant, Jeremy Bentham, John Stuart Mill, John Dewey, and John Rawls. Their canonical works define sets of *principles* for human action from which rules for behavior might be derived. For instance, Bentham's utilitarian principle states that "it is the greatest happiness of the greatest number that is the measure of right and wrong."[6] Making decisions about how to act in the world, then, depends on making a calculation of which action would lead to the "greatest happiness."

Kleinman argues that such principles and the reasoning they entail are often highly abstract. Bioethics often attempts to strip away religion, prejudice, and any other extraneous cultural baggage and consider how a perfectly rational individual might act in a given situation. For instance, bioethics often imagine a Martian—a rational creature landing on the earth and having to decide what to do based on principles and deduction alone. This may certainly produce a well-reasoned position about a particular issue, but is this the best way for humans to actually decide what to do? Kleinman is not so sure:

> Happily or unhappily, there are no Martians; there are unfortunately, many, many humans on our planet who are faced with desperate choices in situations in which the concrete details of historical circumstances, social structural constraints like limited education and income, interpersonal pressure, and a calamity in the household or the workplace are at the core of what a dire ethical dilemma is all about. Thus there is a deeply troubling question in the philosophical formulation of an ethical problem as a rational choice among abstract principles.[7]

Kleinman is pointing out that the particularity of situations *does* matter. When we have to make a life and death decision, religion, culture, experience, and life story are likely to matter *a lot* to us. The kind of principled reasoning that bioethics provides is not likely to be very helpful to us, or at least not very relevant to the practical task of making a "good" choice.

For Kleinman, and other critics, bioethics has become based on a "view from nowhere." Its thinking needs to be supplemented with a view of moral issues from "inside of experience." Principles and reasoning alone will not suffice. Anthropology, biography, social history, and literature—the kinds of writing that help us to understand how people actually experience making decisions in traumatic situations—are what is required to arrive at a socially responsible biomedical ethics.

6. Jeremy Bentham, *A Fragment on Government* [1776] (Cambridge: Cambridge University Press, 1988).

7. Arthur Kleinman, *Writing at the Margin: Discourse between Medicine and Anthropology* (Berkeley: University of California Press, 1995), 49.

Bioethics has also become very closely associated with legal reasoning and the legal system. As bioethics became more secular, religious and moral language was increasing replaced with the language of rights and duties. Theologians were replaced by lawyers. Indeed, during the 1970s and '80s, it became increasingly clear that some bioethical debates would ultimately be played out in the courtroom. In 1975, for example, the family of Karen Ann Quinlan successfully appealed to the New Jersey Supreme Court for the right to remove her from the artificial ventilator that they believed was keeping her alive. Quinlan had fallen into a persistent vegetative state after consuming drugs and alcohol at a party. Her devoutly Catholic parents petitioned the court on the grounds that the Church doctrine did not require medical professionals to employ "extraordinary means" to prolong a patient's life. Ultimately, however, Quinlan's fate became a matter of law, not theology.

The secularization of bioethics resulted from the need to increase its legitimacy and applicability in a pluralistic society. The law seems to provide a shared set of principles and language for agreement on how to act. However, some critics have argued that reducing bioethical reasoning to legal reasoning makes bioethics impersonal, suspicious of emotions, and distant from the particularities of human experience (in other words, the reliance on the law has an effect similar to the reliance on abstract philosophical principles, described above). The bioethicist Daniel Callahan argues that collapsing bioethics into the law,

> leaves us . . . too heavily dependent upon the law as the working source of morality. The language of the courts and legislatures becomes our only shared means of discourse. That leaves a great number fearful of the law (as seems the case with many physicians) or dependent on the law to determine the rightness of actions, which it can rarely do since it tells us better what is forbidden or acceptable than what is commendable or right.[8]

The critique of legalistic reasoning in bioethics draws attention to the discipline's attempts to find a universal set of principles and means of reasoning. The effect is to create a form of public morality that may shut out or silence relevant and important aspects of people's private lives. Callahan and others are calling for a wider framework for evaluating right and wrong.

---

8. Daniel Callahan, "Religion and the Secularization of Bioethics," *Hastings Center Report* 20, no. 4 (1990): 2–4. Quotation p. 4.

ETHNOCENTRISM

Fourth, bioethics has been faulted for its ethnocentrism. Again, it is important to pay attention to the philosophical roots of bioethics: the list of canonical authors given above traces a distinctly *Western* tradition. This may exacerbate the problems with principles—not only might the principles be irrelevant to a patient, but they might also be entirely foreign. The principles of Western philosophy might entail a worldview or set of values that is completely incompatible with those of a particular person.

More specifically, Western philosophy places a high value on the individual and individual rights. The Anglo-American tradition in law and philosophy, in particular, is centered on individual rights to autonomy, self-determination, and privacy. Other traditions—particularly in East, South, and Southeast Asia—place more emphasis on social obligations, family responsibility, and communal loyalty. In a particular situation, such ties may outweigh personal autonomy. Confucianism, for instance, emphasizes an individual's obligations and duties within prescribed networks of relationships (rather than an individual's autonomy).

The ethnocentrism of bioethics is not only a problem when it is applied in non-Western nations. Western nations too are increasingly composed of diverse and multicultural groups, each of whom may hold different values and worldviews and understand their experience in fundamentally different ways. Bioethics does not easily accommodate this diversity of experience and culture.

CONCLUSIONS: ENTANGLEMENTS

These four critiques all draw attention to the historical, social, and cultural circumstances that affect moral decisions. They show that there is a messy uncertainty to people's lives and experience and that the details of this messiness often make a great deal of practical difference to what people should or should not do. All this suggests that moral and bioethical dilemmas don't necessarily have definitive answers: rather, answers will depend on context and culture in complicated ways. In others words, it suggests that "should we or shouldn't we?" is often a messy and contingent question. These critiques direct us towards all kinds of other important questions, many of which need to be answered *before* we can legitimately ask "should we or shouldn't we?" Such questions include these: Who is making the decision? On whose behalf are they deciding? Who is benefiting? Who is losing out? And are we using the most appropriate or socially useful categories?

Again, despite these criticisms, the point of this chapter is not to disparage all bioethics or to suggest that it is useless. Bioethics does have much to offer in helping society to make decisions about biotechnology. The aim here has been to suggest that bioethics offers only a partial perspective on the complexity of the problems that biotechnologies and biomedicine raise. As such, we need to read and evaluate it carefully and critically; bioethics should not always have the last word on telling us what to do with biotechnologies. The complexity and contingency of biomedicine, embedded as it is with culture and politics, suggests that "should we or shouldn't we?" often requires complex and contingent answers.

This point can be illustrated by examining the consequences of the entanglement between contemporary biomedicine and corporations. A patient who is to be subjected to a new drug or therapy sits within a complex web of competing interests that includes doctors, hospitals, medical schools, university labs, insurance companies, Big Pharma, biotech companies, instrument manufacturers, and test providers. A moral decision about "what is good for the patient" depends on understanding and unpacking this web of overlapping interests, power, and financial dependencies. Until we know more about how this system works (Who is making decisions for whom? For what reasons? Who is benefiting from what?), it is going to remain difficult to make responsible decisions about what might be right or wrong, good or bad.

FURTHER READING

The history of bioethics can be fairly sharply divided between accounts written by bioethicists and accounts written by historians. For the former see Nancy S. Jecker, Albert R. Jonsen, and Robert A. Perlman, eds., *Bioethics: An Introduction to the History, Methods, and Practice* (Sudbury, MA: Jones and Bartlett, 2007), Albert R. Jonsen, *The Birth of Bioethics* (New York: Oxford University Press, 1998), and Albert R. Jonsen, *A Short History of Medical Ethics* (New York: Oxford University Press, 1999).

For historians' accounts see especially John H. Evans, *Playing God? Human Genetic Engineering and the Rationalization of Public Bioethical Debate* (Chicago: University of Chicago Press, 2002); and the essays by Charles Rosenberg and by John H. Warner in Robert B. Baker, Arthur L. Caplan, Linda L. Emanuel, and Stephen R. Latham, eds., *The American Medical Ethics Revolution: How the AMA's Code of Ethics Has Transformed Physicians' Relationships to Patients, Professionals, and Society* (Baltimore: Johns Hopkins University Press, 1999). A work that examines bioethics in the broader context of changing medical practice is David J. Rothmann, *Strangers at the Bedside: A History of How Law and Bioethics Transformed Medical Decision Making* (New York: Basic Books, 1991). For an extremely critical history see Tom Koch, *Thieves of Virtue: When Bioethics Stole Medicine* (Cambridge, MA: MIT Press, 2012). On the history of institutional review boards in particular see Laura Stark, *Behind Closed Doors: IRBs and the Making of Ethical Research* (Chicago: University of Chicago Press, 2011).

The important foundational texts by Van Rensselaer Potter can be found as Van Rensselaer Potter, "Bioethics, The Science of Survival," *Perspectives on Biology and Medicine* 14 (1970): 127–52 and Van Rensselaer Potter, *Bioethics: Bridge to the Future* (Englewood Cliffs, NJ: Prentice Hall, 1971).

Literature critical of bioethics comes from history of medicine, anthropology of medicine, and bioethics itself. The work of Arthur Kleinman is of particular importance and a good overview can be found in this essay: Arthur Kleinman, "Anthropology of Bioethics," in *Writing at the Margin: Discourse between Anthropology and Medicine* (Berkeley: University of California Press, 1995), 41–55. Kleinman also coedited a special issue of the journal *Daedalus* on bioethics; see Arthur Kleinman, Renée C. Fox, and Allan M. Brandt, eds., "Bioethics and Beyond," *Daedalus* (Special Issue) 128, no. 4 (1999). Here especially see essays by Charles E. Rosenberg, "Meanings, Policies, and Medicine: On the Bioethical Enterprise and History," pp. 27–46; Kleinman himself, "Moral Experience and Ethical Reflection: Can Ethnography Reconcile Them? A Quandary for 'The New Bioethics,'" pp. 69–97; Mary-Jo DelVecchio Good, Esther Mwaikambo, Erastus Amayo, and James M'Imunya Machoki, "Clinical Realities and Moral Dilemmas: Contrasting Perspectives from Academic Medicine in Kenya, Tanzania, and America," pp. 167–196; and chapter 6 of Carl Elliot, *White Coat, Black Hat: Adventures on the Dark Side of Medicine* (Boston: Beacon Press, 2010).

More specifically, on the secularization of bioethics see Daniel Callahan, "Religion and the Secularization of Bioethics," *Hastings Center Report* 20, no. 4 (1990): 2–4. For more information about the case of Karen Ann Quinlan see *In Re Quinlan* 355 A.2d 647 (NJ 1976). For more about the way bioethics applies across cultures and ethnicities see Christoph Rehmann-Sutter, Marcus Düwell, and Dietmar Mieth, eds., *Bioethics in Cultural Contexts: Reflections on Methods and Finitude* (Dordrecht, Netherlands: Springer, 2006). A good example of an attempt to forge a more culturally inclusive bioethics is Heiner Roetz, ed., *Cross-Cultural Issues in Bioethics: The Example of Human Cloning* (New York: Editions Rodopi, 2006).

PART VIII
VIRGIN BIRTHS

FROM THE PILL TO IVF

INTRODUCTION

Many of the biotechnologies we have discussed so far have aimed to control our bodies by intervening in the processes of disease. This chapter begins a discussion of biotechnologies that intervene in the processes of *reproduction*. This includes technologies aimed at stopping reproduction (contraceptives) as well as those aimed at "helping" or enhancing reproduction. The latter, which includes the techniques of in-vitro fertilization (IVF), are sometimes known as assisted reproductive technologies (ARTs).

In this chapter, we will examine the histories that led to development of the contraceptive pill (in the 1950s) and IVF (in the 1970s). The research on animal and human reproduction that led to these breakthroughs was closely linked to the development of reproductive cloning and human embryonic stem cell lines (both discoveries of the 1990s). The cloning of Dolly the sheep (discussed in chapter 16) and the emergence of stem cell science (chapter 17) form part of the same story about the endeavors of biologists to understand and gain increasing levels of control over reproduction.

A BRIEF HISTORY OF CONTRACEPTION

The history of contraception is almost as long as the history of human sexuality itself. Ancient technologies, used in Europe, China, and India, included barrier methods such as caps (made of substances including intestines, waxed paper, tortoise shell, and animal horn) fitted or tied onto the tip of the penis during sexual intercourse. In Europe, such devices evolved into condoms made out of linen, intestines, bladders, and leather. The 1839 invention of vulcanization allowed the development of condoms made of rubber. In some cases, such devices were intended to prevent infection by sexually transmitted diseases (such as syphilis), as well as to prevent pregnancy.

In many cultures and in many periods of history the prevention of pregnancy was considered women's responsibility. Women used materials such as sponges, wool, and cotton as cervical caps or diaphragms (all intended to block the entrance of sperm into the uterus). Probably even more common was the use of plant and mineral mixtures inserted into the vagina and intended to act as spermicides. Such concoctions could either be applied directly or used in a pessary—a pouch or container designed for insertion into the vagina. Substances used at various places and times included acacia gum, honey mixed with sodium carbonate, pomegranate, pennyroyal, myrrh, palm leaf mixed with red chalk, and salt.

Women have also long tried to control reproduction by inducing abortions in cases of unwanted pregnancies. Again, various plants, herbs, and concoctions (of various efficacy) were available for this purpose. In Europe, for instance, silphium, ergot of rye, pennyroyal, slippery elm, nutmeg, and savin (a form of juniper) were used as abortifacients. Most of these achieved their effects—if they were effective at all—by poisoning the pregnant woman enough to induce a miscarriage.

The history of contraception is linked to developments in technology and medicine. But it is also closely linked to the power of women in society: in cultures where women were able to gain access to knowledge and resources they were also able to gain increased control over their own bodies.

THE CONTEXTS OF THE CONTRACEPTIVE PILL

The development of the contraceptive pill marked a significant breakthrough in both biotechnology and women's rights. This pharmaceutical product allowed women to gain an unprecedented level of control over their own fertility and reproductive capacity. We might not usually think of pills or other drugs as biotechnologies. But an increasing number of pharmaceutical products intervene in our bodies on a molecular level and allow us a high level of control over various aspects of our lives. The pill was one of the first such drugs; some others will be considered in chapter 19 (along with a more elaborate justification of why we should consider drugs biotechnologies).

The contraceptive pill was the result of several converging histories and lines of research: women's rights, eugenics, hormone research, and the emergence of the modern pharmaceutical industry. Understanding the pill's cultural, political, and technological significance requires considering all of these strands at once.

The pill was an outcome of the early twentieth-century women's rights movement in the United States. In particular, the work of Katherine McCormick (1875–1967) and Margaret Sanger (1879–1966) generated both the social and financial capital that supported contraception research. McCormick had married into a rich family but early on in her marriage her husband was diagnosed with schizophrenia. She was determined not to have children since they would potentially inherit the disease from their father. From 1909,

McCormick participated in the women's suffrage movement where she met birth control activists, including Sanger.

Sanger worked as a nurse in New York City, often witnessing the catastrophic social, mental, and medical effects of unwanted pregnancies (including the death of women from illegally performed back-alley abortions). She started the first birth control clinic in the United States in 1916 and began to distribute information about birth control. Such actions were not only distasteful to most Americans, but also illegal. The Comstock Act, passed in 1873, criminalized the distribution of material deemed to be "obscene, lewd, or lascivious"—this was taken to include information about contraception as well as contraceptive devices themselves. Sanger was arrested for distributing information about contraception in 1918. She was acquitted of the charges on appeal, and the trial brought a large amount of public attention to the issue of birth control.

After women in the United States received the right to vote in 1920, McCormick and Sanger devoted themselves wholly to work on promoting contraception. In 1921, Sanger formed the American Birth Control League (which eventually became Planned Parenthood Federation of America) and in 1923 established the Clinical Research Bureau, staffed by female doctors who could prescribe birth control devices to women. McCormick assisted these efforts by smuggling diaphragms from Europe into the United States and providing financial backing. By 1929, they were lobbying state and federal governments for the legalization of birth control in the United States. Alongside these immediate goals, however, Sanger and McCormick were also looking for new methods of birth control that would be more effective, be easier to use, and provide greater control for women.

Although many Americans still considered birth control immoral, dangerous, and taboo, Sanger and McCormick's efforts were supported by eugenicists. Eugenicists believed that the health of the population depended on reproduction by "fit" individuals only (see chapter 11). Birth control offered a means by which the reproduction of "unfit" individuals might be stopped or at least curtailed. Even after World War II, when eugenic thinking had been largely discredited, many people still worried that unchecked reproduction would ultimately lead to the exhaustion of food and other resources and have disastrous consequences for the whole human population. During the Cold War, Westerners saw explosive population growth in Asia and Africa as a threat to their own dominance. In particular, the increasing numbers of poor and starving people provided conditions ripe for communist revolution. Cheaper and better contraception, some reasoned, could contribute to a solution. Such eugenic and "population" thinking form an important second context for the development of birth control.

The third context is the scientific research on hormones that expanded during the 1920s. Chapter 2 described the excitement around endocrinology in the early twentieth century. The newly discovered sex hormones, in particular, were considered to be hugely powerful (especially for "rejuvenating" older men). McCormick believed that a hormonal imbalance might be responsible for her husband's schizophrenia and so she began to fund research in endocrinology (she established the Neuroendocrine Research Foundation at Harvard Medical School). However, hormones were very hard to obtain—they usually had to be extracted from the glands of animals—and consequently were very expensive. A cheap, reliable source of hormones would be extremely useful and lucrative.

By the late 1930s, chemists had worked out how to synthesize progesterone (a pregnancy hormone), testosterone (a male sex hormone), and estradiol (a female sex hormone) from cholesterol. But the substances all remained expensive. During World War II, the Allies came to believe that the Luftwaffe (the German air force) were using hormones extracted from the adrenal glands on their pilots to counteract the effects of high-altitude flying. This prompted a major research effort to produce such hormones in quantity in the United States. The Germans, it turned out, were not using hormones and the war ended before the Allied research program could produce any.

But the wartime effort did spur further research into hormones and hormonal synthesis. In 1942, Russell Marker (1902–1995), a chemist from Pennsylvania State College, collected a large Mexican yam called *cabeza de negro*. Marker showed that the yam contained large amounts of a substance called diosgenin and discovered methods of chemically converting this into testosterone, estradiol, and progesterone. By 1944, he had set up his own company—called Syntex—in Mexico to produce progesterone. (This is a good example of the kind of bioprospecting that will be discussed in chapter 22.)

In 1948, the biochemist Edward Kendall (1886–1972) discovered that a hormone that he had isolated from the adrenal glands in the 1930s had a remarkable therapeutic effect on arthritis patients. Cortisone quickly became a wonder drug. Merck & Co. first produced it commercially in 1949. The chemist Percy Julian worked out how to synthesize cortisone from soybeans. This allowed the price of cortisone to drop from $200 per gram in 1949 to $10 per gram in 1951. The cost of cortisone was reduced even further when it was discovered that cortisone could be produced from progesterone made from Marker's yams.

The final context for understanding the emergence of the contraceptive pill, then, is the growth of the pharmaceutical industry in the 1950s. Before World War II, the US and European pharmaceutical industries had consisted of relatively small and local enterprises. From the 1940s onward, firms such as Merck, Eli Lilly, Schering, Wyeth, G. D. Searle, Upjohn, La Roche, and Glaxo began to invest in scientific methods of drug discovery and industrial manufacturing processes. Cortisone, in particular, had demonstrated the immense profits that could be made from a mass-market pharmaceutical. It was in this rapidly expanding and competitive commercial context that the contraceptive pill was developed.

THE ARRIVAL OF THE PILL

The final parts of the story emerge from yet another direction: research on mammalian reproduction. Gregory Pincus (1903–1967) had begun working on reproduction and hormonal biology at Cambridge University in the late 1920s. During the 1930s, then at Harvard, he continued to experiment on the fertilization of rabbit eggs outside the body. As will be discussed later in this chapter, this research formed a basis for the discoveries that led to human IVF. But in the 1930s, studying reproduction was considered low-status science; for many it was even embarrassing or immoral work. Such problems of prestige and reputation no doubt played a role in Harvard's decision not to grant Pincus tenure as professor in 1937. Realizing he needed to strike out on his own, in 1944, Pincus set up a private scientific research institute to study reproduction called the Worcester Foundation for Experimental Research (in Worcester, Massachusetts).

Pincus' research at the Foundation was funded partly by pharmaceutical companies, including G. D. Searle. But he had also been commissioned by Sanger to be on the lookout for any compounds with contraceptive potential. In the early 1950s, more hormonal compounds were derived from Mexican yams. Most were quickly patented by pharmaceutical companies. Two of the new substances seemed to have contraceptive potential: Norethynodrel (patented by Searle) and Norethisterone (patented by Syntex). In 1951, now funded by Sanger, Planned Parenthood, and McCormick, Pincus began experimenting with these compounds. He showed that they could prevent pregnancy in rats and rabbits.

By 1955, Pincus was looking for human subjects on which to test his drug. Since he didn't have any patients of his own, he needed to find a collaborator who did. John Rock, a gynecologist at Harvard Medical School, specialized in the treatment of infertility (Rock, too, will enter our story again as a key figure in the development of IVF). Rock believed that female infertility could sometimes be caused by a sort of mixing-up of the cycle of ovulation and menstruation. If the cycle could be stopped for a time, he thought, maybe it would reset itself and fertility would be restored. Based on this theory, Rock allowed Pincus to test his new compounds on his patients as part of their treatment for infertility. The trials showed that the orally administered pills suppressed ovulation in almost all women.

But, of course, to see whether the drug really worked as a contraceptive Pincus needed to test it on fertile women. In 1956, Pincus organized more trials to take place on the island of Puerto Rico, a territory of the United States. This allowed Pincus to avoid media scrutiny and to avoid the problems involved in testing a contraceptive in a nation where birth control remained illegal in many states (including in Massachusetts). In other words, the marginal racial and citizen status of women in Puerto Rico made it easier for Pincus to conduct his experiment on them.

The trials proved the drug safe and effective and the FDA approved the pill in 1957 as a treatment for gynecological disorders, infertility, habitual miscarriage, and excessive menstruation. It was not designated for use as a contraceptive (this would have been far too controversial). However, women soon realized that physicians could prescribe the pill in any way they saw fit (so-called off-label use—a legal and widespread practice in the United States and other countries). The pill (first sold under the label Enovid) was an immediate and massive success, making huge profits for G. D. Searle. By late 1959, already half a million women were on the pill; by 1965, over 6.5 million married women were taking it (plus an unknown number of unmarried women). The FDA approved the application to market the pill as a contraceptive in 1961.

The cultural impact of oral contraceptives was immediate and immense. For one thing, the pill marked a turning point in how people consumed drugs: it was a medication that women took not to get well, but to allow them to change their lifestyle and behavior. This contributed to shifting people's ideas about what drugs could be *for* (this will be discussed further in chapter 19). But more importantly, the pill played a significant role in the cultural transformations of the 1960s. The counterculture fostered, amongst other things, changing attitudes toward sex, sexuality, and women's rights. In particular, women began to demand greater equality in sexual relationships and assert greater control over their bodies. By providing women control over their reproductive capacity, the pill played a significant role in changing attitudes and behaviors and in allowing women greater freedom of sexual expression.

IN-VITRO FERTILIZATION

The contraceptive pill gave us the ability to *stop* reproduction. Since the 1970s, biological knowledge has also been applied to the *starting* or restarting of reproduction. In vitro fertilization—sometimes coupled with other biotechnologies—allows reproduction in circumstances where it was previously impossible (older women, infertile couples, and same-sex couples, for instance).

Like contraception, treatments for infertility have a history that dates to ancient times. Different cultures have deployed a variety of herbal and dietetic means to promote fertility and virility. Artificial insemination has also been used as a means of overcoming infertility: it can allow fertilization in cases where sexual intercourse is not possible (impotence, for example) or where cervical scarring or mucus impedes conception. By obviating sexual intercourse, this relatively simple procedure makes it possible for reproduction to be partially disconnected from the body and from the presence of a male. The 1950s invention of means of freezing sperm for long periods (discussed in chapter 9) made it possible to further disconnect reproduction from the male body (since sperm could come from an anonymous donor who may have deposited the sample many years before the conception). In vitro fertilization is a means of further dissociating sex from reproduction.

The history of in-vitro fertilization is closely connected to the history of animal husbandry. As noted above, in the early twentieth century the biology of human reproduction was a taboo topic and a low-status subject for scientific inquiry. The study of animal reproduction, however, was far better developed, better funded, and more lucrative. Creating bigger and better livestock was a high priority for governments interested in feeding populations and companies interested in profits. And making superior animals meant controlling their reproduction as closely as possible.

The pioneer of this work was Walter Heape (1855–1929), a Cambridge University embryologist. In the 1890s, Heape performed a series of experiments in which he managed to transfer fertilized embryos from the uterus of one rabbit to that of another. In the 1920s, this work was taken up again at Cambridge by Gregory Pincus. Pincus had received his doctoral degree from Harvard, but then spent a year abroad in Cambridge and Berlin. There he began experiments in which he attempted to fertilize rabbit eggs in vitro. Following his return to Harvard in 1931, he published a paper reporting the successful birth of rabbits from eggs fertilized outside the body.[1] Pincus' work received significant popular attention throughout the 1930s—he was criticized for "playing God" and for opening up the possibility of allowing reproduction without men.

1. This claim was later called into question once the nature of sperm and egg maturation was fully understood.

Despite these controversies, many physicians and biologists also saw that this work could lead to a "cure" for human infertility. John Rock, the Harvard Medical School gynecologist, continued Pincus' work, now working with human oocytes (eggs) and sperm. Between the late 1930s and the early 1950s, Rock and his assistants removed hundreds of eggs from women undergoing surgeries (mostly hysterectomies) and exposed many of these to sperm. They observed a few divide into two and then three cells. But overall, the results were inconsistent and disappointing. The main difficulties with achieving IVF were that sperm and egg had to be united at the right stage in their development in order for fertilization to take place. Sperm had to be "capacitated" after ejaculation; eggs had to be extracted at a particular stage of maturity. Moreover, researchers had to find a culture medium that would support the growth and development of oocytes outside the body.

During the 1960s, Robert Edwards, working at Cambridge University, solved many of these problems. Edwards was trained as a physiologist and his research on human embryos sprang from his interest in studying the abnormalities of developing oocytes. Around 1960, Edwards turned his attention to the problems of fertilization: studying developing oocytes meant fertilizing and growing them outside the body. He and his coworkers refined the techniques for using hormones (called gonadotrophins) to stimulate the production of "ripe" oocytes and for culturing them in the lab. In the late 1960s, Edwards, needing a consistent source of eggs for his experiments, began collaborating with Patrick Steptoe, an infertility expert at Oldham General Hospital in Manchester. Steptoe was an expert in the use of the laparoscope — a new instrument that was used to remove eggs from the ovaries. By 1968, Edwards' team had managed to fertilize a human egg in vitro.

By the early 1970s, most of the knowledge that was needed to perform IVF was in place. Such work, however, remained controversial and many scientists doubted that it could ever lead to a normal human birth. In 1971, the UK's Medical Research Council rejected Steptoe and Edwards' application for a large grant for their research. Over the next seven years, Steptoe and Edwards (and other research teams elsewhere, particularly at Monash University in Australia) worked out the means by which the growing multicell embryos could be successfully transferred back to the mother's uterus. In 1973, the Australian team successfully implanted an eight-cell embryo, but the pregnancy ended after only eight days. Steptoe and Edwards' team also faced similar setbacks — only the right balance of hormones in the mother's body would produce a pregnancy. Finally, on November 10, 1977, Steptoe and Edwards performed what turned out to be the first successful IVF procedure. Louise Brown, a normal and healthy girl, was born approximately nine months later on July 25, 1978.

Since then, IVF has become a ubiquitous part of medical practice in developed countries. It is estimated that over five million "test tube babies" have been born around the globe. In the United States, roughly 50,000 babies are born every year using IVF techniques, a figure representing just over 1% of total births. IVF continues to raise controversies, especially within some religious communities (the Catholic Church, for instance, opposes IVF). These controversies center on IVF's potential to allow pregnancy past menopause, its potential to permit impregnation by anonymous male donors, and its role in facilitating reproduction for same-sex, unmarried, and single persons. IVF has also been used in conjunction with genetic testing for preimplantation genetic diagnosis or "screening" of embryos (see chapter 18).

The high cost of IVF has also become a subject of controversy: in many countries, the procedure is available only to the upper economic strata of society. IVF has contributed to the "pathologizing" of infertility—increasingly, infertility has come to be seen as a disease or disability, and reproduction as a fundamental right. The notion of IVF as a *treatment* suggests that infertility is a disease that now has a cure. As a right—or as a problem of health—reproduction should be equally available to all people, regardless of economic status, many argue.

The growing use and acceptance of IVF has caused a gradual expansion of the "right" to reproduce. Not only infertile heterosexual couples, but also postmenopausal women, single women, and same-sex couples are able (in some places) to utilize IVF. This has possibilities for refiguring notions of kinship—that is, how we think about families and our relationships with those around us. Along with sperm donation and surrogacy, IVF has created the potential for new kinds of families, often decoupling biological relationships from social ones (for example, an IVF-conceived embryo may be carried by a surrogate who delivers the child to be raised by the original owners of the egg and sperm). In the short term, this may create social and identity problems for those directly involved ("Who are my 'real' parents?"). In the longer term, it is likely to reshape social expectations and norms regarding families.

IVF also affects how we think about embryos. If an embryo is made outside the body it may come to be seen as something independent of the body. In a traditional pregnancy, the embryo was inseparable from a woman's body, wholly dependent on the uterus. Medical technologies keep prematurely born babies alive outside the body. IVF means that the body is not even necessary for the beginning of new life. The idea that the embryo and the fetus exist independent of the mother's body may reinforce the argument that embryos, too, have their own legal and moral rights.

Our own bodies too, take on different roles in the world of ARTs. Both IVF and the contraceptive pill have contributed to the decoupling of sex from reproduction and the decoupling of reproduction from human biological time. Before the 1950s, the process of reproduction was usually something that took place completely *inside* the body. This process was dependent on bodily rhythms and cycles (such as menstruation-ovulation, menarche-menopause). The pill intervened in such internal processes, while IVF allowed these processes to be replaced with actions that take place partially *outside* the body.

ARTs allow us to "outsource" reproduction to the laboratory, augmenting our bodies with doctors, clinics, sophisticated instruments, test tubes, culture mediums, and laboratory freezers. Many people now regularly augment their brains with information technologies such as computers that help them to think and remember. This changes the kinds of things we can do with our brains and the ways we use them. Likewise, we have begun to augment our reproduction with biotechnologies. This changes the ways we use our bodies and what we can do with them: sex can take on different meanings and significance and we become far less dependent on (or constrained by) our body's own physical and temporal capacities.

CONCLUSIONS: BIOPOWER, REPRODUCTIVE FREEDOM, AND SOCIAL CONTROL

Debates about ARTs are often framed in terms of rights and freedoms: reproductive freedom, freedom for women, the right to choose, and so on. But we need to mindful that providing technologies does not automatically generate choice and promote freedoms. Renate Klein has argued that ARTs may actually limit women's choices about reproduction: by magnifying infertility into a "life crisis," women are increasingly expected and even pressured to undergo expensive medical procedures such as IVF. "In western cultures the stigma of infertility often leads to internalised guilt and inferiority complexes: women's worth is still measured predominantly through motherhood."[2] In a society where motherhood is the norm, the "choice" associated with ARTs may be illusory.

The philosopher Michel Foucault uses the term "biopower" as a label for the various ways in which governments exercise control over bodies and populations. Governments exert power over bodies through direct means

2. Renate Klein, *The Ultimate Colonisation: Reproductive and Genetic Engineering* (Dublin: Attic Press, 1992), 12.

such as violence, imprisonment, and execution; but they also do so in more subtle ways through scientific, educational, and medical institutions. The regulation of sex and reproduction—Foucault argues—is of particular concern to governments since it forms the means by which its own subjects are produced. Following Klein's argument, we might see ARTs as a potentially powerful means of state intervention into reproduction: it allows science and medicine to exert greater control over bodies and populations. That is, through biomedical practice and biotechnology, the state gains greater control over who reproduces and when. Klein sees IVF as "the ultimate colonization"—the invasion of the (usually male) medical practitioner and his technology into women's bodies. Noting the close connection between research on ARTs and research on contraception, Klein argues that ARTs are more about social and population control than about the freedom to choose.

For some readers, all this may seem an extreme view. Nevertheless, it should be acknowledged that ARTs do not automatically provide greater reproductive freedom or choice. Especially as ARTs are coupled to other biotechnologies such as genetic engineering and cloning, we must continue to ask questions about whose interests they are serving.

FURTHER READING

For long histories of contraception see John M. Riddle, *Contraception and Abortion from the Ancient World to the Renaissance* (Cambridge, MA: Harvard University Press, 1994), Angus McLaren, *A History of Contraception: From Antiquity to the Present Day* (Hoboken, NJ: Wiley Blackwell, 1992), and Robert Jütte, *Contraception: A History* (Cambridge: Polity, 2008). On birth control in the United States see Andrea Tone, ed., *Controlling Reproduction: An American History* (Lanham, MD: Scholarly Resources Books, 1997), Andrea Tone, *Devices and Desires: A History of Contraceptives in America* (New York: Hill & Wang, 2001), and Linda Gordon, *The Moral Property of Women: A History of Birth Control Politics in America* (Urbana: University of Illinois Press, 2002). The history of birth control in other parts of the world is less comprehensively documented, although for Victorian England see Angus McLaren, *Birth Control in Nineteenth-Century England* (Teaneck, NJ: Holmes & Meier, 1978) and for mid-twentieth-century Germany see Atina Grossmann, *Reforming Sex: The German Movement for Birth Control and Abortion Reform, 1920–1950* (New York: Oxford University Press, 1997). For a more global perspectives see Judith R. Seltzer, *The Origins and Evolution of Family Planning in Developing Countries* (Santa Monica, CA: RAND, 2002); and Andrew Russell, Elisa J. Sobo, and Mary S. Thompson, eds., *Contraception across Cultures: Technologies, Choices, Constraints* (Oxford: Berg, 2000).

For work more specifically on the history of the contraceptive pill see Lara V. Marks, *Sexual Chemistry: A History of the Contraceptive Pill* (New Haven, CT: Yale University Press, 2001); Jonathan Eig, *The Birth of the Pill: How Four Crusaders Reinvented Sex and Launched a Revolution* (New York: W. W. Norton, 2014); and (for a personal account by one of those involved) Carl Djerassi, *Carl Djerassi: Steroids Made It Possible* (Washington, DC: American Chemical Society, 1990). For Russell Marker's work see American Chemical Society, *The "Marker Degradation" and the Creation of the Mexican Steroid Hormone Industry, 1938–1945* (Washington, DC: American Chemical Society, 1999; http://www.acs.org/content/acs/en/education/whatischemistry/landmarks /progesteronesynthesis.html). On the cultural impact of the pill in the United States see Elaine Tyler May, *America and the Pill: A History of Promise, Peril, and Liberation* (New York: Basic Books, 2011).

To my knowledge there is no comprehensive account of the history of IVF. The following sources tell parts of the story: John D. Biggers, "Walter Heape, FRS: A Pioneer in Reproductive Biology," *Journal of Reproduction and Fertility* 93 (1991): 173–186; John D. Biggers, "IVF and Embryo Transfer: Historical Origin and Development," *Reproductive BioMedicine Online* 25 (2012): 118–127; M. H. Johnson, "Robert Edwards: The Path to IVF," *Reproductive BioMedicine Online* 23 (2011): 245–262; M. H. Johnson, S. B. Franklin, M. Cottingham, N. Hopwood, "Why the Medical Research Council Refused Robert Edwards and Patrick Steptoe Support for Research on Human Conception in 1971," *Human Reproduction* 25 (2010): 2157–2174; and Malcolm Gladwell, "John Rock's Error," *New Yorker*, March 10, 2000. Unfortunately one of the best "long" histories of IVF that places it into the context of agricultural experimentation is available only in German: Christine Schreiber, *Natürlich künstliche Befruchtung? Eine geschichte der In-vitro Fertilisation von 1878 bis 1950* [Natural Artificial Insemination? A History of IVF from 1878 to 1950] (Göttingen: Vandenhoeck & Ruprecht, 2007).

In my discussion of the ethical and social consequences of IVF I have drawn especially on Renate Klein, *The Ultimate Colonisation: Reproductive and Genetic Engineering* (Dublin: Attic Press, 1992). There is also a large bioethics literature related to assisted reproductive therapies, focused especially on issues related to embryo destruction and access. Important works here include Susan Markens, *Surrogate Motherhood and the Politics of Reproduction* (Berkeley: University of California Press, 2007); President's Council on Bioethics, *Reproduction and Responsibility: The Regulation of New Technologies* (Washington, DC: March 2004, available at: https://bioethicsarchive.george town.edu/pcbe/reports/reproductionandresponsibility/); and Anne T. Fidler and Judith Bernstein, "Infertility: From a Personal to a Public Health Problem," *Public Health Reports* 114, no. 6 (1999): 494–511. On the debate over provision of ARTs to same-sex couples see Jacob M. Appel, "May Doctors Refuse Infertility Treatments to Gay Patients?" *Hastings Center Reports* 36, no. 4 (2006): 20–21.

For more advanced readers, there are also several accounts of assisted reproductive technologies from medical and feminist anthropological perspectives. See Sarah Franklin, *Biological Relatives: IVF, Stem Cells, and the Future of Kinship* (Durham, NC: Duke University Press, 2013); Sarah Franklin, *Embodied Progress: A Cultural Account of Assisted Conception* (New York: Routledge, 1997); and Elizabeth Ann Kaplan and Suran Merrill Squier, eds., *Playing Dolly: Technocultural Formations, Fantasies, and Fictions of Assisted Reproduction* (New Brunswick, NJ: Rutgers University Press, 1999).

16   CLONING

INTRODUCTION

Cloning is one of the most widely discussed and controversial areas of biotechnology. The contraceptive pill and IVF are technologies that have become an almost everyday part of Western society's reproductive practices. Cloning, however, has been met with alarm and resistance. In order to describe how cloning became so significant and divisive, this chapter will take an unusual approach. It will analyze the history of cloning through a discussion of sheep and their relationship to humans. For many people, Dolly the sheep, born in Scotland in 1996, represents cloning. Usually, the fact that Dolly was a sheep is not given much thought—it is assumed to be more or less insignificant. But this chapter argues that the Dolly's being a sheep was, first, not coincidental, and second, very significant for how we understand cloning.

By situating Dolly—and cloning—within the context of sheep breeding we can begin to see some of the continuities between older agricultural practices and newer biotechnological practices. We will also see how our cultural and economic relationships to animals makes us particularly wary and worried about reproductive biotechnologies like cloning.

ROBERT BAKEWELL AND THE HISTORY OF SHEEP-BREEDING

Humans have a long relationship with sheep. Several of the earliest civilizations, especially those in the Middle East, cultivated sheep. These societies were dependent on their flocks for milk, meat, and wool. Some historians have argued that the sheep has had a greater impact on human history than any other animal (the dog is also of great importance).

In the Judeo-Christian religions that emerged from these civilizations, sheep play important symbolic roles. Lambs are symbols of purity, innocence, and righteousness. Sheep appear as markers of wealth, exchange value, ownership, and inheritance. Jesus is often depicted as a shepherd, tending his flock; but he is also described as the "lamb of God." In many Biblical stories sheep become representatives of humans, standing in for them—most significantly in Exodus the Jews in Egypt sacrifice lambs in place of their first-born sons.

In Europe, sheep became more and more important throughout the medieval and early modern periods. Wool, used especially for clothing, was the basis of many early forms of trade and became vital to the development of European economies, especially in England and Scotland. By the eighteenth century, European kingdoms were vying to produce wool of the highest quality that would give them an advantage in the wool trade. As populations grew, especially in cities, sheep were also considered to be an important source of meat. Producing sheep with both finer wool and more meat required attention to animal husbandry and the development of new sheep breeds.

One of the most renowned sheep breeders of the eighteenth century was the English agriculturalist Robert Bakewell (1725–1795). On his farm in Leicestershire, Bakewell developed extraordinary new breeding methods. The most important result of these was his New Leicester (or Dishley Ram—figure 16.1). This hornless, square-bodied, and short-legged sheep variety was specially designed by Bakewell to provide "mutton for the masses." The large body was especially meaty and hence this new breed of sheep was especially well suited to provide food for growing populations (it was exported to colonies in Australia and North America for this purpose).

Bakewell's work required a high level of knowledge and skill about sheep, breeding techniques, and farming practices. Bakewell knew that the value of his work was not limited only to individual sheep that he produced. Rather, when he sold one of his carefully bred sheep, or when he put it out to stud (selling its reproductive powers), he knew he was really selling a kind of template or design for producing animals of the same type. Bakewell and his contemporaries did not have the language of genes, but they knew that breeding could create a kind of type that had value that could be passed on from sheep to sheep (the historian Harriet Ritvo has called this "breed wealth"). In other words, Bakewell was well aware of the economic significance of his breeding work. He usually kept his methods a secret, attempting to protect his intellectual property. For Bakewell, controlling the fertility and reproduction of sheep meant controlling capital.

This example shows how, in the eighteenth century, sheep reproduction was already intertwined with economics, power, and politics—generating breed wealth allowed countries to feed their populations and maintain a prosperous trade in wool.

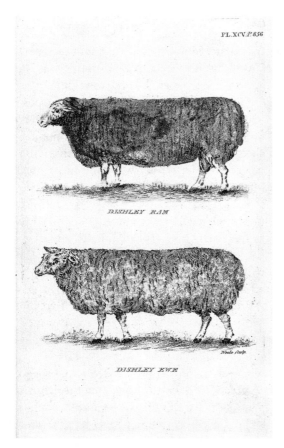

16.1 Dishley Ram. Drawing by Richard Phillips, 1807–1808. Robert Bakewell's Dishley Ram or New Leicester. The large-bodied and stocky sheep was able to produce more "mutton for the masses." Source: Wellcome Library, London. Used under CC BY 4.0 license.

THE ROSLIN INSTITUTE AND PHARM ANIMALS

This economic and political importance of sheep and sheep breeding continued into the twentieth century. In the wake of World War II, the British economy was severely strained. Importing food from overseas was expensive and the British government was looking for ways to reduce the nation's dependence on overseas agriculture. If more grains and meat could be produced at home, economic conditions could be improved. To this end, the British government aimed to establish a set of publicly funded research organizations to improve the output of British farms. Scientific research on animal genetics and animal breeding, the government reasoned, would lead to greater farm productivity and, ultimately, more food.

Although Scotland possessed an industrialized economy, it retained an important agricultural sector (especially sheep) and the region had a strong record of research on animals and agriculture. The University of Edinburgh had established an Institute of Animal Genetics in 1919. In 1947, the UK government tapped this expertise to create the Poultry Research Centre and the Animal Breeding Research Organization, both closely affiliated with the University. In 1985, these organizations were merged to form the Institute of Animal Physiology and Genetics Research, with a research station at Roslin, near Edinburgh. In the 1990s, further restructuring led to the creation of the Roslin Institute at first as a part of the Biotechnology and Biological Sciences Research Council, and later (in 1995) as an independent company.

This complicated institutional trajectory shows how scientific research on animal breeding developed from government-funded efforts to increase food supplies to private ventures into biotechnology. But throughout its history, Roslin's research has remained focused on animals. Even today, the Institute's modern laboratory facilities are surrounded by large pastures where sheep graze happily on the Scottish grass. The aim has always been to use science to harness and increase the productivity of animals.

The project that eventually gave rise to Dolly began as an attempt to redirect the productivity of animals beyond food (or wool). Scientists at Roslin wanted to transform mere farm animals into so-called *pharm animals*. These animals, as the word play suggests, can produce pharmaceutical products. More specifically, Roslin researchers aimed to genetically engineer sheep so that pharmaceutically valuable proteins could be expressed in their milk. For instance, if the gene for human insulin could be inserted into an appropriate place into a sheep genome, that sheep might make human insulin that would be excreted in their milk. The pure drug could then be obtained simply by separating it from the rest of the milk. For this work, Roslin scientists were funded by a biotech company, PPL Therapeutics; in exchange for its money, PPL would gain exclusive rights to the license for the new drug-production methods. Such a project was certainly biotechnological, but it was also part of a British tradition of sheep breeding and agricultural experimentation that stretched back centuries.

SOMATIC-CELL NUCLEAR TRANSFER

Making pharm animals seems a far cry from cloning. How did an attempt to make drugs end up making sheep? In fact, the creation of Dolly was a by-product of the pharm animal project—she was the means to the end of making reliable livestock drug-factories.

The crucial step in making pharm animals requires getting human genes (such as the gene for human insulin) inside animal cells. Animals, like humans, are made up of billions of individual cells—it would be impossible to insert the human gene into each cell individually. Even if the human gene needed to be inside only the mammary cells (since they are associated with milk production), this would still somehow require getting human DNA inside millions of animal cells. But, of course, all animal cells initially grow from one single cell — the fertilized egg. If it was possible to insert the human DNA into the egg, then it might be possible that this DNA would be copied and passed on to all the cells in the adult animal body (in the same way that the animal's own DNA is copied and passed on).

This would work, however, only if the human DNA was somehow incorporated into the animal's DNA — it would have to be taken up into the animal's own genome (so that it could be copied along with it). At first, during the 1980s, the scientists at Roslin just hoped for the best. They took the human gene they wanted to insert and just injected it into the animal egg cell (this technique is called pronuclear microinjection). After many attempts on different eggs, some did incorporate the human genes. These first experiments led to the birth of a sheep named Tracy in 1990. A transgenic organism, Tracy could produce human proteins in her milk, just as planned.[1]

However, the microinjection method was hit and miss — it had taken hundreds of attempts to get the gene into an egg and the results (in terms of getting proteins to express in the milk) were highly variable. Ian Wilmut, the lead scientist on the project, wanted a more reliable and systematic method that could be used in a commercial pharmaceutical setting. In in-vitro fertilization, eggs are extracted from a female, fertilized outside the body, allowed to develop into an embryo, and reinserted into a uterus to grow into a fetus (see chapter 15). The pharm animal project was far more ambitious. It required, first, reproducing, growing, and manipulating extracted eggs in the lab *without* allowing them to transform into embryos. And then, once the lab work and genetic manipulation was complete, it required taking those petri-grown cells and making them into viable embryos.

Wilmut and his team imagined being able to manipulate an egg in the lab

---

1. Tracy was the most successful of the five transgenic sheep produced by Roslin, producing 35 grams of human protein per liter of milk (the protein was in fact alpha-1-antitrypsin [AAT], a protein normally produced in the human liver and that can be used in the treatment of cystic fibrosis). PPL established commercial flocks of transgenic animals from Tracy by sexual reproduction. In 1996, PPL was floated on the London Stock Exchange and valued at GBP 120 million. PPL signed an agreement with the German pharmaceutical company Bayer and sheep-produced AAT entered clinical trials. PPL went out of business when this collaboration broke down in 2003.

just like genetic engineers manipulated single-celled organisms like bacteria. But eggs were difficult to work with — injecting DNA was a difficult task, they came in a limited supply (they had to be extracted one by one from female sheep), and, once fertilized, they quickly turned into embryos. Single-celled organisms, on the other hand, could be conveniently laid out in a petri dish where DNA could be inserted easily; also, they could be multiplied thousands or millions of times in the lab. Moreover, there were well-known techniques for turning off genes in single-celled organisms (as well as inserting new ones). How might it be possible to get egg cells to behave more like bacteria?

One possible approach was to manipulate non-egg sheep cells and then turn them *back* into eggs later on. Colonies of non-egg sheep cells (especially those taken from growing sheep embryos and fetuses) grown in the lab *could* be treated more like bacteria. The trick was to be able to turn them back into sheep. Initially, Wilmut attempted to use embryonic stem cells. Collaborating with Matthew Kaufman and Martin Hooper at nearby Edinburgh University (who worked mostly with mouse cells), Wilmut attempted to grow whole sheep from cultured embryonic stem cells. These efforts were unsuccessful — it didn't seem possible to make the stem cells back into viable embryos.

The Roslin scientists moved on to investigate alternative techniques. Between 1990 and 1996, Wilmut, now joined by Keith Campbell, worked to create sheep from different kinds of cells (not stem cells) grown in the lab. The key to their success was the finding that the stage of the cell cycle mattered a great deal in controlling the development of the cells. In particular, they found that cells in the part of their cycle called G0 had special abilities to become different kinds of cells. By intervening during the G0 phase, Wilmut and Campbell could turn lab-grown cells back into cells that could become sheep.

Megan and Morag, two lambs born in 1995, were produced from embryonic cells that had been passed through up to thirteen passages in the lab (this means that the cells had been through as many as thirteen cycles of growth and division outside the body). The team then planned a further series of experiments to create sheep from nine-day-old embryonic cells (these lambs were called Cedric, Cecil, Cyril, and Tuppence) and twenty-six-day-old fetal fibroblast cells (these lambs were called Taffy and Tweed; fibroblast cells are the main components of connective tissue). These experiments all showed that living sheep could be made from lab-grown cells.

However, all these sheep were derived from embryonic and fetal cells. What about other kinds of cells? Wilmut and Campbell's technique would be more powerful if it could be applied to any sheep cell. This would allow them to begin any modification or manipulation in the lab starting from any existing sheep rather than having to go back to the embryonic stage for each new experiment. This, finally, is where we get to Dolly. The aim of Wilmut and Campbell's next experiments was to move beyond producing sheep from embryonic and fetal cells and instead to produce sheep from cultured *adult* sheep cells. The technique they invented to achieve this they named *somatic-cell nuclear transfer*. Somatic cells make up most of the cells in our body — they are distinguished from *germ cells* (sperm and eggs) that usually pass on their DNA from generation to generation. Nuclear transfer refers to the process whereby the nucleus (including the DNA) of one cell is transferred into a different cell.

Somatic-cell nuclear transfer is illustrated in figure 16.2. The process involves three female sheep, labeled A, B, and C. Ewe A is the source of the egg. An egg is extracted from A and the nucleus (containing her DNA) is removed. Ewe C is Dolly's genetic mother — she is the source of the genetic material from which Dolly is made. At the time of Dolly's birth, ewe C had been dead for some time. During her lifetime, some mammary tissues had been extracted from C and stored in a lab freezer. To make Dolly, the breast tissues were defrosted and the nucleus extracted from one of the cells — this nucleus was then fused with the de-nucleated egg cell taken from A.[2] This fusion was the most difficult part of the somatic-cell nuclear transfer procedure. The result was an egg cell taken from A with genetic material taken entirely from the adult cell of C. Wilmut and Campbell's prior work meant that they knew how to transform this fused cell into a growing embryo. At the appropriate time, this embryo was transferred into the uterus of ewe B, making her pregnant. The lamb that was born from this pregnancy was Dolly.

In Wilmut and Campbell's original experiment, both ewes A and B were black-faced sheep, while C was a white-faced sheep. Dolly was a white-faced sheep born to a black-faced mother — this offered visual evidence that Dolly's genetic material came from C, and not from A or B.

Dolly was born in 1996. However, she was not a transgenic animal. The aim of Roslin's pharm animal program was only fully realized in the following year. The somatic-cell nuclear transfer technique provided Wilmut and Campbell with a way to genetically manipulate any animal cells in the lab and then turn them back into a live sheep embryo, just as they had wished.

---

2. Dolly's mammary origin provided the inspiration for her name: she was named for Dolly Parton, a singer and song writer noted for her large breasts. Although no doubt intended as a joke, this certainly raises interesting questions about the psychosexual motivations of Dolly's creators.

Somatic Cell Nuclear Transfer

Adult cells extracted from mammary gland

Unfertilized egg cell from donor
Cytoplasm
Nucleus
DNA

Tissue Donor
Finn Dorset Ewe C

Egg Donor
Scottish Blackface Ewe A

Nucleus
DNA
Cytoplasm

Micropipette

Nucleus removed

Donor cells grown and stored in lab

Enucleated egg cell

Electrical pulses

Cell fusion

Electrical pulse induced cell division

Surrogate Mother
Scottish Blackface Ewe B

Embryo implanted

Cell division

Dolly
Finn Dorset lamb

16.2 Somatic cell nuclear transfer. A donor cell is taken from the mammary tissue of ewe C. The nucleus (including all the DNA) is extracted from this cell. An egg cell is extracted from ewe A. This egg cell is "enucleated"—that is, the nucleus is removed. The nucleus from C and the enucleated egg from A are fused using electric pulses. The result is an egg cell with DNA coming entirely from ewe C. This egg is cultured in the lab and directed to grow into an embryo. At the appropriate time, this embryo is implanted into the uterus of ewe B, the surrogate mother. The lamb that is born is Dolly. Source: Illustration by Jerry Teo.

Returning to figure 16.2, imagine now that the defrosted cell taken from B is not immediately fused with the egg cell. Instead, the cell is cultured in the lab. During this process, genes may be added or turned off—it can be genetically engineered just like single-celled organisms. After this process is complete, the nucleus of this manipulated cell (or one cell from the colony that has been grown in the lab) can be fused with the egg from A and transformed into an embryo. Since all of the genetic material from the resulting sheep comes from this manipulated cell, the animal will be wholly transgenic. This is what the Roslin Institute achieved with the birth of another sheep, Polly, in 1997.

DOLLY MANIA

So far, we haven't mentioned cloning. As far and Wilmut and Campbell's experimental program was concerned, the fact that Dolly was a clone was almost incidental to their central goal (creating transgenic animals). A clone is an identical copy. In biology, cloning can refer to several different processes. In chapter 3, we encountered the process of recombining DNA by splicing it into a plasmid and making copies using bacterial reproduction. Since this involves making millions of identical copies of a piece of DNA, this process is sometimes called *molecular cloning*. It is important to distinguish this from the kinds of cloning we are talking about here.

With plants and animals, clones refer to two or more individuals that are *genetically* identical. This doesn't mean that the individuals are wholly identical in appearance or other characteristics—it just means that they have the same genome. This occurs often in both plants and bacteria, both of which can reproduce asexually, just by making genetic copies of themselves. Insects such as worker bees and soldier ants are also clones of one another.

Artificially cloned animals, too, were produced long before Dolly. In the 1950s, biologists wanted to know what happened to genes during the development from an egg to a whole organism. Did the genes change in order to produce the different types of cells in the body? In 1952, in order to answer this question, Robert Briggs and Thomas King conducted an experiment in which they moved the nuclei of developing frog cells into an embryonic cell with its nucleus removed (since the nucleus contained the genes, they were effectively transplanting all the genes from one cell to another). This laid the groundwork for the work of John Gurdon at the University of Oxford. In 1958, Gurdon transplanted the nucleus of cells taken from the intestines of tadpoles into an enucleated frog embryo. These embryos grew into fully developed *Xenopus* frogs. Not only had Gurdon cloned tadpoles, but he had also showed that genes did not disappear or change during development (the genes taken from the intestines must somehow retain the instructions for making a whole organism; we now know that this occurs through certain genes being turned on or off in different cells).

Following Gurdon's work, a carp was cloned in China in 1963 and a mouse was cloned in Russia in 1986. And, of course, the experiments at Roslin produced numerous sheep before Dolly (Cedric, Cecil, Cyril, Tuppence, Megan, and Morag) that were all cloned from embryonic and fetal cells. All these clones were made in the same way: by opening up an egg cell, replacing its genes with those of another cell, and then patching up the egg in such a way as to allow it to develop into a fully grown organism.

So Dolly was certainly neither the first cloned animal, nor the first artificially cloned animal, nor the first cloned mammal, nor even the first cloned sheep. Yet it was Dolly that caught headlines around the world, even making the cover of *Time* magazine. Something about Dolly caught the public imagination and made everyone suddenly pay attention to cloning. What was so special about Dolly, then? One important factor was that Dolly was the first animal to be cloned from *adult* cells. This seemed to turn the usual processes of birth, reproduction, and death on its head. Dolly signaled that it was now possible to take a single bodily cell, reprogram it, and remake it into a living being that could be born, live, and die all over again. Looking a little more closely, Dolly seemed not only to give us a great deal more control over biology, but also to alter the process of reproduction, and to rework notions of kinship and biological time. The rest of this section will explore each of these ideas—control, reproduction, kinship, and time—in more detail.

Many scientists around the world were quick to realize that Dolly promised significant extensions to their biotechnological powers. Somatic-cell nuclear transfer offered new possibilities and tools for genetic manipulation. In particular, many biologists believed that new techniques could be combined with genetic engineering and stem cells to form powerful new ways of building and manipulating animal tissue. In Wilmut and Campbell's book about cloning, *The Second Creation*, they wrote:

> As the decades and centuries pass, the science of cloning and the technologies that flow from it will affect all aspects of human life—the things that people can do, the way we live, even, if we choose, the kinds of people we are. Those future technologies will offer our successors a degree of control over life's processes that will come effectively to seem absolute . . . that expression, biologically impossible, seems to have lost all meaning. In the twenty-first century and beyond, human ambition will be bound only by

the laws of physics, the rules of logic, and our descendants' own sense of right and wrong.[3]

Suddenly, nothing seemed out of reach—cloning made anything possible. In fact, Wilmut and Campbell downplayed the reproductive implications of their work since they feared that human cloning would be far too controversial. Instead, they described a future in which somatic-cell nuclear transfer could be used to create stem cells to repair the human body (so-called *therapeutic cloning*). Rather than using somatic-cell nuclear transfer to produce whole copies of organisms (as with Dolly), the technique could be repurposed to grow human organs or tissues that could be used as spare parts for our bodies. Since the organs would be derived from a patient's own cells, they would be perfect genetic matches and not suffer the problems of immune rejection that plagued other kinds of transplants. This "new biotechnology," as they called it, captured the imagination by offering the potential for vast new powers to manipulate bodies.

But whether Wilmut and Campbell liked it or not, cloning also garnered attention because it rewrote the rules for reproduction. For one thing, cloning transforms the act of reproduction too. Like IVF, cloning refigures a bodily act (sexual intercourse) into a laboratory act. The anthropologist Sarah Franklin has suggested that cloning has disassembled sex, rearranged it, reversed it, redeployed it, and redirected it. By taking reproduction outside the body and using it for other purposes (for making new tissues or organs, for example), sex takes on new possibilities and new meanings. Ironically, one piece of evidence that Dolly was a "normal" sheep was she was able to sexually reproduce herself (she gave birth to a healthy lamb in 1998 and five more subsequently). The need to demonstrate that Dolly could reproduce "normally" suggests some degree of discomfort with these new reproductive modes.

New modes of reproduction create new possibilities for kinship. Who were Dolly's parents? Ewe A, B, or C? (figure 16.2) A cell line living in a lab? A piece of tissue from a freezer? In a society where reproductive cloning is possible new kinds of family units and social relations also become possible. Should you consider your reproductive clone to be your offspring or your sibling? The meanings of such common concepts as "father," "mother," "sister," and "cousin" are at stake. In a society in which nuclear families are still the norm and in which nontraditional family units (such as same-sex parents) still pro-

3. Ian Wilmut, Keith Campbell, and Colin Tudge, *The Second Creation: Dolly and the Age of Biological Control* (Cambridge, MA: Harvard University Press, 2001), 5.

voke controversy, the potential for the disruption of kinship is surely another reason for our fascination and unease with Dolly.

Finally, the new level of biological control and the re-formation of reproduction disrupts our normal sense of biological time. Human lives are oriented and grounded by biographical narratives that begin with birth and proceed to growth, aging, and eventually, death. Our biology provides a kind of arrow pointing from cradle to grave. Biologists long thought that once a cell had assumed its adult form, it could not go back: you could not make an egg cell out of a liver cell. Somatic-cell nuclear transfer showed this to be wrong. The introduction of cloning opens up new possibilities: adult cells could be "recapacitated" into embryonic cells. This seemed to defy, in a way, the process of aging—old cells could be made young again and adult individuals could be reborn as clones. This creates the possibility of different kinds of biographical stories—the straightforward birth to death narrative is rerouted, lengthened, or shortened. In the future, as the technological possibilities of cloning are explored, this may allow us to imagine our lives and our identities in vastly different ways.

Dolly's death generated further controversy over whether she had lived a "normal" life. Dolly was euthanized in 2003 due to complications from lung cancer caused by a retrovirus. The virus is quite common in sheep, but Dolly's death at six years amounted to only about half the life expectancy of Finn Dorset sheep (Dolly's breed). Many wondered: Did the fact that she was created by cloning contribute to her premature death? (In particular, scientists noted that the sheep from which she was cloned had been six years old at the time the tissue was extracted, giving Dolly a "genetic age" of six at birth.) This controversy indicates not only a concern with the fate of any future cloned human being, but also an unease about the potential of cloning to disrupt normal life stories.

CONCLUSIONS: WHY ARE WE SO SHEEPISH?

The story I have told here is only one way of exploring the history and meanings of cloning. I might, for instance, have focused far more on the careers and work of the scientists involved or on the possibilities for therapy generated by somatic-cell nuclear transfer. This narrative connects cloning to the long history of agricultural and animal manipulation that leads from Bakewell to Roslin. It also places the emphasis on sheep. I have done this because the fact that Dolly was a sheep can help us to explain her fame. It can help us to understand why Dolly, rather than a tadpole or a carp, came to represent cloning.

Our culture places a high value on originality. We have a fear of copies: illegitimate copies of books, movies, clothes, or music are made illegal and governments go to great lengths to prevent their distribution; plagiarism is the cardinal sin for academic work. Copies are disruptive of the proper order of things and are seen as illegitimate and subordinate. We value uniqueness and authenticity in objects and also in ourselves—our individuality and autonomy as persons are closely guarded.

Cloning taps into our worries about copying. By offering the (future) possibility of duplicating ourselves, it threatens our uniqueness and our identity as human agents. Copying anything—especially a human—is a disruptive and unsettling idea in our society.

Sheep raise this problem in a particularly acute way. When you think about sheep, how do you usually describe them? Unless you happen to be a vet or a farmer, probably every sheep looks the same to you: they are more or less identical (counting sheep to go to sleep can be useful only because they are so boringly identical). Also, most people think of sheep as rather stupid—even comically so. They follow each other around in a flock, unable to think for themselves, and are usually very timid. Sheep, to say the least, are not given very much credit.

But perhaps we also see something of ourselves in sheep. After all, sheep often serve as representatives of humans (at least in Judeo-Christian culture). At times, we too can be stupid, conformist, and timid creatures. In a mass society, we even seem to become more similar to one another. And our own sheepishness scares us: we fear becoming less unique and more like everyone else. Some of our concerns about cloning—and some of the notoriety of Dolly—should be attributed to this fear of becoming more sheepish. Cloning, especially the cloning of sheep, raises the idea that we may become more conformist, less unique, and more objectified. Like sheep, we may become devalued as individuals and commodified as products of biotechnology.

FURTHER READING

The approach used in this chapter borrows extensively from Sarah Franklin, *Dolly Mixtures: The Remaking of Genealogy* (Durham, NC: Duke University Press, 2007). For Franklin's work on kinship see also Sarah Franklin, "Biologization Revisited: Kinship Theory in the Context of the New Biologies," in *Relative Values: Reconfiguring Kinship Studies*, ed. Sarah Franklin and Susan McKinnon (Durham, NC: Duke University Press, 2001), 302–328. In my account of the significance of cloning for our thinking about our bodies and ourselves I also drew on Susan M. Squier, *Liminal Lives: Imagining the Human at the Frontiers of Biomedicine* (Durham, NC: Duke University Press, 2004). Another approach to the history and significance of cloning can be found in Jane Maienschein, *Whose View of Life? Embryos, Cloning, and Stem Cells* (Cambridge, MA: Harvard University Press, 2003). On the history of the technique itself see Nick Campbell, "Milestone 5 (1952): Turning Back Time," *Nature Milestones: Development* (2004), http://www.nature.com /milestones/development/milestones/full/milestone5.html; and on the work of the Roslin Institute see Wilmut and Campbell's own account: Ian Wilmut, Keith Campbell, and Colin Tudge, *The Second Creation: Dolly and the Age of Biological Control* (Cambridge, MA: Harvard University Press, 2001).

On the history and significance of animals, especially livestock, in Britain see Robert Trow-Smith, *A History of British Livestock Husbandry, to 1700* (London: Routledge and Kegan Paul, 2006) and Harriet Ritvo, *The Animal Estate: The English and Other Creatures in the Victorian Age* (Cambridge, MA: Harvard University Press, 1989).

17 STEM CELLS

INTRODUCTION

Like cloning, stem cells are one of the most widely discussed and controversial scientific subjects. Like the last chapter, this chapter will take an unconventional approach to this complex topic. Just as we discussed cloning by discussing sheep, here we will discuss stem cells by discussing single-celled organisms. Amoebas, bacteria, and yeast—all single-celled—live differently, reproduce differently, and die differently from humans. But stem cells are single cells that have some of the properties of single-celled organisms but also have the potential to generate whole human beings. This capacity forces us to think about our lives and our bodies in radically different ways. Like the designer babies discussed in chapter 18, stem cells offer remarkable new ways of manipulating and altering our bodies.

We usually think of amoebas (and their single-celled cousins) as organisms far removed from humans, in terms of both complexity and evolution. Nevertheless, all organisms must solve many of the same problems in order to survive. In particular, all species must find ways of reproducing themselves. Stem cells, like cloning, offer new opportunities for human reproduction that suggest we may be more like amoeba after all. Exploring this insight can help us to understand some of our cultural and social responses to stem cells.

AUGUST WEISMANN AND THE PROBLEM OF DEATH

To understand the relationship between stem cells and amoeba, we must look back at the thinking of a German biologist from the nineteenth century. August Weismann (1834–1914) became well known for his contributions to understanding Charles Darwin's theory of evolution. But Weismann was also obsessed with death. Why should animals have to die, he wondered, and why do some live longer than others?

Even more puzzlingly, some simple organisms didn't seem to die at all. In single-celled organisms reproduction takes place by cell division: a single individual copies its genetic material and divides into two equal parts. As Weismann wrote, "The origin of new individuals is not connected with the death of the old; but increase by division takes place in such a way that the two parts into which an organism separates are exactly equivalent to one another, and neither of them is older or younger than the other."[1] When an amoeba or bacteria (protozoa) divided into two, some of its parts went into both—it was not clear which was the "parent" and which was the "offspring." Since these divisions could occur over and over again, it seemed to Weismann that protozoa could not really be said to die.

Of course, individual protozoa do die—they may be exposed to a poison or deprived of nutrients, water, etc. But Weismann's point was that this was not necessarily an older generation dying off to be replaced by a younger generation—in fact, as long as *some* protozoa survived, some parts of the originating individuals would always persist. This is quite different from multicellular organisms (called metazoans by biologists). With birds, insects, or mammals, for instance, an older generation produces a younger generation of offspring; the bodies of the older generation rot away, their bodily cells are (mostly) not passed on to their offspring. Weisman expressed this rather dramatically by arguing that humans (and other metazoans) had somehow lost the "power of unending life." How and why had this occurred?

Weismann's answer to this question was the basis of his *germ plasm theory* (the idea for which he is best remembered). First, Weismann noticed that some special types of cells in the metazoan body *were* passed on from generation to generation—these were the sperm and egg cells, which he called the *germ* cells. A human egg cell, for instance, passes on some of its physical parts into the cells of the offspring; some of these new cells become the germ cells of the next generation; in turn they also pass on their parts. In other words, germ cells are immortal in the same sense that protozoans are immortal. What had happened with metazoans like humans, Weismann argued, was that a division of labor had arisen between germ cells and bodily cells. Germ cells took charge of reproduction and were immortal, while somatic cells (the rest of the cells in the body) took care of other functions (such as energy production, metabolism, and respiration).

The germ plasm theory implied that animals consisted of collections of individual cells working together for mutual benefit. The "death" of the body was just the breaking-up of this collection—in fact some cells, the germ cells, lived on. "The dissolution of a cell-colony into its component living elements," Weismann argued, "can only be called death in the most figurative sense, and can have nothing to do with the real death of individuals; it only consists in a

---

1. August Weismann, "Life and Death," in *Essays Upon Heredity and Kindred Biological Problems*, trans. Selmar Schönland and Arthur E. Shipley (Oxford: Clarendon Press, 1889). Quotation p. 111.

change from a higher to a lower stage of individuality."[2] This is a radical view of animal life. Rather than thinking of humans as coherent, singular entities, it reimagines them as large, cooperating *colonies* of cells.

THE ORIGINS OF STEM CELL SCIENCE

Weismann's distinction between germ cells and somatic cells proved extremely influential in twentieth-century biology. It inspired other biologists, first in Germany, to closely investigate the development of cells in animal embryos. In particular, these biologists wanted to know how the embryonic cells generated the variety of different types of cells in the adult body (this is called *differentiation*). The embryologists Theodor Boveri (1862–1915) and Valentin Häcker (1864–1927) attempted to identify the cells in the embryo that gave rise to Weismann's germ cells. They called these cells *Stammzelle* (figure 17.1). By the end of the nineteenth century, the idea of "stem cells"—as some kind of original cells that get passed on from generation to generation—had spread to the United States through the work of Edmund B. Wilson (1856–1939).

Rather than working forward from embryos, other biologists attempted to work backwards from mature bodily cells. In the early twentieth century, hematologists attempted to identify the origins of all the different types of blood cells (red blood cells, white blood cells, platelets). The Russian-born biologist Alexander A. Maximow (1874–1928) proposed that all these cells developed from a single common precursor type. These precursor blood cells also became known as hematopoietic (or blood-generating) "stem cells." Maximow's experiments showed that some embryo-derived cells had remarkable properties of survival and an ability to transform into different types of cells. However, the difficulty of culturing cells outside the body (see chapter 9) made it impossible to pursue the investigation of these cells much further.

By the 1960s, however, new techniques of cell culture had been developed. Stem cells had also captured the attention of biologists studying both mammalian reproduction and cancer. At Cambridge University, Robert Edwards' research on in-vitro fertilization (see chapter 15) allowed him to grow mammalian embryos in his lab, providing a source of cells for experimentation. Edwards and his students conducted a series of experiments on blastocysts

2. August Weismann, "Life and Death," in *Essays Upon Heredity and Kindred Biological Problems*, trans. Selmar Schönland and Arthur E. Shipley (Oxford: Clarendon Press, 1889). Quotation p. 126.

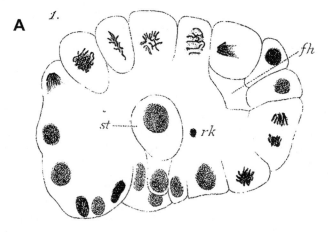

A

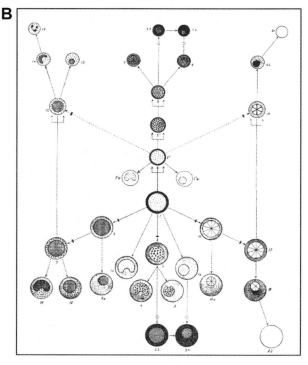

B

17.1 Early twentieth-century stem cells. Artur Pappenheim's sketch of hematopoiesis (the generation of blood cells) from 1905. The cell in the center of the lower diagram is the hypothesized progenitor of all the different blood cell types. Already at the beginning of the twentieth century, biologists had an idea that some special cells could transform into many of the different types of cells in the body. Source: Miguel Ramalho-Santos and Holger Willenbring, "On the Origin of the Term 'Stem Cell,'" *Cell Stem Cell* 1, no. 7 (2007): 35–38, figure 1. Used by permission.

(embryos 5–8 days after fertilization). They demonstrated that cells taken from rabbit blastocysts could transform into many independent cell lines comprising different tissues. This work contributed to the successful culturing of mouse embryonic stem cells by Martin Evans and Matthew Kaufman in 1981.

The "discovery" of stem cells, however, is usually attributed to the cancer researchers Ernest McCulloch and James Till. Oncologists wanted to know more about the process through which radiation killed cancer cells. Bone marrow cells were of particular interest because of their seemingly remarkable ability to regenerate blood cells after a bone marrow transplant (often required for leukemia treatments when bone marrow cells had been killed by radiation). McCulloch and Till's work sought to understand the response of bone marrow cells to radiation. In their experiments, conducted at the Ontario Cancer Institute in the early 1960s, they irradiated mice so that their own bone marrow was inactive. They then injected bone marrow cells from other mice into the irradiated mice to see whether they could survive. They usually could not. However, autopsies of their dead mice showed that they had developed unusual nodules on their spleens that seemed to be differentiating into different types of blood cells (red, white, and platelets). Following up on these experiments, they showed that the cells in these nodules derived from a single type of cell that could renew itself indefinitely and also gave rise to all the different types of blood cells. This proved Maximow's notion that blood cells all derived from a single progenitor. But it also showed that the efficacy of bone marrow transplants depended on the powers of one particular kind of cell.

Despite this long history, stem cells came to popular attention only in 1998 when James Thomson, at the University of Wisconsin–Madison, succeeded in producing a *human* embryonic stem cell line. Growing human stem cells in the lab was a difficult feat and Thomson's success seemed to indicate that medicine would soon be able to harness the regenerative properties of stem cells. Scientists and the media quickly imagined that stem cells could be used to grow all kinds of human tissues that could be used for therapies. However, we should be careful not to read this present-day concern backwards into history. Stem cells emerged from research in embryology, development, hematology, mammalian reproduction, cell culture, and cancer. Understanding the power and the problems of stem cells requires that we remember their multiplicity of roles and capacities.

Stem cells can be classified in several different ways. As we have already seen, the term *stem cell* has been used in a variety of ways too. For Häcker, stem cells were cells from which the germ line cells were derived; for Maximow they were the precursors for different types of blood cells. In fact, there is a whole hierarchy of different cells within an animal body that have different abilities to reproduce themselves and other cells. However, what all these notions have in common is that stem cells are the generators of other cells: they are the cells from which other cells derive. If we were to draw a family tree of cell types showing how one type is produced from another (figure 17.2) then stem cells are those nearest the root or stem of the tree.

To be more precise, we can define stem cells as having three particular properties:

1) Stem cells are capable of renewing themselves over long periods of time and able to divide almost indefinitely. This property is sometimes called *immortality*. Regular tissue cells (say skin cells) can be taken from the human body and grown in a lab. However, such cells are capable of dividing and renewing themselves only a fixed number of times (around 40 to 60 times for human cells; this is known as the Hayflick limit). This is also true of most cells growing in our body—they eventually die. Stem cells, however, are not limited in this way and can divide and grow forever.
2) Stem cells are unspecialized cells. They do not perform specific functions within the body except to regenerate themselves and to produce other types of cells.
3) Stem cells can differentiate into other types of cells. That is, stem cells have the potential to become several different types of cells in the body. For example, neural stem cells can become neurons, astrocytes, or dendrocytes.

Different types of stem cells are usually classified according to their "potency." The more different types of cells that a stem cell can differentiate into, the more potent it is. In figure 17.2, the more potent stem cells are those closest to the center. *Totipotent* (or omnipotent) stem cells can differentiate into any of the different kinds of cells needed to form an embryo, including the placenta. *Pluripotent* cells can form any of the bodily cells needed to make an adult animal. *Multipotent* cells have the ability to develop into a limited number of types of cells. For instance, hematopoietic stem cells are multi-

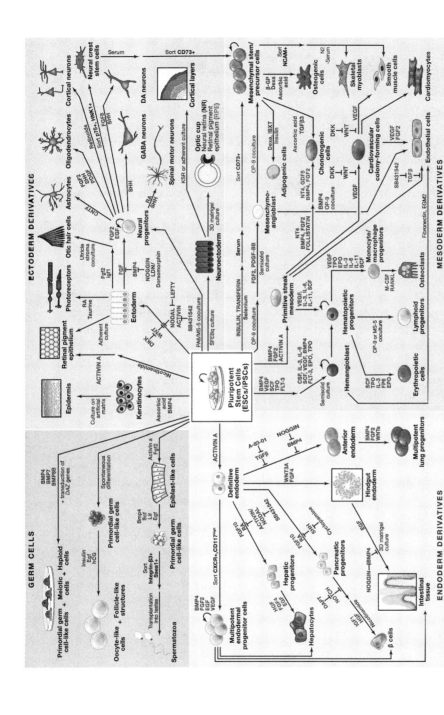

17.2 Stem cell family tree. Stem cells give rise to more and more specialized types of cells. Cells at the center of the diagram are pluripotent and can give rise to many types of cells in the body. Cells further toward the edge of the figure are less potent since they can give rise to only some types of cells. At the far edges of the figure we find the individual types of specialized cells that make up our organs and tissues. Source: Williams et al., "Snapshot: Directed Differentiation of ESCs and iPSCs," *Cell* 149 (2012). Used by permission.

potent since they can give rise to all the different kinds of blood cells, but not non-blood-type cells such as brain or muscle cells. *Oligopotent* stem cells have the ability to differentiate into just a few types of cells. For instance, a lymphoid cell can give rise to B and T cells, but not other types of blood cells.

Stem cells can also be classified according to where they are found. Embryonic stem cells (which are usually also pluripotent) can be found only in embryos. Adult stem cells, on the other hand, can be found in various places in the adult body (the bone marrow, for instance, contains hematopoietic stem cells). Adult stem cells are usually multipotent. In addition to the bone marrow, blood vessels, brain, skeletal muscle, skin, and numerous other organs each contain their own adult stem cells.

As bodies develop from embryos, the cells within them gradually lose potency. Pluripotent embryonic cells become multipotent adult stem cells, which in turn become oligopotent cells that then develop into all the specialized types of cells in the body. The specific developmental path taken by particular stem cells (that is, when it differentiates and what kind of cell it turns into) is a complex process governed by genetic, epigenetic, intercellular, and environmental signals. However, in 2006 Shinya Yamanaka at Kyoto University showed that it was possible to *reverse* these pathways. Yamanaka managed to take adult somatic cells (not adult stem cells) and "reprogram" them by turning on a particular set of genes (now known as the Yamanaka factors). These cells then became pluripotent, having similar features to embryonic stem cells. Such cells are known as *induced pluripotent stem cells* (iPSC).

THE POLITICS OF STEM CELLS

The discovery of induced pluripotent stem cells created a renewed enthusiasm for stem cell–based therapies. Since iPSC can be derived from readily available adult cells (such as adipose [fat] cells), rather than from embryos, they appeared to offer the greatest potential for creating stem cell therapies. In the decade prior to the iPSC discovery, stem cell science had become an increasingly divisive political issue, especially in the United States. James Thomson's 1998 experiments brought embryonic stem cells wide public attention. The enthusiasm for potential treatments that might arise from stem cells was quickly tempered by concerns about the sources of these tissues.

Embryonic stem cells, as their name suggests, must be extracted from embryos—that is, from a growing egg, a few days after fertilization. Removing the stem cells effectively destroys the embryo. Once the cells are removed, the embryo no longer has the potential to develop into a human being in the usual way. Most of the embryos used for embryonic stem cell research were those left over from IVF procedures. During a standard IVF treatment, many eggs are extracted from the woman's ovaries and fertilized in vitro. Only some of these growing embryos are selected for reimplantation in the uterus. The rest are usually frozen and stored in the lab or clinic for potential future use (for example, if the pregnancy fails or if the woman wants another child). Eventually, the unused embryos are discarded. Stem cell biologists (with the permission of the man and woman whose eggs and sperm were used to create the embryo) arranged to use some of these discarded embryos for extracting stem cells for their research.

However, some individuals—many guided by their religious convictions—believe that human life begins with the fertilization of the egg. Whether the embryo is to be discarded or not, embryonic stem cell research involves actions that destroy the embryo and therefore destroy its human potential. On this view, embryonic stem cell research amounts to the taking of a human life (that is, it is equivalent to murder).

In the United States, this dilemma became especially acute because of its links to the debates over abortion. The landmark Supreme Court case Roe v. Wade (1973) upheld the right to obtain an abortion. Nevertheless, abortion has remained a highly contentious political and social issue. The central argument of those opposed to abortion was that human life begins at conception, that abortion amounted to taking that life, and that therefore it is morally wrong. Anti-abortion activists used this argument not only to oppose abortion, but also to attempt to restrict various forms of scientific research that utilized embryonic tissues. These tissues, they argued, were derived from the destruction of a human life. From the mid-1970s onwards, legislators, bioethics committees, the media, and the public debated the merits of various forms of embryo-based research. Box 17.1 lists some of the outcomes of these discussions: it shows how this scientific work endured constant reversals of fortune and funding.

The emergence of human embryonic stem cell research in the late 1990s re-enlivened these debates. Those opposed to abortion believed that they had to oppose stem cell research too. On the other hand, those that supported abortion rights generally believed that human life begins some time after fertilization (for instance, at the time of implantation in the uterus, or at the first signs of brain activity, or even at birth). On that view, the use of discarded embryos for scientific researched poses no special moral problems.

The intensity and highly politicized nature of the abortion debate made embryonic stem cell research a highly charged issue. In 1999, President Bill Clinton's National Bioethics Advisory Commission recommended permitting research to proceed on "supernumerary" (extra, leftover) embryos. But

**Box 17.1 Timeline of Regulations on Human Embryonic
Tissue Research in the United States**

1975—National Commission for Protection of Human Subjects approves fetal research.

1984—Moratorium is imposed on fetal tissue research.

1993—National Institutes of Health Revitalization Act reverses 1984 moratorium.

1994—National Institutes of Health Embryo Research Panel recommends federal funding for embryo research.

1995—Dickey-Wicker Amendment prohibits federal funding in cases where an embryo is destroyed or specifically created for research.

1995—President Clinton appoints National Bioethics Advisory Commission to report on morality of embryo research.

1999—National Bioethics Advisory Commission recommends permitting research on supernumerary embryos.

2001—President Bush announces restriction of federal funds for embryo research (except on approved and existing lines); Bush appoints President's Council on Bioethics to study the issue.

2005—Congress passes Stem Cell Research Enhancement Act to reverse restrictions on stem cell research but it is vetoed by President Bush.

2007—Congress again passes the Stem Cell Research Enhancement Act but it is again vetoed by Bush.

2009—President Obama rescinds Bush's restrictions on federal funding for stem cell research.

the new political landscape that resulted from the 2000 Presidential and Congressional elections changed the fate of stem cell research once again. On August 9, 2001, President George W. Bush announced in a live national television address that he was restricting the use of federal funds for embryo-based research. The moratorium did not affect the use of private or state funds, and did not apply to stem cell lines that had already been created from discarded embryos. Nevertheless, many biologists were outraged.

Since 2001, stem cell science has continued to be a contentious political issue (some of the political turning points after 2001 are detailed in Box 17.1). The discovery of iPSCs opened up lines of research not dependent on embryos. To some extent, this has lessened the urgency of the issue. However, many scientists still claimed that understanding and unlocking the full potential of stem cells would require research on embryos. Meanwhile, other nations saw the American restrictions on stem cell science as an opportunity. In particular, nations trying to build up their capacity in bioscience and biotechnology began to invest in stem cell science. For instance, Singapore set up "Biopolis" in 2003 and South Korea established the "World Stem Cell Hub" in 2005. Both nations hoped to attract world-leading scientists to generate the next big breakthrough in stem cell science.

ALIVENESS, DEADNESS, AND HUMAN POTENTIAL

Why have the debates over stem cells proved so divisive (at least in the United States)? One answer to this question can be found in returning to Weismann and his view of bodies as colonies. For those opposed to stem cell science, the central issue is the notion of the destruction of "human potential"—destroying an embryo destroys the potential of that embryo becoming a fully grown human being.

But stem cells hold all kinds of other "potential." In particular, they hold the potential to make, remake, or regenerate all sorts of human parts. So far, the only stem cell–based therapy that has been shown to be effective is bone-marrow transplantation. This procedure has been performed since the 1960s, even before the role and nature of stem cells was fully understood. Nevertheless, some scientists, journalists, and other stem-cell advocates point to many other possible uses of stem cells as "regenerative" therapies. The notion is that cells could be extracted from a patient's body, induced into pluripotency, and then grown to make any number of bodily tissues (brain, bone, tooth, heart, muscle, spleen, liver, etc.—see box 17.2). These tissues could then be transplanted back into the patient's body. Since these tissues would be derived from the patient's own cells, such procedures would avoid any problems of immune-system rejection. Stem cells have been called an *all-purpose spare parts system* for the human body.

Stem cells, like other biotechnologies, are promissory objects. Stem cell science is driven by the hype surrounding their possible future uses, and stem cells themselves are exciting because they promise the hope of future cures. This hype and promise drive careers, generate investments, and create controversy. They have even inspired some patients to travel from the United States and Europe to seek out experimental stem cell treatments in South America, South Korea, and China (this phenomenon is known as *stem cell tourism*).

This promise and potential also provide the main argument in favor of pursuing stem cell science. Advocates of stem cells argue that the potential of stem cells for doing good, through therapy, makes it morally wrong *not* to undertake the research. The quadriplegic and former star of 1970s and 1980s *Superman* movies, Christopher Reeve, believed that stem cells offered him the hope of a cure for his spinal cord injuries. He strongly advocated loosen-

Box 17.2 Potential Stem Cell Therapies (2014)

- Cancers
  - Brain cancer
  - Testicular cancer
  - Non-Hodgkin lymphoma
  - Ovarian cancer
  - Neuroblastoma
- Autoimmune diseases
  - Type I diabetes
  - Crohn disease
  - Multiple sclerosis
  - Rheumatoid arthritis
- Cardiovascular diseases
  - Acute heart damage
- Ocular conditions
  - Corneal regeneration
- Neural degenerative diseases and injuries
  - Parkinson disease
  - Spinal cord injury
  - Stroke damage
- Anemias and blood conditions
  - Sickle-cell anemia
  - Thalassemia
- Wounds and injuries
  - Limb gangrene
  - Surface wound healing
  - Jawbone replacement
  - Skull bone repair
- Metabolic disorders
  - Osteopetrosis
  - Hurler syndrome
- Liver disease
  - Chronic liver failure
  - Liver cirrhosis

Source: http://www.explorestemcells.co.uk/UsesForStemCellsCategory.html

ing of federal restrictions and raised large amounts of money for research (Reeve died from complications of his condition in 2004).

For Reeve and other stem cell supporters, stem cells represented a different kind of potential. Rather than focusing on the potential of the embryo to become a human being, they pointed out that stem cells had the potential to become all kinds of other human parts. They have the potential to enhance, repair, regenerate, and redirect human lives. Rather than destroying the potential of an embryo, many argue that stem cells *harness* this very potential, transforming and reshaping it in ways that may profoundly affect human lives. Taking a stem cell from an embryo and growing it outside the body may close off the potential that it becomes a whole human being. However, if the stem cell grows into an organ or a piece of brain tissue, it opens up all kinds of other "potential" in those cells.

Seeing this human potential of stem cells involves just the kind of idea about the human body that Weismann advocated over a century ago. If we think about humans as colonies of cells working together (rather than as wholly singular and cohesive individuals), then it is easier to think of stem cells growing in a petri dish as just a different kind of human colony. The embryonic colony of cells can be directed to become a colony growing in a lab, which can in turn be directed to become part of another colony forming another human body. The anthropologists Catherine Waldby and Susan Squier argue that stem cell lines are actually "human" inasmuch as humans are considered "integrated colonies of amoeboid beings."

Weismann's ideas can help us understand why debates over stem cells have proved so polemical and persistent: the view of each side rests on a radically different view of human bodies and human potential. Advocates of stem cells perceive human bodies and human potential not merely in the singular egg cell growing into the singular body, but in the possibilities of thousands of cells that may be redirected and proliferated into multiple forms and parts. Weismann's view of life and death suggests an alternative view of human potential that places less emphasis on the embryo as a the singular source of human life. This suggests that coming to terms with a more "colonial" view of the human body might be a starting point for making progress in debates over stem cells.

CONCLUSIONS: GREY GOO

Inspired by single-celled organisms like bacteria, Weismann suggested an alternative way of thinking about life and death: single-celled organisms just keep on dividing forever, never really dying. Applying this idea to multicellular organisms, we could say that *some* cells in our body (the germ cells) continue on in the same way, passed on from generation to generation without dying. The collection of cells that makes up our body might cease functioning and disintegrate, but some cells persist into the next generation, forming new bodies around them. The "colony" changes, but the cell persists. However, we usually think of our bodies as more than colonies and we associate the end of our bodies with the end of our life. The Weismannian view, however, takes a cell's point of view: life continues on beyond the body.

Stem cell technologies, like Weismann, locate the power of life in the cell, not the body. Do clumps of cells living in a petri dish count as human? Usually we would say "no." But what if these cells could regenerate parts of human beings, or even whole human beings? In chapter 9, we saw that new technologies of tissue culture have transformed how we think about the boundaries of bodily life. Cultured stem cells may have an even more dramatic effect on how we think about what counts as human and not human, alive and dead. Stem cells, with their uncanny power to regenerate themselves ("immortality"), are alive and persist in their aliveness beyond our bodies. In this respect, they should perhaps be considered *even more* alive than our bodies. Stem cells suggest a Weismannian view of life in which aliveness (and death) does not depend on bodies, which are mere colonies built to support the immortal germ cells.

One of the best explorations of the possibilities and potential of cell colonies comes from science fiction. In Greg Bear's *Blood Music* (1985), a molecular biology experiment leads to the creation of colonies of human cells that become self aware. The cells manage to escape from the lab, first taking over individual human bodies, but quickly multiplying and managing to subsume and merge all humans into a collective of super-intelligent cells. The earth is overtaken not by aliens, but by an integrated, highly intelligent cooperative consciousness. This being or colony looks essentially like "grey goo"—a formless mass of cells covering the surface of the earth.

This novel suggests how biotechnologies like stem cells are forcing us to come face to face with alternative modes of existence and being. The grey goo suggests how aliveness and human existence is not necessarily tethered to bodies. Like Weismann, we are forced to think differently about living and dying. Stem cell technologies, like grey goo, suggest the power and possibilities of living beyond the body. In *Blood Music*, the triumph of the grey goo is partly celebrated as a release from the travails of lonely, individual existence; but at the same time, the narrative raises deep fears about loss of identity, individuality, bodily integrity, and autonomy. Our social and cultural concerns about stem cells are partly driven by similar worries: if a human can be regenerated from a single cell, what and where are "we" within the colonies of cells that make up our bodies? Stem cells confront us with a radically different view of life, death, human existence, and identity.

FURTHER READING

For August Weismann's thinking about the immortality of cellular life see "Life and Death," in *Essays Upon Heredity and Kindred Biological Problems*, trans. Selmar Schönland and Arthur E. Shipley (Oxford: Clarendon Press, 1889). There is not a great deal of scholarship on Weismann in English, but some information can be found in F. B. Churchill, "August Weismann and a Break from Tradition," *Journal of the History of Biology* 1 (1968): 91–112; and F. B. Churchill, "Hertwig, Weismann, and the Meaning of the Reduction Division, Circa 1890," *Isis* 61 (1970): 429–457.

On the "long" history of stem cell science see Miguel Ramalho-Santos and Holger Willenbring, "On the Origin of the Term 'Stem Cell,'" *Cell Stem Cell* 1, no. 1 (2007): 35–38; Robert G. Edwards, "IVF and the History of Stem Cells," *Nature* 413 (2001): 349–351; Sara Shostak, "(Re)defining Stem Cells," *Bioessays* 28, no. 3 (2006): 301–308; and Evelyn Strauss, "Albert Lasker Basic Medical Research Award: Ernest McCulloch and James Till" (2005), http://www.laskerfoundation.org/awards/2005_b_description.htm.

The debates over stem cell research have given rise to a small industry of social science commentary on this topic, especially in the American context. An excellent starting point is Jane Maienschein, *Whose View of Life? Embryos, Cloning, and Stem Cells* (Cambridge, MA: Harvard University Press, 2003). See also Charis Thompson, *Good Science: The Ethical Choreography of Stem Cell Research* (Cambridge, MA: MIT Press, 2013); J. Benjamin Hurlbut and Jason S. Robert, "Stem Cells, Science, and Public Reason," *Journal of Policy Analysis and Management* 31 (2012): 707–714; and Adam Keiper, ed., "The Stem Cell Debates: Lessons for Science and Politics," *New Atlantis* (Special Report) 34 (2012). For a comparative perspective on stem cell policies in the United States, United Kingdom, and Germany see Sheila Jasanoff, "In the Democracies of DNA: Ontological Uncertainty and Political Order in Three States," *New Genetics and Society* 24, no. 2 (2005): 139–155; and also Jasanoff's wider account of biotechnology regulation in democratic societies in Sheila Jasanoff, *Designs on Nature: Science and Democracy in Europe and the United States* (Princeton, NJ: Princeton University Press, 2005).

Beyond the United States and Europe, the significance of stem cells has been explored by Catherine Waldby, "Singapore Biopolis: Bare Life in the City-State," *East Asian Science, Technology and Society* 3, no. 2–3 (2009): 367–383; Charis Thomson, "Asian Regeneration? Nationalism and Internationalism in Stem Cell Research in South Korea and Singapore," in *Asian Biotech:*

Ethics and Communities of Fate, ed. Aihwa Ong and Nancy N. Chen (Durham, NC: Duke University Press, 2010), 95–117; and Jennifer A. Liu, "Asian Regeneration? Technohybridity in Taiwan's Biotech?" *East Asian Science, Technology and Society* 6, no. 3 (2012): 401–414.

The implications of thinking about our bodies as colonies of cells are explored by Nik Brown, Alison Kraft, and Paul Martin, "The Promissory Pasts of Blood Stem Cells," *Biosocieties* 1, no. 3 (2006): 329–348; Catherine Waldby and Susan M. Squier, "Ontogeny, Ontology, and Phylogeny: Embryonic Life and Stem Cell Technologies," *Configurations* 11, no. 1 (2003): 27–46; and in the science fiction novel by Greg Bear titled *Blood Music* (Westminster, MD: Arbor House, 1985). For broader reflections (by anthropologists) on the refiguring of kinship through stem cells see also Sarah Franklin, *Biological Relatives: IVF, Stem Cells, and the Future of Kinship* (Durham, NC: Duke University Press, 2013) and Sarah Franklin and Margaret Lock, eds., *Remaking Life and Death: Toward an Anthropology of the Biosciences* (Santa Fe, NM: School of American Research Press, 2003).

In the 1997 film *Gattaca* (directed by Andrew Niccol) the action takes place in an imagined future society in which genetic engineering and discrimination have become the norm. The protagonist, wishing to become an elite astronaut, must adopt a false identity in order to avoid the prejudices against "natural born" or "in-valid" humans (that is, those who are not genetically selected). The combination of genetic engineering and biometric surveillance creates not only personal hardship, but also an oppressive, rigid, and conformist social and political order.

Although it is (science) fiction, the film is a powerful critique of genetic engineering. By exploring the possible consequences of rampant genetic discrimination *Gattaca* warns us about the eugenic possibilities that emerge from attempts to achieve human perfection through biology. Some elements of *Gattaca*'s future remain far-fetched. Nevertheless, such technological imaginaries play an important role in our understanding of biotechnologies, particularly in creating fears and expectations about the future. As such, it is worthwhile exploring whether *Gattaca* offers a realistic depiction of the future and what other kinds of risks may be present in building "designer babies."

This chapter explores the consequences of coupling IVF (discussed in chapter 15) with the kinds of biological knowledge and testing technology that has emerged from the Human Genome Project (chapters 12 and 13). What is emerging from this is not only the ability to select the children we want, but also a significant industry and market that deal in the creating, buying, and selling of babies.

HAVING A BABY?

In chapter 13, we described the vast number of genetic tests now available for pregnant women. Amniocentesis, chorionic villus sampling, and umbilical blood sampling can all be used to provide material for the genetic testing of the fetus. In some cases, such tests will show risk factors for traits such as autism or heart disease. In others, it will show evidence of debilitating or fatal genetic disorders (such as neural tube defects). In the latter cases, the pregnancy may be terminated.

With the growth of IVF-assisted births, however, another opportunity for genetic testing presents itself. During the IVF procedure, once the eggs are fertilized by the sperm, they are grown in the laboratory for a period of time before they are re-implanted into the mother's uterus. During this time, it is possible to perform genetic testing on cells in the new embryo—embryos that test positive for genetic disorders (or show high disease risk) may be discarded. This "screening" allows the "best" embryos to be selected for implantation into the uterus. Since it occurs prior to implantation, this process is known as preimplantation genetic diagnosis (PGD). As more genetic tests for more diseases and traits become available, PGD will allow finer and finer control over the selection of embryos (and hence children), while avoiding the need for abortions.

THE REPRODUCTIVE-INDUSTRIAL COMPLEX

As the use of IVF expands, the use of PGD is also growing. Many nations restrict the use of PGD to screening for recognized medical conditions. In the United Kingdom, for instance, PGD is not permitted for "social or psychological characteristics, normal physical variations, or any other conditions which are not associated with disability or a serious medical condition."[1] This restriction extends to using PGD for sex selection. The laws regarding PGD are even stricter in Germany. The United States, on the other hand, has no specific federal or state laws regulating PGD.

However, such restrictions depend on the definition of *recognized medical conditions*. As we saw in chapter 13, such definitions change over time and depend strongly on social and political ideas about what counts as "normal." This means that the distinction between using PGD simply to ensure that an offspring is free of genetic disease and the use of PGD for "enhancement" is not clear-cut.

Despite the restrictions, individuals have found ways to use PGD for creating offspring with tailored characteristics. In 2002, for instance, British parents traveled to the United States to avoid UK laws preventing PGD. The Grahams' first child, Saskia, had been born with a rare form of leukemia. A bone marrow transplant from an individual with a human leukocyte antigen (HLA) match for Saskia could donate hematopoietic stem cells and save her life. Unfortunately, neither of the Graham parents matched. However, it was

1. Human Fertilization and Embryology Act 1990 (UK).

possible that if the Grahams conceived another child, that child could be an HLA match for Saskia. Along with their doctors and genetic counselors, the Grahams realized that it would be possible to screen embryos using PGD techniques to create a child with an HLA match for Saskia. The Grahams' second child, Imogen, was conceived in this manner and her bone marrow was harvested to save her sister. Both children returned to a normal and healthy childhood.

The birth of so-called savior or sacrifice siblings such as Imogen raises serious ethical questions. The savior is brought into being for the distinct purpose of saving the life of another. Is it morally acceptable to create a child not for its own sake, but for the sake of another? And, will the child be psychologically affected by the knowledge that he or she only came into being to save someone else? On the other hand, is it morally right to let a child die from leukemia when a treatment is possible?

As IVF and PGD become cheaper and more ubiquitous, some policymakers have argued that imposing bans and restrictions is the wrong approach. Some people, they argue, are desperate to have children; others, like the Grahams, are desperate to save their existing children. People will continue to want the best possible or healthiest possible babies and be willing to pay for the privilege, regardless of any attempts to halt the use of this technology. One proponent of this view is Debora Spar, who outlines such a view in her book *The Baby Business: How Money, Science, and Politics Drive the Commerce of Conception* (2006). Spar argues that there is a *market* in babies—there is a *demand* created by would-be parents and a ready *supply* in the form of IVF clinics, sperm banks, and potential surrogates. Governments may choose to place heavy restrictions on ARTs, but this will only create a *black market*, Spar argues.

Describing children and reproduction in terms of a market may seem distasteful since it casts babies as commodities to be bought, sold, and traded. But, this view does offer an alternative to a discussion of ARTs that is framed in purely ethical terms. While the moral debate may lead to a set of "rights" and "wrongs," the market view opens up questions about how such a market works, how it might be regulated, how it relates to other markets, and who is gaining and losing in the trade. Moral questions will always remain crucial, but answering them fully may require understanding the market in babies first.

DESIGNING BABIES, DESIGNING SOCIETY

This section will try to understand the possible effects of designer babies by understanding reproduction as a market. This market includes not only IVF and PGD, but also sperm donation, surrogacy, and genetic testing and counseling.

Surrogacy has a long heritage—in many societies, parents with many children would often give away some of their offspring to childless couples. This was usually not a directly financial transaction, although it was understood that children were both a source of economic burden (they had to be fed and housed) and, later, economic value (they could be put to work for the family). Artificial insemination of a surrogate made it possible for infertile couples to adopt a child that was genetically related to the father (but not the mother). The commercialization of artificial insemination in the 1980s made it possible for would-be parents to contract with a surrogate who was then impregnated with the father's sperm. At the end of the pregnancy, the surrogate would give the baby over for adoption.

IVF allows couples to take the next step: their own egg and sperm can be combined in vitro and implanted into the surrogate. For a fee, the surrogate then (assuming all goes well) gives birth to a baby that is genetically unrelated to her; she becomes merely a womb for hire. Many nations (including Canada, Israel, the United Kingdom, Australia, and some US states) now have bans on such commercial arrangements. This has caused couples from developed nations to seek surrogates in poorer countries such as Guatemala and India. There, the amount of money on offer (between $20,000 and $120,000) is a strong incentive for women to engage in such "baby farming."

In addition to "gestational surrogacy," there is an increasing market for sperm and eggs. Since the 1970s, sperm banks have collected and frozen sperm for distribution to single women and couples. At first, both the collection and use of the sperm remained relatively informal. Banks paid and collected only small fees for their services, which were mostly directed at the treatment of male infertility. Increasingly, however, sperm banks have evolved into specialized services paying high fees for high-value sperm and marketing it as a specialized product (such as that from Nobel Prize–winners or Olympic athletes, for example) or to particular clients (single women or lesbian couples, for example). A typical sperm bank in the United States charges $250–$400 per sample for sperm and pays the donor around $75 per specimen.

Eggs can fetch even higher prices. Once again, this market began informally and was aimed at medical treatment: women who had damaged eggs or ovaries could become pregnant by obtaining eggs from friends or family.

During the 1990s, the demand for eggs grew, and some clinic began to offer remuneration for egg donation. Since then, prices have soared. Advertisements promise $50,000 for eggs harvested from certain women (usually specified as young, healthy, athletic, tall, white, and with a high SAT score). Although the sale of eggs is banned in some countries, it is unregulated in the United States.

Rather than something absolutely new, PGD represents only an extension of these practices; rather than selecting and buying suitable eggs, sperm, or surrogates, PGD allows consumers to tailor their own eggs and sperm. In all cases, what is purchased is a level of increased control over reproduction and offspring. The market here acts as a kind of substitute for biology: if you are unable to reproduce (due to infertility, or due to being a single person, or due to being a same-sex couple) you can purchase the necessary means. Likewise, if you are unable to produce the *kind* of offspring you desire or you are unwilling to endure pregnancy, the market can also provide a solution.

This suggests that the market itself is increasingly determining how families are made in the twenty-first century. Parent-offspring kinships are made not necessarily via family and marriage relationships, but rather through market relationships. It has become possible to design families and kinship in the ways we want, freed of the constraints of biology. The availability of sperm, eggs, surrogacy, and PGD offers us the ability not only to design our children, but also to design new kinds of social units. Designer babies are often thought to entail a future inhabited by superlatively gifted individuals. But the future may equally be transformed by fundamental changes to the sorts of families and kinship relations produced by the buying and selling of reproductive abilities.

MARKETS, PROPERTY, AND REGULATION

Considering ARTs as a market also allows us to think about how such commerce should be regulated. Spar argues that the baby business is currently a *grey market*, existing on the edges of legality and legitimacy. Grey markets, partly because they lack regulation, present numerous opportunities for exploitation and inequity.

To work effectively, and to be regulated effectively, markets need clear notions of *property*. We can use the example of the market for land, or real estate. If I buy a piece of land, I know that this entails a very clear set of rights: the land has certain boundaries that are carefully recorded by local and state authorities, and I usually have the right to construct a building on that land. My rights might be limited in all sorts of ways—I might not be able to build a skyscraper, or I may not be able to cut down particular trees on the site — but for the most part such restrictions are well defined. In any case, the level of control I have over that space is determined by legal principles.

In the baby business, on the other hand, there is no clear notion of who owns what. If a couple use their sperm and egg to implant an embryo in a surrogate, who owns that embryo? Who has rights to control it? Who gets to determine what can and cannot be done with it? Do such rights belong to the mother, the father, the surrogate, or some other party such as a clinic or laboratory?[2] As Spar argues, such rights are being worked out in practice anyway: couples draw up contracts with surrogates that lay out specific rights and responsibilities. But such ad hoc arrangements can lead to exploitation: typically the contracting couple has more resources than the surrogate, allowing them to hire a lawyer to write a contract in their own favor. If property rights were determined by laws instead of one-time contracts, the potential for uncertainty, cheating, and exploitation would diminish.

The problem with most unregulated markets is that they generate inequalities — haves and have-nots. Some people are able to take advantage of the market and enrich themselves, while others may be exploited by it. By regulating markets, governments can intervene to smooth out this process and reign in excesses. Most governments, for instance, do not permit cartels or monopolies to dominate markets and exploit consumers; this would be an unfair skewing of the power in the market towards large companies. In a similar way, governments could intervene to ensure that reproductive markets are not too far skewed towards any of the particular groups involved (ART providers, prospective parents, or surrogates).

One of the ways to do this would be to clarify rules regarding the ownership of the means of reproduction (eggs, sperm, embryos, fetuses, babies). By making reproductive transactions more predictable and secure, property laws would likely increase the usage of ARTs. However, it would offer greater legal protection to all those involved (including the resulting children). By

---

2. Also, who bears responsibility if things go wrong? This issue was raised by a 2013 case known as "Baby Gammy" in which an Australian couple contracted with a Thai surrogate. An ultrasound revealed that the surrogate was carrying twins, one with Down Syndrome. The Australian couple requested an abortion, which the surrogate refused. When the twins were born, the Australians left Thailand with the healthy child, leaving "Baby Gammy" in Thailand with his birth mother. Based on this case, Thailand subsequently banned all paid surrogacy. See Jonathan Pearlman, "Thailand Bans Surrogate Babies from Leaving after Baby Gammy Controversy," *Telegraph*, August 15, 2014.

adjusting ownership rules, governments would have a subtle tool for guiding market actors towards ethically and socially acceptable outcomes.

CONCLUSIONS: CONSUMERISM AND COMMODIFICATION

The danger of this kind of reasoning is that it leads to the commodification of the products in question—in this case, babies. That is, it allows us to think of sperm, eggs, and embryos as things to be bought and sold. But, as we have seen elsewhere in this book, such commodification is not limited to reproduction—cells, organs, DNA, and other parts of bodies are increasingly up for sale.

Indeed, commodification can be understood as part of a much broader trend in twentieth- and twenty-first century history. The historian Lizabeth Cohen has called America a "consumer's republic," suggesting the importance of consumerism to modern US politics and culture. This drive for consumerism is a central part of the American dream—that is, the notion that you can spend your way to happiness. We have moved from designer jeans to designer babies; but it is part of the same quest for happiness that has created the market demand for perfectible offspring. The almost endless choice offered by modern consumer markets has led to the belief that we should also have almost endless choice when it comes to children.

If we want to halt PGD and the baby business, then outright bans and legal restrictions will not be enough (and, in fact, may make the situation worse by driving markets further underground). Rather, we will need to create a culture that is less dedicated to the form of consumerism that fuels a belief in self-expression and self-worth through unlimited purchasing power.

FURTHER READING

Significant parts of this chapter rely on the arguments of Debora Spar, *The Baby Business: How Money, Science, and Politics Drive the Commerce of Conception* (Cambridge, MA: Harvard University Press, 2006). On the relationship between designer babies, biotechnology, and capitalism more generally see Melinda Cooper, *Life as Surplus: Biotechnology and Capitalism in the Neoliberal Era* (Seattle: University of Washington Press, 2008). Lizabeth Cohen's history of the United States in the twentieth century as a "consumer's republic" is Lizabeth Cohen, *A Consumer's Republic: The Politics of Mass Consumption in Postwar America* (New York: Vintage, 2003).

There is also an extensive bioethics-rooted literature concerning designer babies: see Nicolas Agar, *Liberal Eugenics: In Defence of Human Enhancement* (Oxford: Blackwell, 2004); J. Glover, *Choosing Children: Genes, Disability, and*

Design (New York: Oxford University Press, 2006); R. M. Green, *Babies by Design: The Ethics of Genetic Choice* (New Haven, CT: Yale University Press, 2007); T. H. Murray, *The Worth of a Child* (Berkeley: University of California Press, 1996); M. Sandel, *The Case Against Perfection: Ethics in the Age of Genetic Engineering* (Cambridge, MA: Harvard University Press, 2008); Julian Savulescu, "Deaf Lesbian, 'Designer Disability,' and the Future of Medicine," *British Medical Journal* 325, no. 7367 (2002): 771–773; Brandon Keim, "Designer Babies: A Right to Choose?" *Wired*, August 3, 2009; and Richard Alleyne, "Genetically Engineering 'Ethical' Babies Is a Moral Obligation, Says Oxford Professor," *Telegraph*, August 16, 2012.

PART X
MINDING YOUR OWN BIOLOGICAL BUSINESS

19   DRUGS AND DESIGNER BODIES

INTRODUCTION: DRUGS AS BIOTECHNOLOGIES

This chapter considers pharmaceutical drugs as biotechnologies. Why should they be included here? The definition of biotechnology offered in chapter 1 suggested that biotechnologies (1) are sociotechnical systems that include active biological processes and (2) are aimed at control over these processes at the molecular level. Pharmaceutical products are certainly part of a sociotechnical system (one that includes pharmaceutical companies, physicians, hospitals, pharmacies, patent laws, and so on) and the action of drugs intervenes in biological processes within the human body. The intervention is certainly at a molecular level: drugs manipulate proteins, DNA, RNA, immunoglobulins, and other small molecules inside our cells. So, at least by our definition, drugs do seem to be a kind of biotechnology.

But what might we gain by talking about drugs in this way? What is the payoff from highlighting the similarities between pharmaceuticals and other, more familiar, biotechnologies such as cloning, GMFs, and tissue culture? For one thing, biotechnology companies participate in the research and development leading to pharmaceutical products. Understanding the consequences of biotech requires seeing the effects of these products when they leave labs and enter hospitals, clinics, and bodies. Beyond this, though, this chapter will illustrate how drugs, like other biotechnologies, are having important social, political, and economic effects. Like other biotechnologies, they help us to control and redesign human lives as we see fit. Along with consumer genetics services discussed in the next chapter, drugs are allowing us to increasingly tailor our bodies and minds according to our desires. Understanding drugs as biotechnologies helps us to see that drugs are now far more than pills that move us from illness to health, from pathological to normal. Rather, they are implicated in redefining what counts as "healthy" or "normal," what counts as part of our body or part of our identity, and what kind of life we can expect to live. This chapter will explore these issues primarily through three case studies: the contraceptive pill, Prozac, and Viagra.

THREE CASE STUDIES

For most of human history people have been ingesting herbs, potions, elixirs, compounds, and mixtures intended to heal. Such *materia medica* were often produced and sold by individuals—apothecaries, alchemists, herbalists—who had little or no connection to physicians. Although some such concoctions were effective, most were developed by trial and error, or on notions of "sympathetic healing" (the notion that resemblances between objects provided clues to their healing properties—a walnut shell could help heal a skull fracture, for instance), rather than medical theories.

From the seventeenth century onward, both well-meaning and unscrupulous individuals developed a wide range of remedies that were claimed to contain exotic ingredients and to produce remarkable healing effects on the body. The proliferation of these "patent medicines"—heavily advertised but usually ineffective—provided the justification for closer government scrutiny over drugs. In 1906, the United States passed the Pure Food and Drug Act, which attempted to regulate the labeling of foods and drugs (that is, to make sure the labels were not false or misleading) and to ban poisonous patent medicines. By 1938, this law was replaced by the more comprehensive Federal Food, Drug, and Cosmetic Act that gave power to the US Food and Drug Administration (FDA) to oversee the approval of medicines.

During the early decades of the twentieth century, research in physiology, microbiology, and biochemistry led to the discovery of several medically important compounds. Vitamins were shown to be crucial chemical factors in food; in the early 1920s insulin was purified from pancreatic cells and found to be effective in the treatment of diabetes; and the antibiotic penicillin was discovered in 1928 and entered mass production during World War II. These research-derived drugs formed the basis for the modern pharmaceutical industry that was to flourish from the 1950s on.

I: THE CONTRACEPTIVE PILL

The history of the contraceptive pill is discussed in detail in chapter 17. Here, we will just highlight some of the features of the pill that are particularly relevant to wider trends in the development of pharmaceutical products in the second half of the twentieth century.

First, the pill was a drug that was to be taken by perfectly healthy women. Rather than curing a disease, the pill was used to prevent unwanted conception and pregnancy. Chapter 17 noted the dramatic cultural effects that this had in 1960s America. Second, the pill was a drug that—in order to achieve its desired effects—needed to be taken every day for many years (potentially for several decades). Since the pill was not used to transform sick bodies into healthy ones, there was no "end" to the regimen (except menopause or the desire to become pregnant). Increasingly, women in developed countries could expect to take the pill every day for a large part of their fertile lives.

Third, it is important to note that the development of the contraceptive pill was a thoroughly commercial undertaking. The drug was invented, developed, and brought to market through pharmaceutical-industry research and development (rather than academic laboratory work). Finally, the pill is a drug that gives the user increased control over her own biology, directly intervening in the hormones that control fertility and ovulation. This control over biology allows an increased control over one's life: the ability to self-regulate fertility (turn it on and off almost at will) permits greater freedom in sexual and social behavior. It allows women (and to some extent men too) to remake their lives in particular ways. In this sense, the pill might be called the first *lifestyle drug*.

II: PROZAC

Beginning in the 1950s, pharmaceutical companies began to market a new generation of minor-tranquilizer drugs that aimed to control anxiety. Initially, the most popular of these was Miltown (Meprobamate, introduced 1955). In the 1960s, this drug was superseded by the sedative and muscle relaxant Valium (Diazepam, introduced 1962). Used to treat anxiety, panic attacks, and insomnia, by the 1970s in the United States, Valium was viewed by physicians and psychiatrists as a drug that could remake the patient's lifestyle. In particular, Valium could be used to counteract the stressful effects of modern life, helping patients (especially women) to cope and readjust their personalities to fast-paced American society. A headline from a magazine advertisement for the drug (directed at physicians) described the typical patient: "35, single, and psychoneurotic." "You probably see many such Jans in your practice," the copy continued, "The unmarrieds with low self-esteem. Jan never found a man to measure up to her father. Now she realizes she's in a losing pattern—and that she may *never* marry." Valium was the solution for such a "tense, over anxious patient who has a neurotic sense of failure, guilt or loss"[1] (figure 19.1).

Physicians increasingly understood the brain and personality as chemical-molecular systems. Depression and anxiety were no longer due to repressed desires or unconscious fears (as a Freudian might have seen them), but rather psychological states were increasingly seen to depend on chemical balances within the brain. Drugs such as Valium became popular because they were

1. Quotations from advertisement in physicians' journal for Valium, 1970.

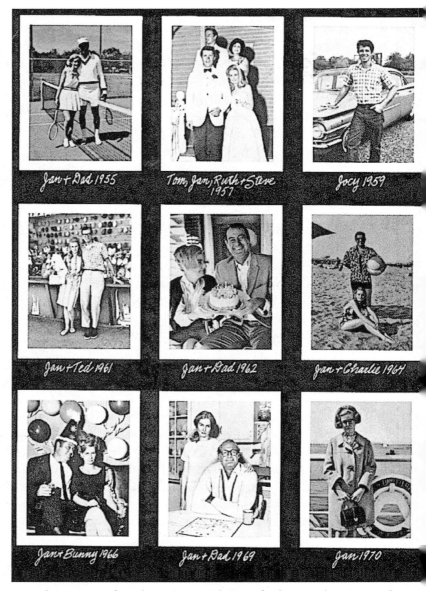

19.1 Advertisement for Valium, 1970. Marketing of Valium to physicians in the 1970s suggested that it could help patients adjust to the stresses and conflicting demands of modern life, remaking them as calmer and better people. Source: ProCon website. http://prescriptiondrugs.procon.org/view.resource.php?resource ID=005637

35, single and psychoneurotic

The purser on her cruise ship took the last snapshot of Jan. You probably see many such Jans in your practice. The unmarrieds with low self-esteem. Jan never found a man to measure up to her father. Now she realizes she's in a losing pattern—and that she may *never* marry.

Valium (diazepam) can be a useful adjunct in the therapy of the tense, over anxious patient who has a neurotic sense of failure, guilt or loss. Over the years, Valium has proven its value in the relief of psychoneurotic states—anxiety, apprehension, agitation, alone or with depressive symptoms.

Valium 10-mg tablets help relieve the emotional "storms" of psychoneurotic tension and the depressive symptoms that can go hand-in-hand with it. Valium 2-mg or 5-mg tablets, *t.i.d.* or *q.i.d.*, are usually sufficient for milder tension and anxiety states. An *h.s.* dose added to the *t.i.d.* dosage often facilitates a good night's rest.

Valium® (diazepam)

for psychoneurotic states manifested by psychic tension and depressive symptoms

Roche LABORATORIES
Division of Hoffmann-La Roche Inc
Nutley, New Jersey 07110

means through which this balance might be manipulated, adjusting molecules in order to modify and control personality and lifestyle.

In the 1990s, Prozac replaced Valium as the most popular "pill for the mind." Developed by the pharmaceutical company Eli Lilly in the 1970s, Prozac (Fluoxetine) was approved by the FDA in 1987. A serotonin re-uptake inhibitor, Prozac became widely used in the treatment of depression, obsessive-compulsive disorder, bulimia, anorexia, and panic disorder (serotonin is a molecule that is involved in the regulation of mood, appetite, and sleep). By 2002, it was prescribed to over 33 million patients per year in the United State alone. By 2008, Prozac helped to make antidepressants the third most common type of prescription drug taken in the United States.

The sale of Prozac was driven by the near-miraculous power that the drug had in improving people's lives. In 1994, Elizabeth Wurtzel described (in her book *Prozac Nation*) how the drug acted to almost magically recreate her identity:

> And then something just changed in me. . . . It happened just like that. . . . It was as if the miasma of depression had lifted off me . . . in the same way the fog in San Francisco rises as the day wears on. . . . Was it the Prozac? No doubt. . . . It took a long time for me to get used to my contentedness. It was so hard for me to formulate a way of being and thinking in which the starting point was not depression.[2]

Many other patients reported similar feelings: Prozac enabled them to remake themselves and take on a new personality, free of depressive feelings.

But the huge market for Prozac was created in part by changing notions of depression. Historians of medicine have shown how the boundaries between "disease" and "normality" are malleable—they are negotiated through interactions between doctors, patients, hospitals, drug companies, governments, and other people or institutions. Our understanding of depression (or anxiety) can change in response to the availability and characteristics of new treatments (such as Valium or Prozac). In the 1970s, the parameters of "anxiety" depended in part on the characteristics of Valium, while in the 1990s, Prozac helped to shape how people understood depression. To be "anxious" meant, in part, to possess symptoms that Valium could alleviate; to be depressed meant, in part, to have symptoms that Prozac could alleviate.

Prozac's effectiveness on a variety of symptoms meant that more and more individuals came to take the drug. This changed and expanded the definition of depression to include many of the people taking Prozac. As ideas about

2. Elizabeth Wurtzel, *Prozac Nation* (New York: Riverhead, 1994). Quotation p. 329.

depression shifted, more and more people needed and took the drug to make themselves as "normal" (that is, as "not depressed" according to Prozac's own standard). It is not the case that people taking Prozac were not really depressed or that they didn't really need Prozac. Quite the opposite: the shifting boundaries of "normal" and "diseased" made their suffering all the more real and Prozac provided a real solution. For many, Prozac became a constant fixture of life: like the contraceptive pill, it was not something to be taken until you were "well," but something that could be taken more or less indefinitely. That is, the drug could become a part of one's "normal" personality, permanently.

III: VIAGRA

The now well-known effects of the drug sildenafil were discovered by accident. In the early 1990s, the pharmaceutical company Pfizer had developed sildenafil citrate as a treatment for the heart condition angina pectoris. During the clinical trials, some patients reported that they were having "better" erections. Pfizer rearranged its clinical trials to test sildenafil as a treatment for "erectile dysfunction" (ED). Sildenafil sped through the trial process and became available for use under the trade name Viagra in 1998. In its first three months on sale, the drug was prescribed to 2.7 million men (generating sales of over $400 million for Pfizer). It is estimated that over 20 million men worldwide have received prescriptions for Viagra.

As the first oral treatment for ED, Viagra has played a significant role in reshaping medical and popular understanding of this condition. In the early part of the twentieth century, "impotence" was considered part of the normal aging process—men were usually advised to just accept the condition and adjust to it (although some treatments were available that claimed to stop or reverse the aging process, especially using hormones—see the section on "Sex, Hormones, and Rejuvenation" in chapter 2). By the middle of the twentieth century, impotence was increasingly treated as a psychological problem: the cause was considered to be anxiety over the loss of sexual function, or fears of loss of potency. Removing "psychological blocks" through sex therapy could cure impotence.

The emergence of Viagra signaled the molecularization of impotence. Research in the 1980s and 1990s, largely by urologists, discovered physiological and chemical causes for the lack of erections. Impotence, which was caused by psychological factors, was gradually replaced by ED, which was caused by breakdown in the cyclic guanosine monophosphate (cGMP) pathway (cGMP is a messenger chemical that produces smooth-muscle relaxation). ED became understood as a thoroughly chemical or molecular problem. Although these ideas had begun to emerge before the discovery of sildenafil, Viagra fit directly into this new model of ED: it provided a "cure" that matched the new understanding of a new disease.

The emergence of ED was also a medicalization of impotence. The failure to achieve an erection went from being a normal part of the aging process, to a psychological block, to a dysfunction that could be treated with a pill. ED became a medical problem with a pharmaceutical solution. Viagra marketing strives to portray sexual activity at an advanced age not only as possible, but also as a normal part of a healthy lifestyle. Advertisements showing vigorous, healthy, and physically active seniors enjoying their lives attempt to destigmatize and sanitize ED: no longer a sign of psychosexual dysfunction to be embarrassed about, ED is just a chemical imbalance that can be addressed with a pill.

Pfizer is selling not only a cure, but also a lifestyle. In particular, Viagra promises the extension of a fully functioning, happy, and active retirement to an aging population. Coupled with other drugs that extend life and health, Viagra extends and augments our bodily capacities, promising an enhanced control over our own lives and behaviors.

WHAT IS A DRUG?

These three case studies all suggest that we need to think about drugs not just as pills for curing sickness. They are implicated in definitions of disease, patterns of consumption, bodily control, and lifestyle choice. These four themes (explored in turn below) suggest that drugs, like other biotechnologies, are as much products of society and culture as they are products of science and technology.

DEFINITIONS OF DISEASE

We have seen how pharmaceuticals like Prozac and Viagra play a crucial role in creating the diseases or conditions for which they then become the cure. As we have learned elsewhere in this book (especially in chapters 13 and 14), how we think about disease changes over time and is historically and culturally specific—what we think of as illness at one place and time may not count as illness in another place and time (hysteria, homosexuality, and attention deficit hyperactivity disorder are all good examples of diseases that have either disappeared or recently emerged). Pharmaceuticals have come to play a significant role in the process through which diseases are made and unmade.

This is most obvious when we examine a drug like Paxil (paroxetine). Paxil is used to treat (among other things) "social anxiety disorder" (or "social phobia"), a condition defined by fear of situations that involve scrutiny or judgment by others (such as parties or other social events). Probably almost everyone has felt anxious in a social situation at some point in his or her life; certainly, some people feel more anxious than others. But where is the boundary between normal feelings of anxiety and pathological anxiety? Paxil allows us to group together certain sets of symptoms and certain kinds of feelings (those which Paxil can alleviate) and label them a *disorder*. In other words, the drug helps to define some kinds of anxiety as normal and others as disordered.

Viagra has achieved something similar: it has taken a problem that people have to cope with in their lives (impotence) and medicalized it into a disorder with a cure. Prozac too has helped to redefine the boundaries of depression, making us think differently about what kinds of feelings and symptoms belong in this category and which belong outside of it. The drugs create a new sense of normalcy that the drug itself helps us to conform to.

PRODUCTS

It is obvious, of course, that pharmaceuticals are products—remedies and medicines of all kinds have been bought and sold for a long time. Increasingly, though, pharmaceutical products have become consumer items not only for the sick, but also for the healthy. Pharmaceutical marketing for drugs such as Viagra, the contraceptive pill, and Nexium (for heartburn) promote the notion that drugs are necessary for the full enjoyment of one's life.

Like a stronger laundry detergent or better motor oil, pharmaceuticals should be constantly available to the consumer. For prescription-only drugs, advertisements encourage us simply to "just ask your doctor." The historian of medicine Carl Elliott has coined the phrase "better than well" to describe the aims of the modern pharmaceutical consumer. For pharmaceutical companies, selling more products means not just catering to the sick, but also convincing well individuals that they deserve to live longer and better lives (see the list of top-selling pharmaceuticals in box 19.1).

BODILY CONTROL

Being "better than well" includes having enhanced control over your own body. Whether this is the power to control reproduction (as with the contraceptive pill) or the power to have an erection on demand (as with Viagra), drugs put us in control of our own biological functions.

Once again, drug company advertising plays to this theme: "disordered"

Box 19.1 Top-Selling Pharmaceutical Products (2011)

Lipitor
- Company: Pfizer
- Purpose: lowers cholesterol, reduces risk of heart attack and stroke
- FDA approved: 1996
- Sales in 2011: $7.7 billion

Plavix
- Company: Bristol-Myers Squibb, Sanofi Aventis
- Purpose: prevents blood clots, reduces risk of heart attack and stroke
- FDA approved: 1997
- Sales in 2011: $6.8 billion

Nexium
- Company: AstraZeneca
- Purpose: treats acid reflux
- FDA approved: 2000
- Sales in 2011: $6.2 billion

Abilify
- Company: Otsuka, Bristol-Myers Squibb
- Purpose: antipsychotic and antidepressive
- FDA approved: 2002
- Sales in 2011: $5.2 billion

Advair
- Company: GlaxoSmithKline
- Purpose: management of chronic asthma
- FDA approved: 2000
- Sales in 2011: $4.6 billion

lives are lives out of control. "When you suffer from restless legs syndrome," a commercial for Mirapex (pramipexole) announces, "your legs don't want to stop, even when you do." The drug allows you to regain control not only over your legs, but also over your life: "Your legs feel better, and you feel better."[3] Drugs not only make you feel better but also become empowering agents.

We have seen in this book how many other kinds of biotechnologies aim to control biological systems and processes at the molecular level. Pharmaceuticals provide another powerful way in which we can manipulate and control living things, making them plastic and shaping them to suit our needs.

3. Mirapex television advertisement, 2007. http://www.youtube.com/watch?v=IcCq CYuS-Q0.

Seroquel
- Company: AstraZeneca
- Purpose: treats schizophrenia, bipolar disorder, depression
- FDA approved: 1997
- Sales in 2011: $4.6 billion

Singulair
- Company: Merck
- Purpose: treats asthma and seasonal allergies
- FDA approved: 1998
- Sales in 2011: $4.6 billion

Crestor
- Company: Shionogi, AstraZeneca
- Purpose: reduces cholesterol, prevents heart disease
- FDA approved: 1996
- Sales in 2011: $4.4 billion

Cymbalta
- Company: Eli Lilly
- Purpose: treats depression and anxiety
- FDA approved: 2004
- Sales in 2011: $3.7 billion

Humira
- Company: Abbott Laboratories
- Purpose: anti-inflammatory used for rheumatoid arthritis and Crohn disease
- FDA approved: 1998
- Sales in 2011: $3.5 billion

DESIGNING LIFE

The ability to consume pharmaceuticals at will and to use them to control our bodies implies a further step: that we can use them to redesign our lives. I have already suggested how selling Viagra is about marketing not only a drug, but also a particular lifestyle. Many other pharmaceutical products promote similar lifestyle choices: the ability to live longer, happier, more actively, and without the problems usually associated with aging. The decision to take a drug is not necessarily one about health, but rather one about living life in a certain way.

In the case of psychoactive drugs, such as Prozac, the psychiatrist Peter Kramer has used the term "cosmetic psychopharmacology." Cosmetic surgeries are performed voluntarily in order to give ourselves more desirable or highly rewarded physical features. Likewise, drugs like Prozac move our psychological states towards those which others find normal or attractive. Like cosmetic surgery patients, those who undergo cosmetic psychopharmacology may be perfectly healthy—the aim is not a cure. Rather, drugs are used for enhancement—that is, for remaking ourselves (including our personalities) as we desire.

CONCLUSIONS: THE PHARMACEUTICAL PERSON

Drugs, like other biotechnologies, force us to confront questions about the plasticity of life and the boundaries of our bodies. They are complicated technological, social, and economic objects that play an increasingly powerful role in determining who we are, what we want, and how we live.

Some psychiatrists have argued that we are on the brink of a revolution in neuro-manipulation. Soon we may be able to pick and choose drugs that enhance our memories, increase our attention spans, alter our moods, and augment our reflexes. Whether or not such a revolution eventuates, this chapter suggests the power of existing and widely available drugs to build and rebuild persons and personalities. Doctors and psychopharmacologists concoct powerful drug cocktails, tailored to particular individuals, making their personalities to order. What is most disturbing about such scenarios is that we increasingly think about taking such pills not for enhancement (to make us better), but to make us *who we really are*. Without the drugs, we are shy, depressed, or manic; with the drugs we can become our true selves. We may become "pharmaceutical persons"—individuals who can be who they are only with the assistance of chemicals. The drugs themselves become part of our very identity.

FURTHER READING

For the general histories of pharmaceuticals, psychiatry, and psychopharmaceuticals see David Healy, *The Creation of Psychopharmacology* (Cambridge, MA: Harvard University Press, 2002), Edward Shorter, *A History of Psychiatry: From the Era of the Asylum to the Age of Prozac* (New York: Wiley, 1998), and Jeremy Greene, *Prescribing by Numbers: Drugs and the Definition of Disease* (Baltimore: Johns Hopkins University Press, 2008).

My analysis of Prozac draws on David Healy, *Let Them Eat Prozac: The Unhealthy Relationship between the Pharmaceutical Industry and Depression* (New York: New York University Press, 2006); Elizabeth Wurtzel, *Prozac Nation: A Memoir* (New York: Riverhead, 1994); Peter D. Kramer, *Listening to Prozac* (New York: Penguin, 1993); and Peter Breggin, *Talking Back to Prozac: What Doctors Aren't Telling You about Today's Most Controversial Drug* (New York: St. Martin's, 1995). A short and useful overview of some of this literature is Siddhartha Mukherjee, "Post-Prozac Nation: The Science and History of Treating Depression," *New York Times Magazine*, April 19, 2012.

On the history of Viagra see Jennifer R. Fishman, "Making Viagra: From Impotence to Erectile Dysfunction," in *Medicating Modern America*, ed. Andrea Tone and Elizabeth Siegal Watkinds (New York: New York University Press, 2006), 229–252. For scholarship on the relationship between drugs and health in American society more generally see Carl Elliott and Peter D. Kramer, *Better than Well: American Medicine Meets the American Dream* (New York: W. W. Norton, 2004) and Joseph Dumit, *Drugs for Life: How Pharmaceutical Companies Define Our Health* (Durham, NC: Duke University Press, 2012). For a more general critique of the pharmaceutical industry see Ben Goldacre, *Bad Pharma: How Drug Companies Mislead Doctors and Harm Patients* (London: Fourth Estate, 2012).

For analyses of drug company marketing and "cosmetic" psychopharmacology see Emily Martin, "The Pharmaceutical Person," *Biosocieties* 1, no. 3 (2006): 273–287; Anjan Chatterjee, "Cosmetic Neurology: The Controversy over Enhancing Movement, Metation, and Mood," *Neurology* 63, no. 6 (2004): 968–974; and Nathan P. Greenslit and Ted J. Kaptchuk, "Antidepressants and Advertising: Psychopharmaceuticals in Crisis," *Yale Journal of Biology and Medicine* 85, no. 1 (2012): 153–158.

INTRODUCTION: AFTER THE GENOME

The Human Genome Project (HGP) officially completed its work in 2003, offering biologists the chance to examine the full DNA readout of the human species (see chapter 12). In many ways the HGP was a triumph of technological and collaborative scientific effort. But in other ways it was a disappointment: the genome revealed the causes of only a handful of diseases, and biologists understood little about how it worked. As the HGP came to an end, a handful of new subdisciplines emerged. These fields—systems biology, proteomics, interactomics, epigenetics, integrative biology—aimed to understand how biological parts (genes, proteins, RNA transcripts) acted together to make cells and organisms work.

Biologists still hoped to understand the relationship of the genome to particular diseases and traits. Now, however, it seemed that these relationships might be much more complicated: many genes might be involved in determining simple traits; genes seemed to work together in complex networks and combinations; epigenetic factors such as methylation and histone modification seemed to play a significant role; genes could be turned on and off by RNA molecules in the form of microRNAs; genes could be spliced together in multiple ways; and environmental signals also proved to have a large impact on gene expression.

Soon after the HGP, a vast amount of "postgenomic" biology was underway. This involved a huge amount of additional DNA and RNA sequencing. By comparing thousands or even millions of fragments of human DNA, biologists hoped that they would be able to discover patterns that revealed the most important sites on the genome that are involved in determining traits and diseases. Such work required matching genomes with data about individuals, especially about their health. The accumulation of sequence data and individual health data has generated a growing stock of information about the relationships between genes and diseases.

By the mid-2000s, some entrepreneurs saw a commercial opportunity for utilizing this information (much of which was published in scientific journals). Genome sequencing (or partial sequencing) could be sold directly to consumers, providing them with a set of information about their probable health risks. The first so-called personal genomics services sprang up in 2007. Although complete genome sequencing was still too expensive, for about $1,000 these companies offered a comprehensive genetic test that showed results for around one hundred diseases. These companies argued that genomic information should not remain in the exclusive domain of scientists or biotech companies, but could rather be used to empower consumers.

This chapter will begin by discussing the technological and scientific background (including new sequencing technologies) that has made personal genomics possible. It will then examine personal genomics in more detail, paying particular attention to the aims, potential benefits, and potential risks of this new field. Like pharmaceuticals, direct-to-consumer genomics promises an increasing personalization of health care and an increasing ability to use biotechnology to tailor medical interventions to our own bodies and our own desires. However, personal genomics also raises difficult questions about who has the right and the proper expertise to use an individual's genomic information.

SEQUENCE, SEQUENCE, SEQUENCE

Personal genomics has been made possible by extraordinary advances in DNA and RNA sequencing technology. Since 2003, the pace of sequencing has rapidly increased. The genomes of dozens of other organisms—including dogs, elephants, zebra fish, kangaroos, mosquitoes, as well as hundred of microorganisms—have been sequenced. Much more human DNA sequence is now available as well. Projects such as the HapMap project and the Genographic Project have sampled DNA from populations across the globe, searching for genetic markers of difference (see chapters 21 and 22). In 2007 and 2008, the first complete genomes belonging to particular individuals were published (they belonged to the biologist-entrepreneur Craig Venter and to James Watson, the co-discoverer of the structure of DNA). In 2012, the "1000 Genomes Project" announced the results of a study of 1,092 completely sequenced genomes.

This massive amount of sequencing was made possible by the development of new DNA and RNA sequencing technologies. Beginning in 2005, several companies—including Illumina, Applied Biosystems, 454 Life Sciences, and Helicos Biosciences—brought to market rapid sequencing machines. The HGP had been completed using automated versions of the Sanger sequencing methods (chain termination with capillary electrophoresis using fluorescent dyes) invented in the 1970s. The so-called next-generation machines used a range of new techniques including polony sequencing, pyrosequencing, ion semiconductor sequencing, and DNA nanoball sequencing (see box 20.1). These methods, as well as the intense competition between these companies,

Box 20.1 Next-Generation Sequencing Methods (2014)

Sequencing by Synthesis
- Sequences by synthesizing DNA using specially designed fluorescent nucleotides stimulated by lasers and imaged with a digital camera.
- First available: 2006
- Approximate cost per megabase: $0.01
- Speed: up to 200,000 megabases per day
- Used in: Illumina HiSeq range (including the HiSeq X Ten)

Pyrosequencing
- Utilizes special light-producing enzyme (luciferase) that emits light when nucleotides are added; light is detected by digital camera.
- First available: 2005
- Approximate cost per megabase: $10
- Speed: up to 700 megabases per day
- Used in: Roche-454 Life Sciences range

Ligation
- Uses fluorescently-active ligase protein; ligase activated when DNA bases match between strands.
- First available: 2006
- Approximate cost per megabase: $0.15
- Speed: up to 20,000 megabases per day
- Used in: Applied Biosystems SOLiD range

Ion semiconductor sequencing
- Uses silicon-semiconductor chip pH meters to detect incorporation of nucleotides during synthesis.
- First available: 2010
- Approximate cost per megabase: $5
- Speed: up to 12,000 megabases per day
- Used in: Ion Torrent Personal Genome Machine

DNA nanoball sequencing
- Fragments of DNA are made into circles and then multiplied into nanoballs; sequencing is then performed by ligation and detected via fluorescence.
- First available: 2009
- Used in: Complete Genomics

Polony sequencing
- Short DNA fragments are attached to beads; DNA sequencing performed by ligation as described above.
- First available: 2005
- Used in: open source technology developed by George Church

Nanopore
- DNA is made to flow through an electrically active nano-scale pore; the change in current across the pore depends on the size, shape, and length of the sequence; each kind of nucleotide produces a distinct signal in the current.
- In development; Oxford Nanopore is testing a portable USB-compatible device called MinION based on this technology.

Mass spectrometry
- Small fragments of DNA are compared by mass rather than by shape or size; small differences in mass between nucleotides are detectable by mass spectrometry.
- In development.

For more detail see: E. R. Mardis, "Next-Generation DNA Sequencing Methods," *Annual Reviews of Human Genetics* 9 (2008): 387–402; and Michael L. Metzker, "Sequencing Technology — The Next Generation," *Nature Reviews Genetics* 11 (2010): 31–46.

caused the price of DNA sequencing to drop rapidly. As shown in figure 20.1, the speed of DNA sequencing has also increased exponentially.[1]

This rapid increase in speed and decrease in cost has caused many in the biotech industries to compare DNA sequencing to Moore's Law. In 1965, Gordon Moore, one of the founders of Intel, predicted that the number of transistors that it would be possible to fit onto a fixed area of silicon wafer would double every eighteen months. Remarkably, this exponential growth in computer chip technology has, so far, continued. DNA sequencing appears to be progressing even faster.[2]

The personal genomics industry has also been driven by improvements in DNA microarray (also known as DNA chip) technologies. Microarrays consist of thousands of short segments of DNA (called probes) attached to a glass, plastic, or silicon slide. Each short piece of DNA can act as a detector, reporting the presence of a complementary piece of DNA. This means that a microarray can test for thousands of DNA mutations simultaneously. Advances in chip-printing technology mean that it became possible to print millions of DNA probes on a single slide, offering "massively parallel" tests

1. In 2013, Illumina's sequencing machines could produce approximately 100 giga-bases of DNA sequence per day (roughly 30 human genomes of sequence or the equivalent of one human genome project per hour).

2. In fact, the cost of DNA sequencing outstripped Moore's Law in 2007 and has accelerated since: http://www.genome.gov/sequencingcosts/.

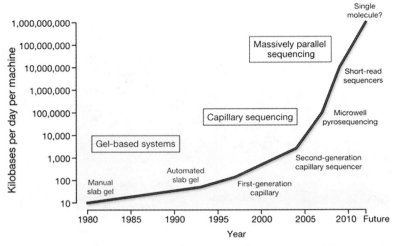

20.1 Increases in sequencing speed. Exponential increases in DNA sequencing speeds between 1980 and 2010. Sequencing technology has evolved from Sanger sequencing on gels (see chapter 12), to highly automated methods using capillary tubes, to a variety of next-generation methods (see box 20.1). Source: Stratton et al., "The Cancer Genome," *Nature* 458 (2009): 719–724, figure 3. Used with permission.

(called such because so many experiments are being conducted at once on one chip).

GENOME-WIDE ASSOCIATION STUDIES

It is one thing to gather huge amounts of sequence, but quite another to know what it all means. What do particular genes or genomic locations do? Or, asking the question the other way around, which genomic locations are involved in cancer, or obesity, or schizophrenia, or height? Since the HGP, biologists have devoted a large amount of work to answering such questions. Genome-Wide Association Studies (GWAS) have become one of the most important means of investigating the relationships between DNA sequence and traits.

GWAS require both DNA sequencing and collecting health information from large numbers of individuals. Such studies compare health data to people's genetic profiles. Importantly, GWAS (despite their name) do not conduct whole-genome sequencing (at the time they started, this was still too expensive). Rather, the studies examined hundreds of thousands of locations on the genome where humans were known to differ significantly from individual to individual (these points are usually called "common variants").

For example, say you wanted to compare two individuals' copies of a gene (let's call it X). It would be possible to sequence the whole of gene X for each person and then compare the gene sequence letter by letter. But most of the letters are going to be identical (because it's the same gene), so much of this sequencing is wasted effort. Let's suppose that we know (from previous sequencing of other individuals) that only three places on gene X usually differ from person to person: the 33rd letter, the 578th letter, and the 812th letter. At the 33rd position, 80% of people might have an A, for instance, and 20% a C and at the 578th letter, 27% might have an A and 5% a T, and 68% a G. If you could just determine the letters at these specific places (33rd, 578th, and 812th position), then you would capture almost all of the important information about an individual's copy of the gene.[3]

GWAS collected information about common variants from thousands of people and aimed to match patterns in these variants to patterns in disease. For instance, let us imagine a study of schizophrenia that involves 1,000 people; 500 have the disease and 500 do not. Information about hundreds of thousands of common variants across the genome are collected from each individual in the study. Simultaneously, health information is collected; in a study of schizophrenia, the most relevant health information is whether or not that individual has the disease. Biologists then use computers and statistical techniques to search for any links between variations in the genome and the occurrence of schizophrenia. Out of the 500 people with schizophrenia, let's suppose 100 share the same mutation at some position (call it Y) in the genome; moreover, out of the 500 non-schizophrenics in the study, only 50 have that mutation at position Y. This would suggest that the mutation at position Y might have something to do with causing schizophrenia since it is significantly overrepresented in schizophrenic individuals.[4] A similar analysis is performed for every sequenced position on the genome—some are shown to be statistically related to schizophrenia and others are not. (For a more detailed explanation of this example see box 20.2 and figure 20.2.)

Of course, not all schizophrenics share the mutation at position Y and some non-schizophrenic people *do* share it, so position Y is not a perfect predictor of the disease. But *statistically*, at least, it seems to be important. Since

---

3. Of course, any particular individual might have another mutation at another point in the gene—say the 35th position. But the argument is that since this variant is doesn't occur very often in the population it is unlikely to be very important in causing diseases or significant traits.

4. In practice, it is a little bit more complicated since other factors such as sex, age, and ethnic background must be taken into account.

Box 20.2 Explaining Genome-Wide Association Studies

A genome-wide association study (GWAS) requires both sequence data and health data from a large number of individuals. These individuals are categorized into two groups: those having the disease (the patients) and those without the disease (the controls).

For the sake of simplicity, we will use an example in which there are 1,000 individuals: 500 patients and 500 controls. In a GWAS, hundreds of thousands of locations on the genome are analyzed. We will just examine two: let's call them Y and Z.

At position Y:
- all the individuals in the trial have either an A or a C;
- amongst the patients, 400 have A and 100 have C;
- amongst the controls, 450 have A and 50 have C.

At position Z:
- all the individuals in the trial have either a G or a C;
- amongst the patients, 395 have G, and 105 have C;
- amongst the controls, 400 have G and 100 have C.

Analysis of position Y:
- both variations occur in the patients and the controls;
- 10% of controls have C, while 20% of patients have a C;
- this suggests that position Y is strongly associated with the disease since there is a significant difference between the percentages of Cs in each group;
- of all individuals with a C, twice as many are in the patient group (100 compared to 50); the odds of an individual with a C being a patient are 2:1;
- of all individuals with an A, for every 8 in the patient group there are 9 in the control group (400 compared to 450); the odds of an individual with an A being a patient are 8:9;
- the effect of C versus A at this position is measured by the *odds ratio*—the ratio between the odds that a patient has a C and the odds that a patient has an A; in this case $2 \div (8/9) = \textbf{2.25}$;
- this indicates that a person with a C at position Y is 2.25 times more likely to suffer from the disease.

Analysis of position Z:
- 20% of controls have C, while 21% of patients have C;
- this suggests that position Z is not strongly associated with the disease since there is no significant difference between those in each group; this difference could be due to chance;
- calculating the odds ratio in the same way as above gives: **1.06**;
- this suggest that a person with a C at position Z is only 1.06 times more likely to suffer from the disease.

Genome-Wide Association Studies (GWAS)

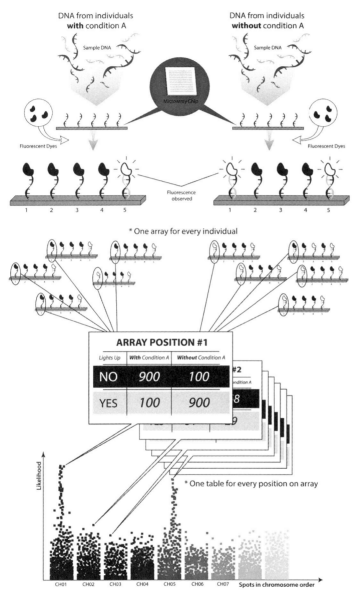

20.2 Genome-wide Association Studies. Refer to box 20.2. Fluorescent dyes on the arrays light up when a mutation is detected. Data from many arrays is used to assemble two-by-two tables from which odds ratios can be calculated. These ratios are summarized as dots on a Manhattan plot. Source: Illustration by Jerry Teo.

2005, GWAS have been conducted on obesity, height, coronary heart disease, age-related macular generation, types I and II diabetes, rheumatoid arthritis, Crohn disease, bipolar disorder, hypertension, and hundreds of other traits. Most of these have found tens or even hundreds of genomic positions to be associated with one particular trait. For instance, height seems to be affected by hundreds of thousands of different places on the genome.[5] It is important to stress that biologists do not understand, in most cases, how or why particular locations affect particular diseases or traits—GWAS discover statistical patterns, not functional pathways.

Interestingly, even all these hundreds of variations don't seem to be able to fully account for the *heritability* of traits. Most traits are determined partly by your genome (passed down from your parents) and partly by environmental factors. For instance, we know that your height depends partly on how tall your parents are, but also on what kind of nutrition you receive when you are growing. By comparing the heights of parents to their offspring, we can estimate what fraction of the trait is heritable and what fraction is environmental. In the case of height, it appears to be about 80% heritable. But adding up the contributions of the genomic variations found in the GWAS doesn't get even close to the 80% figure (in fact, using about 50 locations explains about 5% of variation, but almost 200 locations are needed to explain 10%). The problem is the same for other diseases and traits too.

This suggests that GWAS are missing something important with respect to how our genome works. One possibility is that other variants—apart from the common ones examined by GWAS—are more important than biologists thought. Epigenetic factors and interactions between genes are other possibilities. In any case, biologists still have much to learn about how genomic sequence gives rise to particular traits.

INTRODUCTION TO PERSONAL GENOMICS: AIMS, BENEFITS, AND RISKS

The personal genomics industry dates only to the middle of the 2000s. The company 23andMe (named for the 23 pairs of human chromosomes) was founded in 2006 and offered its first testing services to the public in November 2007. Navigenics was likewise founded in 2006 and began selling services in early 2008. DeCODE Genetics was founded in 1996, but

5. For more information on complex traits see Wenqing Fu, Timothy O'Connor, and Joshua Akey, "Genetic Architecture of Quantitative Traits and Complex Diseases," *Current Opinion in Genetics and Development* 23, no. 6 (2013): 678–683.

began offering personal genomics services (called deCODEme) in 2007. Since then, a handful of other companies—including Pathway Genomics, Existence Genetics, and HelloGene—have entered (and, in most cases, left) the market. The primary service offered by these companies was genotyping (that is, sequencing the common variants covered by GWAS). However, companies developing sequencing technologies—including Illumina, Oxford Nanopore Technologies, Pacific Biosciences, Complete Genomics, and 454 Life Sciences—rapidly brought whole-genome sequencing into the consumer price range.[6] For a brief period, personal genomics was a highly competitive marketplace in which the cost was rapidly decreasing (owing to both improvements in technology and economies of scale as the market expanded).

By 2013, 23andMe had outlasted most of its competitors and it will be the focus of the discussion here. Founded by Linda Avey, Paul Cusenza, and Anne Wojcicki, 23andMe has attracted particular attention owing to connections with Google. At the time the company was started, Wojcicki was engaged to the Google cofounder Sergey Brin (they later married, then separated). Google invested $3.9 million of venture capital into the start-up. In 2008, *Time* magazine named 23andMe's genetic test the Invention of the Year and by 2011, 23andMe had genotyped 100,000 customers. The price of their service decreased from an initial $999 to just $99.

Personal genomics services are marketed and sold directly to consumers (that is, without the involvement of hospitals or physicians); the transaction takes place via the web and the mail (or courier service). To obtain a personal genotype, the customer first makes the purchase on the web using a credit card. The website records personal details, including family and health information, and stores this in its database. The website will also require the customer to create a username and password for secure login. The company then sends the customer (via mail or courier) a kit for retrieving a DNA sample. This usually consists of a sealable plastic container—the customer is instructed to spit into the vessel, seal it, and return it to the company. Within a few weeks, the results of the test are available online. The customer can return to the website, login using his or her username and password, and view the results of the test.

The raw results of the test are a large set of base pairs (As, Gs, Ts, and Cs)

6. Competition here too was fierce leading to rapid turnover: 454 was purchased by Roche in 2007 (which closed down the sequencing business in 2013) and Complete Genomics was bought by BGI (China) in 2013. In 2015, Illumina was the clear market leader.

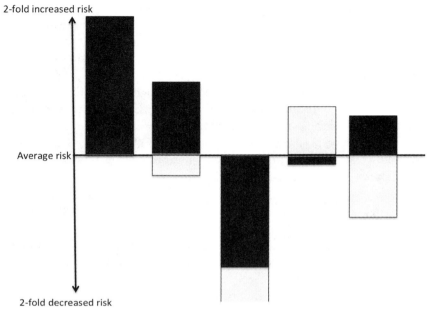

2-fold increased risk

Average risk

2-fold decreased risk

20.3 Representation of an online personal genomics report. Different vertical bars correspond to different genes that affect schizophrenia. The mid-line represents the average risk in the population conferred by a particular gene. Bars above the line indicate that the customer has a higher than average risk for schizophrenia on the basis of that gene. Bars below the line indicate a lower than average risk. The grey areas represent the maximum range of risk for each marker. Since the first bar indicates a risk more than twice the average, this might correspond to a gene containing position Y in box 20.2. Source: Illustration by author.

corresponding to hundreds of thousands of locations on the genome. This is usually not of any use or interest to the customer. Instead, personal genomics companies provide a detailed report based on their analysis of these base pairs. In most cases, these reports provide two kinds of information.

First, they provide information about an individual's risks for a wide variety of diseases and traits. This information is derived from GWAS. For instance, a GWAS study might show that a C nucleotide at a particular position on the genome occurred frequently in people with schizophrenia. If your genotype showed that you had a C at this position, your test would report an elevated risk for schizophrenia on the basis of that information. These risks are usually reported as ratios between your risk of getting the disease and the average risk (for an example see figure 20.3). On average in the whole population only 10% of people might develop schizophrenia.[7] Of those people with the particular mutation to a C nucleotide, however, 20% of people may develop schizophrenia. This would then be reported as a 2.25× risk, since having the C mutation indicates that you are just over twice as likely to get schizophrenia than the average person (for more details on this calculation see box 20.2 and figure 20.2).

However, since many positions on the genome are associated with particular diseases (as we saw in the section above), this information is not always consistent. For instance, there might be seven positions on the genome known to be associated with schizophrenia—a customer's genotype might show elevated risks for three of them, and decreased risks for the other four, for example. Such reports also often show the degree of reliability that can be attributed to particular risk factors—some diseases and traits have been widely studied while other claims may be based on very small study sizes. This raises questions about how such data should be properly interpreted (and by who).

Second, direct-to-consumer genomics often produces reports about customers' ancestry. For instance, genotyping might show that you are of 15% Asian ancestry, 40% European ancestry, and 45% African. Some services include color-coded maps of your chromosomes, indicating which parts are derived from which continents. Different populations (Asian, European, African, Native American, Australia Aborigines, etc.) tend to show different patterns of mutations. For instance, considering a particular location on the genome, people living in Asia might be found to have 10% A, 5% C, 5% G and 80% T at this position; Europeans, on the other hand, might show 50% A, 5% C, 25% G, and 20% T. This would mean that a T at that position makes it likely that this spot on a individual's genome is associated with Asian ancestry. By combining information from hundreds or thousands of such positions, it is possible to statistically infer the ancestry of different parts of a person's genome. This information can then be compiled to give an overall prediction of where the customer's genome came from.

Personal genomics companies also encourage several individuals within the one family to get tested. This makes it possible to see which mutations have been passed on from parents to offspring, and which ones are shared between siblings or more distant family members. Companies have even en-

7. The real number is closer to 1%. The "imaginary" figure of 10% is used here for the purposes of making the mathematics in box 20.2 clearer.

couraged customers to search for relatives through their websites and offer DNA-based versions of social networking in which connections can be made on the basis of matching genotypes.

Personal genomics companies argue that their services offer great benefits both to individuals and to society. Most obviously, they offer increased self-knowledge, especially in relation to one's health. In marketing their services, 23andMe appeal to the ideals of individual autonomy and consumer choice. Knowing that you are at risk for type II diabetes or breast cancer might encourage you to consume less sugar or to make sure you get regular check-ups (or, as in the widely publicized case of Hollywood actress Angelina Jolie, an individual might decide to obtain a double mastectomy in order to reduce the risk of breast cancer). This, the personal genomics industry claims, not only empowers individuals to take responsibility for their own health, but also encourages preventative action that could, in the long run, decrease health care costs (for both the individual and society as a whole). For 23andMe, your genome is your most personal information and you have the right to access, know, and understand it for yourself.

The commercial collection of genomic information is also mobilized in attempts to contribute to the overall store of knowledge about human diseases. 23andMe has close to 750,000 customers who are regularly subjected to surveys about their health—this allowed the company to do their own GWAS-type studies, matching genotypes to diseases amongst their customers (they call this part of their service 23andWe). Since 23andMe remains a private, for-profit company it is unclear how this information will be used, or who it might benefit. Companies such as Google and Facebook take advantage of the fact that consumers are willing to share their private data (emails, photos, etc.) in exchange for their services. In the personal genomics world, consumers are not only paying to share their genomic information, but also voluntary contributing their labor and time to complete surveys that provide valuable data for 23andMe.

Alternative, not-for-profit, versions of personal genomics have also emerged. George Church, a biologist at Harvard Medical School, is using whole-genome sequencing for more explicit public health purposes. Church hopes eventually to enroll 100,000 volunteers into his Personal Genome Project (PGP). These volunteers would be whole-genome sequenced and their medical records would also be made available online. The hope is that this will contribute not only to increased knowledge of human disease, but also to genomically tailored and targeted treatments and pharmaceuticals.

But both the PGP and private personal genomics entail significant risks.

Church requires his volunteers to sign a thirty-five page consent form and to pass a test in genetics and bioethics in order to demonstrate that they understand the potential drawbacks to their participation. Genomic information is identifiable—it can be directly associated with a particular individual. Lapses in security would cause a permanent breach of privacy (you can't change your genome like you might change your password or bank account number). Such information could be used by employers or insurance companies. For instance, particular employers could refuse to employ individuals with an increased risk of schizophrenia. Or, insurance companies could raise the premiums for those with a higher risk of type II diabetes. Since genetic information is relatively new, we do not know what people might do with it, or how they may respond to it.

Some of these concerns have been addressed—at least in the United States—by the creation of the Genetic Information Nondiscrimination Act (GINA). This was signed into law by President George W. Bush in 2008. As its name suggests, the law makes it illegal for employers and insurers to discriminate on the basis of genetic or genomic information. This "civil rights legislation for the genetics age" (as it was called by Senator Edward Kennedy) gave a great boost to the personal genomics industry, offering assurance to customers that their genetic information could not be used against them. However, the law may not provide enough protection. For instance, GINA does not apply to life, disability, or long-term care insurance, causing many people to continue to worry about the consequences of having their DNA sequenced.

However, even if GINA is successfully enforced and genomic information remains private, three important problems remain. First, the information provided by personal genomics may not be very useful. Genetic information is not predictive: tests cannot tell you that you will certainly get schizophrenia, only that your risk is higher or lower than average. In many cases, even the risk is ambiguous—different methods involving genes and different companies and different algorithms that are used to interpret the data lead to different results. It is not clear what all the information means, or who has the ability to interpret it.

Second, personal genomic information could be harmful to customers. This is especially the case when the interpretation of the information is unclear. One possibility is that the knowledge that you have a two-fold increased risk for schizophrenia, for instance, may just cause intense fear and worry. Studies conducted on the effects of personal genomic information have so far shown that it does *not* cause people to worry, nor to change their behavior.[8] Third, personal genomics does not tell us how to act. Even if the evidence from personal genomics is clear, how should a customer (or a physician guiding him or her) respond to it? Personal genomics does not answer that question.

In November 2013, these uncertainties caused the US Food and Drug Administration to intervene. Their "warning letter" asked 23andMe to suspend their personal genomics services on the grounds that the company was offering "diagnosis of disease" without clearance from the FDA. The letter continued:

> Some of the uses for which PGS [personal genome services] are intended are particularly concerning such as assessments for BRCA-related genetic risk and drug responses (e.g., warfarin sensitivity, clopidogrel response, and 5-fluorouracil toxicity) because of the potential health consequences that could result from false positive or false negative assessments for high-risk indications such as these. For instance, if the BRCA-related risk assessment for breast or ovarian cancer reports a false positive, it could lead a patient to undergo prophylactic surgery, chemoprevention, intensive screening, or other morbidity-inducing actions, while a false negative could result in a failure to recognize an actual risk that may exist. Assessments for drug responses carry the risks that patients relying on such tests may begin to self-manage their treatments through dose changes or even abandon certain therapies depending on the outcome of the assessment.[9]

In the FDA's view, the information provided by personal genetic testing was unverified and unreliable and hence could lead to harmful outcomes for consumers. The letter reflected an uncertainly about whether genetic data could really be interpreted into reliable medical guidance (and if so, could a private company be trusted to do so?). A few days later, 23andMe responded by voluntarily suspending their disease-risk services until they could gain FDA clearance (they continue to offer ancestry services). In 2015, the company continued to work towards satisfying the FDA's requirements.

OUR GENOMES, OURSELVES

Personal genomics promises the fulfillment of the vision of the human genome project. According to 23andMe, now you can really see your

8. C. S. Bloss et al., "Impact of Direct-to-Consumer Genomic Testing at Long Term Follow-Up," *Journal of Medical Genetics* 50, no. 6 (2013): 393–400.

9. http://www.fda.gov/iceci/enforcementactions/warningletters/2013/ucm376296.htm.

very own genome for yourself; you can even download it onto your computer, copy it onto a flash drive or CD, and hold it in your hand. The human genome project promised to tell us something about who we are, and personal genomics seeks to deliver on that promise. But some important transformations occurred along the way.

First, the genome is now a fully digitized object. This began with the human genome project itself—biologists uploaded parts of the genome into shared databases. Now, each genome can be uploaded, downloaded, transmitted, compared, shared, and analyzed. All this takes place by and through computers and the Internet. What all our genomic information means—and how we understand who we are—is critically dependent on these information technologies. Genomics has made genomes into data that can flow across the globe. Personal genomics companies have capitalized on and contributed to this vision of a shared, global genome. By linking genomics with the Internet and social media, personal genomics encourages us to think of our genomes as equivalent to other kinds of information that we routinely share and exchange online. If Google's mission is to organize all the world's information, 23andMe's role is to make sure our genomic information is incorporated into this plan.

Second, personal genomics has transformed the genome into a commodity. The vision of personal genomics is that we can download parts of our genome onto our smart phones and tablets, and—using the appropriate app—use it to plan our lives. Perhaps it could help us tailor our daily exercise regime, or we could use it in the supermarket to pick out foods least likely to cause us health problems; it could help our doctor to avoid prescribing drugs that disagree with our genome; and no doubt it could be used to formulate advertising specifically targeted to our DNA. Our genome becomes a product, something that can be bought and sold and that we can use, along with pharmaceuticals, to enhance our lives. It becomes just another consumer item.

Third, genomics has been increasingly linked to the rhetoric of individual rights and consumer rights. Personal genomics companies have promoted their services by marketing access to genomes as liberatory and democratic. Knowing one's genome, and even opening it up to the scrutiny of others, is portrayed as a bold and heroic act. This new way of thinking about our genomes may benefit ourselves and others, but it also opens up new possibilities for the exploitation of personal information.

Finally, personal genomics encourages us to understand ourselves as a set of risk factors. Before genomics we might have identified (if we had these diseases) as a "type II diabetic" or a "schizophrenic." Now, we can be defined by our *risks* for these and a range of other traits. Even though we might not have a disease, we might begin to think of ourselves as "not normal" or "pathological." Large sets of statistics and statistical practices—our chances of being this way or that way—are coming to form an increasingly important part in constructing our identities and our aspirations.

These four transformations suggest the important role that personal genomics is playing in reshaping our ideas about health, disease, and normalcy.

CONCLUSIONS: PERSONALIZED MEDICINE AND PERSONAL RESPONSIBILITY

Personalized genomics is part of a broader program and vision of "personalized medicine." Personalized medicine aims to customize all aspects of health care, tailoring medical decisions, treatments, and practices to individual patients. Such tailoring will rely partly on genotyping and genome sequencing, but may extend to studies of an individual's proteins (proteomics) and metabolism (metabolomics). In the regime of personalized medicine, not only the genome, but also "health" becomes a customizable consumer product. The notion of personalization is used to sell products including drugs, treatments, health insurance policies, health informatics, new technologies, and health services. In this vision, we can now purchase our way to "personal" health.

The idea of personalized medicine also fits with a particular economic and political vision of health care. Namely, a personal, consumer health care comes to seem less like a social, public, or collective responsibility, and more like an individual, personal responsibility. It is up to the individual to take care of their well-being, personally. Personalized health care and personalized genomics are associated with a political and social vision of personal responsibility and individual autonomy. Certainly, personal responsibility and autonomy are important components of a functioning society. But equally, notions of care and collective responsibility for others are critical to a community. In the long term, the emphasis on the individual in the rise of a consumer-driven personal genomics may undermine the notion of health as a public or social good.

FURTHER READING

On next-generation sequencing as its impact see E. L. van Dijk et al., "Ten Years of Next Generation Sequencing Technology," *Trends in Genetics* 30, no. 9 (2014): 418–426; H. P. Buermans and J. T. den Dunnen, "Next Generation Sequencing Technology: Advances and Applications," *Biochimica et Biophysica* 1842, no. 10 (2014): 1932–1941; and Elaine R. Mardis, "The Impact of Next-Generation Sequencing Technology on Genetics," *Annual Reviews of*

Human Genetics 9 (2008): 387–402; and Hallam Stevens, "Dr. Sanger, Meet Mr. Moore: Next-Generation Sequencing Is Driving New Questions and New Modes of Research," *Bioessays* 34, no. 2 (2012): 103–105.

Some of the problems raised by genome-wide association studies are discussed in Allen H. Lango et al., "Hundreds of Variants Clustered in Genomic Loci and Biological Pathways Affect Human Height," *Nature* 467, no. 7317 (2010): 832–838; and Teri A. Manolio et al., "Finding the Missing Heritability of Complex Diseases," *Nature* 461, no. 7265 (2009): 747–753. Also see the introduction to Sarah S. Richardson and Hallam Stevens, eds., *Postgenomics* (Durham, NC: Duke University Press, 2015).

On personal genomics see Michael Fortun, *Promising Genomics: Iceland and deCODE Genetics in a World of Speculation* (Berkeley: University of California Press, 2008), and Nikolas Rose, "Race, Risk, and Medicine in the Age of 'Your Own Personal Genome,'" *Biosocieties* 3, no. 4 (2008): 423–439. On Church's PGP see Misha Angrist, *Here Is a Human Being: At the Dawn of Personal Genomics* (New York: Harper, 2011). On the relationship between genetic testing and race see Alondra Nelson, "Bio Science: Genetic Genealogy Testing and the Pursuit of African Ancestry," *Social Studies of Science* 38, no. 5 (2008): 759–783. For a sense of how personal genomics works and what services are offered it is also worth visiting the websites of 23andMe (http://www.23andme.com) and the Personal Genome Project (http://www.personalgenomes.org/). The FDA's November 2013 warning letter to 23andMe can be found at http://www.fda.gov/iceci/enforcementactions/warningletters/2013/ucm376296.htm. And 23andMe's response is here: http://blog.23andme.com/news/23andme-provides-an-update-regarding-fdas-review/. In 2014, the *New York Times* reported some loopholes in the Genetic Information Nondiscrimination Act: Kira Peikoff, "Fearing Punishment for Bad Genes," *New York Times*, April 7, 2014.

The globalization of biology and the implications of this are explored in Eugene Thacker, *Global Genome: Biotechnology, Politics, Culture* (Cambridge, MA: MIT Press, 2006). Many of the issues raised here will also explored in more detail in Jennifer Reardon's forthcoming volume, *The Postgenomic Condition: Ethics, Justice, and Knowledge after the Genome* (Chicago: University of Chicago Press).

On notions of risk created by personal genomics and the politics of personalized medicine see Nikolas Rose, *The Politics of Life Itself: Biomedicine, Power, and Subjectivity in the Twenty-First Century* (Princeton, NJ: Princeton University Press, 2006), and Nikolas Rose, "Personalized Medicine: Promises, Problems, and Perils of a New Paradigm for Health Care," *Procedia—Social and Behavioral Sciences* 77 (2013): 342–352.

PART XI

BIOTECHNOLOGY AND DIVERSITY

INTRODUCTION

One of the central activities of science has been to divide the world up into categories. Physicists classify stars as red giants and white dwarves; chemists tell us whether substances are acids or bases; and biologists take responsibility for dividing up the plants and animals into kingdoms, phyla, classes, orders, families, genera, species, and varieties. Since biologists are also interested in humans, they have, at various times, also attempted to classify people. We usually call such classifications "races." In chapter 11, the history of eugenics illustrated the close relationship between racism and notions of biological and social improvement. In chapter 13, we explored how regimes of genetic testing could lead to discrimination, including race-based forms of discrimination. Indeed, many aspects of biology, old and new, intersect with questions of race. Tissue culture, xenotransplantation, cloning, and assisted reproductive technologies all have implications for how we think about our relationships to one another.

This chapter focuses on race and biology explicitly. It begins with a brief overview of biological thinking about race in the nineteenth and twentieth century. The main part of the chapter considers how recent advances in biomedicine and biotechnology—in genomics and pharmacology especially—affect our understanding of race. The next chapter describes undertakings such as the HapMap Project and the Genographic Project—major efforts to understand human genetic variation. All these stories suggest that race remains an important lens through which both biologists and ordinary people understand humans. Consequently, race is a critical point of intersection between biotechnology and society. These examples show where and how biological ideas—realized through biotechnology—affect problems of health care, social justice, and equality.

RACE AND SCIENCE IN THE NINETEENTH CENTURY

The most famous classifier of plants and animals—Carl Linnaeus—lived during the eighteenth century (1707–1778). The Swedish botanist, zoologist, and natural historian devised the modern "binomial" (two-name) system of labeling plants (in which humans are *Homo sapiens*, platypus are

Ornithorhynchus anatinus, and the commonly used herb, mint, is *Mentha longifolia*). More importantly, Linnaeus divided and subdivided plants and animals into hierarchical groups (again: kingdoms, phyla, classes, and so on). Linnaeus' classifications were based on observable characteristics of the organisms, usually the sexual organs. The mid-eighteenth century was the time when colonial exploration was revealing a plethora of new and astonishing animals and plants to Europeans. The tables in Linnaeus' *Systema Naturae* (1735) aimed to classify the entirety of the natural world and to bring order to the chaos of life.

Linnaeus also classified humans. According to him, there were four races: *Europaeus albus* (white Europeans), *Americanus rubescens* (red Americans), *Asiaticus fuscus* (brown Asians), and *Africanus niger* (black Africans). (There was also a category—"monstrosus"—for wild humans, monstrous humans, and other humans that didn't fit into Linnaeus' scheme.) These races were also supposed to be associated with various personality traits. Europeans were supposed to be gentle and inventive, Americans stubborn and angry, Asians avaricious and easily distracted, and Africans relaxed and negligent. Therefore ideas about race were associated not only with skin color but also with supposed temperamental qualities. This classification was elaborated in the late eighteenth century by the naturalist and anatomist Johann Friedrich Blumenbach (1752–1840). The third edition of Blumenbach's *On the Natural Varieties of Mankind* (1795) proposed, on the basis of his observations and measurements of skulls (craniometry), five human races: Caucasian, Mongolian, Malayan, Ethiopian, and American.

The Enlightenment, at the end of the eighteenth century, is best known for its efforts to establish sets of "rights" for humans. These liberties would allow knowledge and science to spread and flourish and lead to progress, Enlightenment thinkers believed. But who was entitled to such rights? African slaves in the Americas? Asian peasants under colonial subjugation? The Enlightenment made it crucial to determine who counted as a "human" and who did not in order to clarify who should have access to the full responsibilities of liberty. Consequently, Enlightenment philosophers were prolific in their writings about race as they attempted to show why some people had the capacity for liberty and others did not.[1] Such schemes had the advantage that they also offered a scientific and philosophical justification for European domination over other peoples.

1. For instance, in 1775 Immanuel Kant wrote an essay titled, "On The Different Races of Man" that attempted to scientifically classify races.

Throughout the nineteenth century, the science of race became increasingly technical and sophisticated. Victorians invented ever more elaborate and detailed ways of discovering and describing the differences between groups of people. The use of science meant that race became "naturalized"—that is, it came to seem part of a natural world, rather than a human categorization imposed on the world.

The philosophers Friedrich Hegel (1770–1831) and Auguste Comte (1798–1857) created historical-scientific accounts in which they proposed a linear development of humans from primitive races (hunter-gatherer Africans) to the most civilized (industrial-age Europeans). Other scientists developed techniques for measuring the level of development of different people, especially across races. For instance, measurements of the skull, including its volume, were taken to indicate cognitive and racial characteristics. In the mid-nineteenth century, Adolphe Quetelet (1796–1874) promoted the idea of the "facial angle" to characterize human types. The facial angle was the angle between two imaginary lines drawn on a profile of a human face: the first line went from the middle of the nostril to the earhole, and the second from upper jawbone to the forehead. Europeans were supposed to measure in at around 80°, Africans at around 70°, and orangutans at 58°. This was taken to show the progression from lower to higher types of human beings.

The early nineteenth century also saw phrenology rise to become widely popular (even though it was quickly disowned by physicians). Phrenologists believed that the brain was made up of a large number of different "organs"—a part that controlled love, a part that controlled conscientiousness, a part that controlled secrecy, a part that controlled aggression, and so on. Moreover, phrenologists believed that the function of these parts was directly related to their size and that it was possible to assess the size of these various parts by studying and measuring a person's skull (figure 21.1). In other words, a person who had an enlarged "aggression" organ would be prone to aggression—a skilled phrenologist would be able to discern this by examining a subject's skull. Such schemes easily lent themselves to racial classification: European skulls were said to be distinguishable from Asian or African skulls, reflecting underlying differences in abilities and personality.

Darwin's theories were interpreted to provide even more support for race science. *On the Origin of Species by Means of Natural Selection* (1859) made no specific claims about human races and in general Darwin believed that there were no important biological distinctions between races. However, others (especially Francis Galton) quickly translated Darwin's theories of competition and evolution into social terms. Chapter 11 described how this led to

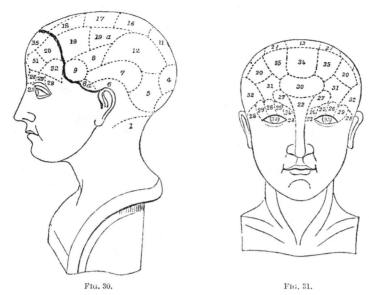

FIG. 30.　　　　　　　　FIG. 31.

21.1 Phrenology. Illustration of the location of the different organs of the head. Source: W. Mattieu Williams, *A Vindication of Phrenology* (London, 1894).

the science of eugenics that dominated thinking about race in the Anglo-American world between about 1880 and 1930. The middle and late nineteenth century saw the birth of many new disciplines including psychology, criminology, sociology, anthropology, and geography. All of these fields drew from and contributed to the discourse on race, finding further evidence that race was a real and important category required to analyze and understand the natural and social world.

RACE AFTER EUGENICS

This section picks up the story of race where chapter 11 ends. That chapter showed how the excesses of Nazi crimes during World War II created a backlash against racism and brought about a widespread discrediting of eugenics. Already in the 1940s, a number of prominent natural and social scientists had begun to speak out against racism. The anthropologist Ashley Montagu (1905–1999) wrote *Man's Most Dangerous Myth: The Fallacy of Race* in 1942. By the 1960s, the era of civil rights in the United States was causing more scientists to distance themselves from the concept of race. In 1964, Frank Livingstone, a physical anthropologist, wrote, "there are no races, only clines"[2] (clines are continuous and gradual changes across a geographic area — in other words, Livingstone was suggesting that there are no clear boundaries between races). Race, as a far as biology was concerned, seemed to be a category on the way to extinction.

For many biologists, Richard Lewontin provided the final nail in the coffin for biologically based race theories. In 1972, Lewontin wrote a short paper called "The Apportionment of Human Diversity." In the paper, Lewontin drew on genetic data from a large number of individuals. A given gene may come in several different varieties — these are called alleles by biologists. Lewontin examined how different alleles were distributed amongst different people. If race was an important biological category, then we would expect to find that most people in one race would share the same alleles and that they would have different alleles from those in other races. In fact, Lewontin found almost the opposite: most of the different alleles (93% of them to be exact) were found *within* races and only a small fraction (the other 7%) were found to differ between races. In other words, there are roughly ten times more genetic differences between people within races than between people of different races. Lewontin concluded:

> It is clear that our perception of relatively large differences between human races and subgroups, as compared to the variation within these groups, is indeed a biased perception and that, based on randomly chosen genetic differences, human races and populations are remarkably similar to each other, with the largest part by far of human variation being accounted for by the differences between individuals.[3]

Partly due to the legacy of eugenics and partly due to findings such as Lewontin's, race was an increasingly illegitimate concept in biological research. But race did not completely disappear either as a social or a scientific category. Psychology was one field in which race has proved particularly persistent. Psychologists have attempted to find scientific grounds to support the notion that different races differ in terms of their cognitive and mental characteristics.

One of the first people to attempt to measure such characteristics was Alfred Binet (1857–1911). Binet developed standardized means of testing chil-

2. Frank B. Livingstone and Theodosius Dobzhansky, "On the Non-Existence of Human Races," *Current Anthropology* 3 (1962): 279–281.

3. Richard D. Lewontin, "The Apportionment of Human Diversity," *Evolutionary Biology* 6 (1973): 381–398. Quotation p. 397.

dren and ranking them with respect to average children of their own age. These methods were adapted at Stanford University by Lewis Terman to create intelligence tests that formed that basis for the "intelligence quotient" (IQ). In the US, intelligence testing quickly took on eugenic connotations. By 1917, the Stanford-Binet test had come into widespread usage by the US Army in order ensure that no "feeble-minded" recruits were enlisted during World War I.

Since then, numerous attempts have been made to refine measurements of intelligence. Many such tests have seemed to show statistically significant differences between races. Some sociologists and psychologists have used such findings to suggest that they point to innate differences in intelligence between races (rather than social, cultural, economic, or other kinds of differences). Even at the end of the twentieth century, this debate continued. In 1994, the psychologist Richard J. Herrnstein and the political scientist Charles Murray published *The Bell Curve: Intelligence and Class Structure in American Life*. Most of this book is devoted to demonstrating the influence of IQ on such factors as job performance, income, and crime. But the authors also suggested that some of the difference in IQ scores between black and white Americans could be attributed to genetics. In other words, it suggested that there was some sort of race-based difference in intelligence. *The Bell Curve* was widely discussed and criticized (but also sometimes supported) in academia and in the popular press. Such debates suggest that despite the discrediting of the blatant scientific racism of the pre–World War II era, race remains a problematic category for biology. Ideas about race still inform some biological thinking and ideas in biology still shape our understandings of race.

BIOTECHNOLOGY AND RACE

So far, this chapter has not touched on biotechnology at all. However, molecular biology, genetics, and genomics all provide us with information about human differences. Such information has been interpreted, misinterpreted, and reinterpreted in racial terms. Genomics research, in particular, has seen a resurgence of the use of race categories. Scientists and politicians worked hard to portray the Human Genome Project as a symbol of human unity and similarity: there was *one* common human genome. However, even before the Human Genome Project ended, a number of efforts began to use the tools of genomics to investigate human differences. The Human Genome Diversity Project, the HapMap Project, the Genographic Project, and the 1000 Genomes Project will be discussed in chapter 22. In these projects, and in genomics more broadly, the use of racial categories became a way of demonstrating the *positive* social role that the research might play. By including minorities in genetic and genomic studies, biologists could show their commitment to extending the benefits of genomic medicine to diverse populations.

Despite this turn to "race-positive" science, however, race remains a problematic concept for biology and biotechnology. This section provides two examples—one from pharmacology and one from genomics—in which race emerges in biotechnological and biomedical work. Many other examples could be added; racial categories and race-based ideas continue to inform biological work in a wide variety of contexts.

CASE STUDY 1: PHARMACOGENOMICS AND BIDIL

Pharmacogenomics refers to attempts to design drugs that are specifically matched to the genome of an individual or a group. From the 1940s onward, there was some recognition that different people responded differently to drugs. For example, a drug that may be toxic to one person may not be toxic to another; or one individual might require a larger dose in order for a drug to be effective. In the 1950s and 1960s, the pharmacologist Werner Kalow (1917–2008) studied these differential drug reactions, arguing that they had an important genetic component. This was plausible on a molecular level. Most drugs cause their effect by binding to enzymes within the human body. The effectiveness and strength of such binding depends heavily on the shape of the molecule (it should be shaped to fit into the enzyme like a key into a lock). Different genes would build differently shaped enzymes, and therefore potentially alter these interactions.

Kalow also maintained that a genetic basis for differences in drug reactions made evolutionary sense. Different groups of people lived in different local environments and were exposed to different kinds of toxins (for example, in their food). These groups would need to evolve distinct enzymes to metabolize such toxins. Kalow arrived at the idea that different groups of people had evolved different responses to drugs. In the postgenomic era, this insight was taken up by pharmaceutical companies who hoped to be able to design "personalized" drugs tailored to groups or individuals. Pharmaceutical companies describe their interest in pharmacogenomics in terms of safety concerns: an adverse drug reaction can kill people, and pharmacogenomics promises a way of avoiding overdoses and poisonings. But it also promises a way for companies to create new markets, protect patents, and increase sales.

The first bespoke drug of this sort was called "BiDil." BiDil underwent clinical trials as a medication for patients with congestive heart failure. In 1997, the FDA rejected the medication since it did not seem to offer any benefit over and above a placebo. However, rather than giving up on the drug,

NitroMed (the company that had developed the drug) reanalyzed the data from their clinical trials. Now, rather than putting the whole test population together, they segmented the data by race. When viewed in this way, the data showed that amongst African-American patients, BiDil did help to reduce the mortality rate and hospitalization rate (compared to a placebo). On the basis of this analysis, the FDA reconsidered and approved the drug for use in African-American patients only in 2005.

Many people immediately voiced objections to the FDA's approval of BiDil, arguing that it amounted to the sanctioning of race-based thinking. Allowing a drug to be approved for one race and not others suggested that there must be real and important biological differences between races. BiDil did have medical benefits for African-Americans at risk of heart attacks (for one thing, it allowed an easier-to-follow dosing regimen). However, it is likely that BiDil would be effective in non-African Americans too—NitroMed just didn't have enough data to say either way. In fact, BiDil is just a combination (in one pill) of two different heart medications (isosorbide dinitrate, a vaso-dilator; and hydralazine, an antihypertensive) that have long been known to be effective in different populations.

So why was BiDil even being tested in the first place? Creating a "new" drug provided a way for NitroMed to market and sell a new product. But, more importantly, it also allowed the company to obtain a new patent on the "new" product (the patent on the original version of the drug would have lasted until 2007, whereas the patent on BiDil now expires in 2020; this is a common practice in the pharmaceutical industry where it is called "ever-greening"). Moreover, NitroMed knew that physicians would continue to prescribe BiDil "off label" to non-African American patients, so the company was not really limiting its market. Rather than merely utilizing race-based categories to provide a targeted drug, NitroMed attempted to exploit those categories in order to expand their own potential profits.[4] By racially segregating their data and extending their patent, NitroMed managed to block the development of a generic, low-cost medication for a racially diverse population that could have benefitted from it. Pharmacogenomics may in fact contribute to, rather than help to alleviate, racial health disparities by making it more expensive for some groups to get access to appropriate medications.

The making of BiDil utilized existing social categories of race (in this case the category "African American"). However, bringing the drug to market also reinforced such categorizations, making them seem more real and granting

4. Ultimately, they were unsuccessful: NitroMed was never profitable and was sold to Deerfield Management in 2009.

them added legitimacy. One important consequence of this is that BiDil reinforces the idea that racial health disparities are caused by genetic differences rather than by the socioeconomic differences between different groups.

CASE STUDY II: MICROCEPHALY AND EVOLUTION

This chapter's second example is taken from genomic studies of brain development. During the 2000s, Bruce Lahn, a geneticist at the University of Chicago, studied a rare brain disorder called *microcephaly*. Microcephaly causes slower-than-normal brain growth in children. This leads to a smaller-than-normal head size as well as a range of cognitive problems. The gene associated with this disorder had been identified as *microcephalin* (abbreviated MCPH1). As well as its role in this disease, MCPH1 is known to be involved in a variety of genetic and metabolic pathways within the body, including many linked to cell division and cell growth. In other words, apart from influencing brain development, MCPH1 appears to play a significant role in controlling bodily growth in general.

In 2005, Lahn and his team published back-to-back papers in *Science* reporting the findings of the genomic analysis of MCPH1 genes. Lahn's team examined DNA samples from a variety of populations around the globe, sequencing and comparing their MCPH1 genes. They found different varieties (different haplotypes) of the MCPH1 gene in sub-Saharan Africa, Europe, and Asia (see figure 21.2). Lahn's claims became controversial, however, when he claimed that his analysis showed that MCPH1 was still "evolving adaptively." This meant that different versions of MCPH1 appeared to be emerging in response to natural selection—some versions of the gene, Lahn claimed, conferred some advantage on the humans that possessed them and therefore persisted in the populations.[5]

What kind of advantage might MCPH1 be producing? At this point Lahn's work began to tread close to speculation. Lahn claimed that since MCPH1 caused microcephaly, it must be more broadly associated with brain size and cognitive development. That is, some versions of the gene caused humans to grow larger brains and therefore to be more intelligent, Lahn argued. In some situations, additional smarts provided larger-brained people with an advan-

---

5. Some critics of Lahn's work have questioned his evidence for this selective advantage. There is only one other case where a gene has been shown to be under active evolutionary selection in humans. This is the well-documented case of lactase persistence: the ability to digest milk as an adult has conferred a selective advantage on Europeans and some African populations where cattle-milk provided plentiful sources of protein and hydration.

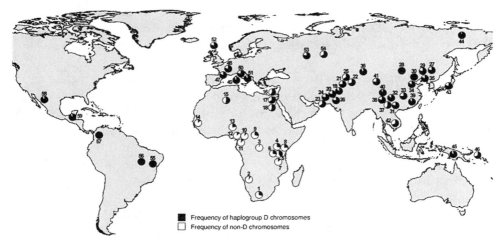

21.2 Global distribution of MCPH1 haplotypes. Lahn's team sequenced MCPH1 genes from different parts of the world. The structure of the MCPH1 gene fell into two broad groups, called here haplotype D (black), and haplotype non-D (white). The small pie charts show the percentage of individuals from various locations with D and non-D haplotypes. This indicates a large proportion of non-D haplotypes in sub-Saharan African and a high proportion of D haplotypes in Europe, Asia, and America. Source: Evans et al., "*Microcephalin*, a Gene Regulating Brain Size, Continues to Evolve Adaptively in Humans," *Science* 309 (2005): 1717–1720, figure 3. Used with permission.

tage over smaller-brained, less-intelligent humans. The differences in intelligence between modern populations, Lahn concluded, could be explained by selection and evolution of the MCPH1 gene.

Lahn's work was both acclaimed and criticized. His papers got published in *Science* and Lahn was awarded tenure and named a Howard Hughes Medical Investigator (a prestigious grant funded by the Howard Hughes Medical Institute). Many biologists have questioned Lahn's findings, challenging both his assumptions and the details of his evidence. Nevertheless, much work in genetics and genomics shares the basic assumption that genes determine intelligence, IQ, or brain size and that some of these differences are congruent with racial divisions. Such thinking shares much in common with nineteenth- and early twentieth-century ideas about race and human difference.

Lahn's work demonstrates these continuities particularly clearly. Basing accounts of intelligence on the size of the brain amounts to a remarkably crude analysis. Do we have any evidence to support the idea that slightly smaller brains are less intelligent brains? Do we know that genes such as

MCPH1 actually control brain size? (Lahn offers no concrete evidence of this.) Do we have any good reasons to think that "intelligence" would be advantageous to some populations (e.g., Europeans) and not to others (e.g., sub-Saharan Africans)?

Such "genomic craniometry" shows clear similarities to the kinds of nineteenth-century science in which racial differences were correlated with skull sizes. Lahn has defended his claims of differences in human intelligence, writing an opinion piece titled "Let's Celebrate Human Genetic Diversity" in *Nature* (with Lanny Ebenstein of the Cato Institute):

> The current moral position is a sort of "biological egalitarianism". This dominant position emerged in recent decades largely to correct grave historical injustices, including genocide, that were committed with the support of pseudoscientific understandings of group diversity. . . . We believe that this position, although well-intentioned, is illogical and even dangerous, as it implies that if significant group diversity were established, discrimination might thereby be justified.[6]

Suggesting that we are all the same, Lahn claims, is bad science and bad policy. But this misses the point. The question is not *whether* we are genetically different from one another (we are) but where those differences lie and what their significance is. Are the differences between broad racial groups ("Europeans," "Asians," "Africans") really the most important differences between humans? Or are old racial categorizations and assumptions still informing our ideas about human differences?

This example from genomics again shows how race remains an important category in twenty-first-century biology. Biomedicine and biotechnology continue to be informed by race-based ideas and continue to reproduce and naturalize race-based thinking. Biotechnology provides us with more and more powerful tools for exploring and explaining human differences. We must be ever more diligent in trying to ensure that such technologies do not reproduce prejudice, entrench differences, or further enact race-based discrimination and injustice.

CONCLUSIONS: RACE AND ENVIRONMENT

Postgenomic biology has struggled with questions of race and difference. Lahn's analysis of MCPH1 certainly provides a fairly blatant example of race-based thinking. Other biologists, however, have begun to develop

6. Bruce Lahn and Lanny Ebenstein, "Let's Celebrate Human Genetic Diversity," *Nature* 461 (2009): 726–728. Quotation p. 726.

more sophisticated approaches. In particular, they have tried to find ways of talking about human differences *without* assuming that these differences fall along traditional racial lines. Such work focuses on "populations" and specific local groups (rather than races). This remains a vexed and complicated issue—biologists and social scientists are still searching for ways of talking and thinking about human differences that allows for constructive science and medicine (for example, greater understanding of human diseases), while eschewing old stereotypes.

A large part of the difficulty here is that "race" itself is a hugely complicated category, representing overlapping ideas about genetics, geography, economics, and social status. It is remarkably easy for these factors to become blurred into something that makes race *look biologically real*. For example, if a particular social group is treated differently by society, they may actually become more different. Political and social divisions may lead to a particular group having different access to health care, a different diet, or exposures to different environments (e.g., toxins and pollutants). If this group is then subjected to tests that measure their response to drugs, or their IQ, or even the expression levels of their genes, the results may, not surprisingly, show significant differences from other social groups or the wider population. Such differences may then be used to further justify treating that particular group differently. Are such differences really attributable to genetics, or to the range of different circumstances of the group?

Such a question may be impossible to answer. Recent evidence shows that "the environment" feeds back to affect genes in all kinds of complicated ways. The environment can cause epigenetic changes—such as histone modification and methylation of genes—that can affect how genes are expressed. These changes can even be passed on from generation to generation. "Genetic" and "environmental" effects may, in practice, be too difficult to separate. This means that asking whether race is a real biological category is a meaningless question—it means that the kinds of inequalities that are built into society may actually make differences into something that is biologically significant. That is, social differences that cause, for instance, some people to live in polluted areas of cities may create persistent and important biological differences between people. Does this mean that racial differences are biologically real or socially produced? Again, the question seems to be an unanswerable one. The categories of social/biological and environmental/genetic don't make any sense in this context.

Given all this, we should remain deeply suspicious of any claims of absolute biological differences between racial groups. No doubt, there are differences between humans, but the examples here show how easily and often they are produced and reproduced by social, political, and economic circumstances.

FURTHER READING

There is a wealth of literature on race science and scientific racism in the eighteenth and nineteenth centuries. For an introduction to the topic see Carolyn Fluehr-Lobban, *Race and Racism: An Introduction* (Lanham, MD: AltaMira Press, 2005). Eminently readable is Stephen Jay Gould's account of the relationship of racism to attempts to quantify human attributes, *The Mismeasure of Man* (New York: W. W. Norton, 1981). On Linnaeus see Lisbet Koerner, *Linnaeus: Nature and Nation* (Cambridge, MA: Harvard University Press, 2001). On Blumenbach see Raj Bhopal, "The Beautiful Skulls and Blumenbach's Errors: The Birth of the Scientific Concept of Race," *British Medical Journal* 335, no. 7633 (2007): 1308–1309.

In the twentieth century much of the race debate has focused on issues of inherent intelligence differences. On the origins of the IQ test see Daniel J. Kevles, "Testing the Army's Intelligence: Psychologists and the Military in World War I," *Journal of American History* 55, no. 3 (1968): 565–581. The controversial work of Herrnstein and Murray is Richard J. Herrnstein and Charles Murray, *The Bell Curve: Intelligence and Class Structure in American Life* (New York: Free Press, 1994). Also important in the debates over race and IQ was the work of Arthur Jensen. See Arthur R. Jensen, "How Much Can We Boost IQ and Scholastic Achievement?" *Harvard Educational Review* 39 (1969): 1–123; and Arthur R. Jensen, *Bias in Mental Testing* (Free Press, 1980). Lewontin's important paper showing the distribution of human genetic diversity within and between races is Richard D. Lewontin, "The Apportionment of Human Diversity," *Evolutionary Biology* 6 (1973): 381–397. For a more recent example of the persistence of attempts to link genetics to race differences see Nicholas Wade, *A Troublesome Inheritance: Genes, Race, and Human History* (New York: Penguin Press, 2014).

The microcephaly example draws primarily on Sarah Richardson, "Race and IQ in the Postgenomic Age: The Microcephaly Case," *Biosocieties* 6, no. 4 (2011): 420–446. The scientific literature here is Evans et al., "*Microcephalin*, a Gene Regulating Brain Size, Continues to Evolve Adaptively in Humans," *Science* 309, no. 5741 (2005): 1717–1720; Nitzan Mekel-Bobrov et al., "Ongoing Adaptive Evolution of *ASPM*, a Brain Size Determinant in *Homo sapiens*," *Science* 309, no. 5741 (2005): 1720–1722; and Bruce Lahn and Lanny Ebenstein, "Let's Celebrate Human Genetic Diversity," *Nature* 461 (2009): 726–728.

On the BiDil case see David S. Jones, "How Personalized Medicine Became Genetic, and Racial: Werner Kalow and the Formations of Pharmacogenomics," *Journal of the History of Medicine and Allied Sciences* 68, no. 1 (2013): 1–48; Pamela Sankar and Jonathan Kahn, "Race Medicine or Race Marketing?" *Health Affairs* (2005, supplementary web exclusive): W5-455–463; Jonathan Kahn, *Race in a Bottle: The Story of BiDil and Racialized Medicine in a Post-Genomic Age* (New York: Columbia University Press, 2012); Howard Brody and Linda Hunt, "BiDil: Assessing a Race-Based Pharmaceutical," *Annals of Family Medicine* 4, no. 6 (2006): 556–560; Sheldon Krimsky, "The Art of Medicine: The Short Life of a Race Drug," *Lancet* 379 (2012): 114–115; Duana Fulwiley, "The Molecularization of Race: Institutionalizing Human Difference in Pharmacogenetics Practice," *Science as Culture* 16, no. 1 (2007): 1; and Anne Pollock, *Medicating Race: Heart Disease and Durable Preoccupations with Difference* (Durham, NC: Duke University Press, 2012). Kahn also has an article-length exposition in *Scientific American*: Jonathan Kahn, "Race in a Bottle," *Scientific American* 279 (2007): 40–45.

For more general reading on the history of race and biomedicine see Troy Duster, "Lessons from History: Why Race and Ethnicity Have Played a Major Role in Biomedical Research," *Journal of Law, Medicine, and Ethics* 34, no. 3 (2006): 487–496; Michael Montoya, *Making the Mexican Diabetic: Race, Science, and the Genetics of Inequality* (Berkeley: University of California Press, 2011); and Keith Wailoo, Alondra Nelson, and Catherine Lee, eds., *Genetics and the Unsettled Past: The Collision of DNA, Race, and History* (New Brunswick, NJ: Rutgers University Press, 2003). For more information on the role of race in postgenomic biology see Catherine Bliss, *Race Decoded: The Genomic Fight for Social Justice* (Palo Alto, CA: Stanford University Press, 2012); and Sandra S. Lee, Barbara A. Koenig, and Sarah S. Richardson, eds., *Revisiting Race in a Genomic Age* (New Brunswick, NJ: Rutgers University Press, 2008).

INTRODUCTION

In his documentary *The Journey of Man: A Genetic Odyssey* (2003), self-styled explorer-scientist Spencer Wells ventured out across Africa, Asia, and the Americas to discover human history. However, Wells is no ordinary explorer—he is not hunting for ruins of ancient civilizations or human fossils. Rather, he is after people's genes. By comparing DNA between people from different parts of the globe, Wells endeavored to reconstruct the route by which early humans left Africa, moved through Asia, and populated the rest of the planet. In Wells' work, a detailed human family tree can also be used as a map of human migration history.

Near the end of his documentary-journey Wells arrives in America—according to his analysis, this was the final frontier for the humans that had emerged from Africa, traveled across Asia, Siberia, and the Bering Strait, and finally to the Americas. In Arizona, Wells takes his film crew into a Navajo reservation, to ask them about their own origin stories: "We have a story of where we came from . . . a very sacred story . . . passed on through generations," they report. "We believe that we were created here. . . . This is where we came up from the ground . . . . We were birthed into this place." In response, Wells tells the Navajo about his findings: "I also have my own sense of what this story might be using science. . . . We're all related to people who lived in Africa about 50,000 years ago. . . . You yourself are essentially African."[1] The Navajo seem to reluctantly accept this, at least on camera—especially when they are shown Wells' "family album" of faces that the Navajo agree look similar enough to their own.

The viewer is supposed to understand this rustic scene as a triumph of modern science over myth—the Navajo are shown the truth about human origins and relatedness. Yet there is something a little unsettling about Wells' interactions. Why should the Navajo accept his story? Why should it be more compelling to them than their own? What might be the political, social, and cultural consequences of his evidence that the Navajo have only settled in this region just a few thousand years ago? And what political and economic cir-

---

1. *Journey of Man* [documentary] Dir. Clive Maltby (PBS, 2003).

cumstance have allowed Wells to generate, gather evidence, and tell his story in the first place?

In particular, it is important to realize that the concept of "native American DNA" on which Wells' scientific studies are premised is itself a consequence of the colonial histories of native American tribes. The way that native Americans in general—and particular tribes such as the Navajo—exist is a result of the way European settlers understood, grouped, managed, rounded up, displaced, and resettled the first Americans from the seventeenth century onwards. Nineteenth-century European ideas about race, purity, and mixing, in particular, played a crucial role in making the category of Native American and the particular subcategories ("tribes") within it. This complicates the idea that DNA analysis can provide an objective and authoritative account of the past—the scientific stories it tells us are inextricably linked to the political and social histories of various groups.

In the previous chapter we explored some of the history of biological thinking about racial difference and examined how this has played out in recent pharmacology and genomics. This chapter takes a slightly different approach, examining narratives about "difference" and "sameness" in the context of colonialism and global history. Biotechnology is not something that just takes place in labs, or gets written up in academic journals, or is put into pills. It is increasingly something that takes place all over the world—in cities, homes, jungles, and oceans. Biotechnology is part of globalization—it is both a result and a cause of increased trade and interaction between people worldwide. These new global power relations also create new opportunities for exploitation and domination. As such, biotechnology should be understood and analyzed against a background of colonialism and colonial history. To what extent does biotechnology reinforce global economic and political divides? To what extent does it recreate colonial patterns of power? Answering questions such as these will help us to understand what role biotechnology is playing in the emerging global order.

MAPPING HUMAN DIVERSITY

During the Human Genome Project (HGP), its proponents worked hard to emphasize the fact that the result of their efforts—*the* human genome—would represent a shared heritage of all the world's peoples. It was valuable, many biologists argued, because it was a "common thread" that connected everyone. At the same time, the public was told that genes—the most important component parts of the genome—were the things that made us unique: tall or short, black or white, smart or stupid, normal or pathological, and so on. DNA forensics could even be used to single out our DNA from the DNA of all the other billions of individuals on the planet.

This seems like a paradox: the genome is supposed to be both something in common and something that makes us unique. The truth is, of course, somewhere in between. The DNA that was used in the public HGP was in fact taken from blood (women) and sperm (men) from many different individuals. Not all of these samples were used, but the identities of the donors are unknown in order to protect their privacy. In the private project run by Celera Genomics, DNA was taken from a few donors of European, African, American, and Asian descent. Craig Venter, the leader of the project, has since revealed that his DNA was amongst that used for the project. The reason that the project was able to mix and match samples in this way is because a large fraction of the genome is indeed shared between all humans: after all, this has a good deal to do with what makes us humans and not gorillas, fruit flies, or petunias.

But it is also true that particular regions or sites on the genome tend to vary between different individuals. These variations partially account for human difference. They come in several different forms, the most straightforward being a single nucleotide polymorphism (SNP, pronounced "snip"). In a SNP, one letter of the genetic code, say an A, is simply substituted for another, say a G. So where I have an A in my genome, you might have a G.[2] On average, SNPs occur roughly one in one thousand nucleotides. Although this sounds like a small fraction (0.1%), if we consider that the human genome is 3 billion base pairs in total, this means that there are millions of SNPs (about 3 million–4 million in one individual). Because of where they occur and how the genetic code works, only a very small number of these SNPs have any effect on how your body looks or behaves. But analysis of SNPs (and other kinds of variations) can be used to track differences between individuals or populations—to examine ancestry or paternity, for instance. The genome is an immensely large and complex object: it can be made to tell stories about both human sameness and human difference, depending on the level at which it is scrutinized.

In the late 1980s, as the Human Genome Project began to take shape, some voices of opposition emerged from the biological community. One of these voices was that of an Italian geneticist, Luigi Cavalli-Sforza, then working at

---

2. A SNP involves the substitution of one nucleotide for another. Other possible differences between sequences include the addition or deletion of one or more nucleotides or the repetition of a group of nucleotides.

Stanford. Cavalli-Sforza had been working on tracking human genetic variation since the 1970s. His argument was that the HGP smacked of Eurocentrism. Funded largely by European and North American institutions, taking place mostly in this trans-Atlantic context, and using DNA gathered for the most part from Caucasians, how could the HGP claim to represent *everyone*? In 1991, Cavalli-Sforza, now joined by other population geneticists, proposed a project that would conduct a large-scale survey of the world's genetic diversity. At first as part of the HGP, and then later independently of it, the Human Genome Diversity Project (HGDP), would collect, analyze, and store blood samples from isolated and indigenous human populations across the globe.

Cavalli-Sforza and his collaborators stressed that this project needed to be done now, before many of these populations "vanished" as they merged into neighboring populations. This would "destroy irrevocably" our ability to reconstruct our evolutionary history. By comparing genes from different populations, Cavalli-Sforza and his collaborators hoped to construct a "human family tree" (figure 22.1) that would describe the relationship between different groups. Such a tree could also be used to infer the patterns of migration that humans had followed out of Africa. The HGDP's proponents also argued that the data they collected could be used to find genetic markers of disease that might ultimately be used to create population-specific tests and treatments.

The HGDP attracted the initial support of the Human Genome Organization, the National Science Foundation, and the National Institutes of Health. Beginning in 1993, however, things rapidly began to unravel. Following some criticism of the project from a group of physical anthropologists, the HGDP came under fire from indigenous leaders around the world. The Third World Network used the burgeoning World Wide Web to raise awareness of the project, arguing that it violated the human rights of indigenous peoples, making them subjects of scientific research and "material for patenting." The 1993 World Congress of Indigenous Peoples labeled the HGDP a "vampire project" and encouraged many indigenous groups to sign petitions and declarations condemning the project. The main concern of indigenous leaders was that the project could lead to a "genetics of race"—it could lead to new forms of discrimination based on genetic testing.

Alongside this, it was not clear how the HGDP would secure the informed consent of populations for extracting their DNA (indeed, it was unclear what truly informed consent would even mean in this cross-cultural context), or how the HGDP would exert control over the data it produced. Wouldn't the data end up leading to patents that would benefit Western corporations?

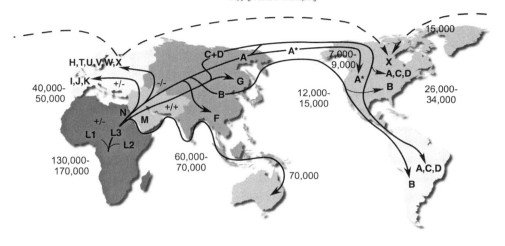

Human mtDNA Migrations

http://www.mitomap.org/pub/MITOMAP/MitomapFigures/WorldMigrations.pdf

Copyright 2002 © Mitomap.org

+/-, +/+, or -/- = Dde I 10394 / Alu I 10397
* = Rsa I 16329

Mutation rate = 2.2 - 2.9 % / MYR
Time estimates are YBP

22.1 Human family tree. Map of the world showing migration of human populations. The earliest transcontinental migrations (70–60 thousand years ago) populated Australia and parts of Asia. A further wave populated Europe some 50–40 thousand years ago, and successive waves of migration populated the more distant parts of Asia and the Americas by approximately 15–12 thousand years ago. Source: Mitomap. Available at: http://www.mitomap.org/pub/MITOMAP/ MitomapFigures/WorldMigrations.pdf. Used under CCA-Share Alike 3.0 license: http://creativecommons.org/licenses/by/3.0/. Modified from color to black and white.

Or, more fancifully, it could even be used to create population-specific bioweapons that could be targeted based on genetics. In any case, it seemed that once indigenous groups had given away their blood, they would have no control over what happened next.

Although the HGDP limped forward through the 1990s, it was essentially crippled by this controversy. However, research on human genetic diversity and variation has hardly ceased. In 2002, the National Human Genome Research Institute launched the $100 million International Haplotype Map Project (commonly known as HapMap). A haplotype is a set of several SNPs arranged consecutively along a chromosome. Tracking the frequency with which different haplotypes occur in different populations is a way of tracking human variation. The HapMap project initially took samples from 269 individuals from four populations (Yorubans in Nigeria, US residents of European ancestry, Japanese from Tokyo, Chinese from Beijing) and tracked millions of SNPs. The HapMap successfully released its data in 2007 and then expanded to include samples from over one thousand individuals. Since the end of the HGP, numerous other projects to map human diversity have sprung up. One of the most prominent is the Genographic Project (begun in 2005), led by Spencer Wells and funded by National Geographic. Moreover in 2008, an international team of geneticists launched the 1000 Genomes Project, aiming to completely sequence the genomes of more than one thousand individuals and create a massive amount of data for studying human variation.

All of these projects, including the HGDP, were undertaken with noble intentions towards indigenous peoples and developing nations. The projects aimed to make genomics more inclusive and to extend the medical benefits of its inventions to a wider and more diverse human population. Yet there have been continued and renewed objections from indigenous leaders. In 2005, the Indigenous Peoples' Council on Biocolonialism released a statement criticizing the Genographic Project and calling for a boycott of IBM, Gateway Computers, and National Geographic (as sponsors of the project):

The Genographic Project is exploitative and unethical because it will use Indigenous peoples as subjects of scientific curiosity in research that provides no benefit to Indigenous peoples, yet subjects them to significant risks. Researchers will take the blood or other bodily tissue samples for their own use in order to further their own speculative theories on human history.

Indigenous peoples are concerned that the Genographic Project will discount Indigenous knowledge, oral histories, and undermine our human rights. In fact, their informed consent form states: "It is possible that some of the findings that result from this study may contradict an oral, written, or other traditional held by you or by members of your group." Despite the speculative nature of genetic research on human histories, these findings could be used to undermine "indigenousness" or "aboriginality" of Indigenous peoples and our rights as the original inhabitants of our territories. Such theories, carrying the weight of Western science, could be used to undermine our human rights to our territories and jeopardize our unique political status. Indigenous peoples oppose this kind of research because our creation stories and languages carry information about our genealogy and ancestors. We do not need genetic testing to tell us where we come from.[3]

The main worry of the Indigenous Peoples' Council on Biocolonialism is that genomics will have important legal, political, and economic consequences. Indigenous peoples' rights to land and special status are based on their cultural beliefs. The alternative stories and histories constructed with genomics, they realized, would contradict their own. The power of science was such that genomic stories could be used to undermine the claims of indigenous people to land and special status.

Yet the truth of these "biological" histories is constructed on the legacy of colonialism. For one thing, projects that analyze indigenous DNA often require the existence and persistence of "pure" populations or groups. Such groups were often constructed by colonial authorities in the first place. By excluding indigenous populations from certain areas, by collecting them on reservations, and by grouping people together for purposes of administration and control, colonial authorities often ensured that groups remained isolated. In other cases, different groups were forced together in order to fit into colonial categories. Groups remained "unmixed" (and thus ripe for genetic analysis) because they were forced to be so.

The results of the DNA analysis are then used to reinforce ideas of difference and distinct origins. This is a form of circular reasoning that shows the difficulty of attempting to separate the biological and the sociopolitical origins of groups. To put this another way, consider the problem of deciding which groups or individuals are selected for study in the first place—what (or who?) defines a tribe (or a group) in the first place? Where are the boundaries of the group? Who is counted as a member of a group and who is not? (And who decides?) You can't use the DNA to answer these questions, since to start out with you don't know which genetic markers define the population—that is what you are trying to find out!

Instead biologists need to fall back on social, political, and historical categorizations (for example, you might use traditional definitions like "if at least 3 of your 4 grandparents belong to X, then you belong to X"). Practically, the

<hr>

3. Global Indigenous Caucus, Buffalo River Dine Nation, International Indian Treaty Council, Indigenous People's Council on Biocolonialism, "Collective Statement of Indigenous Organizations Opposing 'The Genographic Project.'" Fifth session, UN Permanent Forum on Indigenous Issues, New York, May 15–20, 2006. http://www.ipcb.org/issues /human_genetics/htmls/unpf5_collstate.html.

selection and definition of groups often depends on the historical circum-stances in which specific populations were separated or segregated in the colonial context. Yet which groups are selected, and how they are defined, is likely to be critical in determining the outcome of the DNA analysis: choose groups in a different way, and your study is likely to find vastly different re-sults.

The results of DNA testing are inextricably bound up with the social and political histories of the groups being tested. As such, collecting, analyzing, and comparing DNA from human populations are political actions with wide-ranging global consequences. Here genomics affects not only the mean-ing of "race," but also indigenous peoples' rights to their own land, culture, and blood. The dependence of such work on colonial categories and histories makes such science particularly problematic. Biologists' misunderstanding of this wider significance of their work is part of the reason for the continuing difficulties encountered by many of these projects.

FROM COLONIALISM TO BIOCOLONIALISM

The projects to study human diversity are also a powerful reminder of the historical connections between biology and power. The objections of indigenous peoples and developing nations to the HGDP and its successor projects, and the accusations of "piracy" and "vampirism," can only be fully explained by examining the historical relationship between Western science and non-Western lands and cultures. In short, this is a history deeply en-twined with colonialism. The suspicion that existed between indigenous peoples on the one hand and Western scientific institutions on the other was partly caused by colonial exploitation. Why would developing nations now trust institutions that had participated and collaborated in the imperial enter-prise?

To appreciate the significance of this history, it is worthwhile briefly de-scribing the long and close relationship between biology and colonialism. Colonialism might be loosely defined as the exploitation of the resources of one nation by another. In the sixteenth century, the Spanish *conquistadors* in the New World hoped that these resources would be gold and silver. Indeed, some such treasures were unearthed. However, the most valuable resources turned out to be the living ones: the bodies of indigenous people that per-formed labor and were traded as slaves, and the vast botanical resources of the conquered territories. These included new food crops such as maize, toma-toes, potatoes, sweet potatoes, and squash; special medicinal plants such as tobacco, chocolate, cinchona (used to make quinine), and coca; plants for the manufacture of special materials such as rubber, yucca, and agave (for twine); and of course spices (cloves, cinnamon, peppers, cardamom, turmeric).

Making the best use of this flora, however, was no easy task. These plants had to be discovered, improved, adapted to new climates and conditions, and studied in order to find the best methods of cultivation. This was a major part of the colonial enterprise and natural history was the Big Science of colonial times. Botanists were employed to seek out, classify, and study plants for the sake of empire. The primary sites of this work—the greatest laboratories of the day—were the botanical gardens. Upon his arrival at Singapore in 1819, one of Sir Stamford Raffles' first tasks was the founding of a "Botanical and Experimental Garden" on Government Hill. The purpose of the garden was to evaluate possible cultivation crops that might become economically important either for food production, spices, or other raw materials. Other gardens already existed in important colonial centers like St. Vincent, Calcutta, and Rio de Janeiro. These gardens regularly exchanged specimens and information with botanists working in the metropolises of Europe, such as those at Kew near London and the *Jardin des Plantes* in Paris. The result was a colonial botanical network that proved crucial to the maintaining the prosperity of the empire. In 1877, for instance, Henry Ridley, a naturalist working at the Singapore Botanical Gardens, received rubber plant seedlings from Kew. The rubber plant had originated in Brazil, but Ridley's experiments adapted it for cultivation in South-East Asia. During the 1880s, Ridley introduced rubber to large parts of British Malaya, which subsequently became the world's largest producer and exporter of rubber.

The colonial enterprise was fundamentally dependent on biological resources and the use of highly funded and highly organized scientific enterprises to manage and exploit them. Biotechnology also relies on gathering biological resources from around the globe and putting them to economic uses. Does this make biotechnology a neocolonial enterprise?

BIOPROSPECTING AND BIOPIRACY

The blood and DNA of indigenous people are not the only things biologists are after. Many of the products of biotechnology are derived from biological specimens collected across the globe. In the pharmaceutical industry in particular, those searching for new drugs are well aware that the next blockbuster might be lurking in a seed in the Peruvian jungle or in a seaweed off the coast of Borneo. The effort to collect such potentially valuable biological samples is known as *bioprospecting*. Biotech relies heavily on the world's biodiversity. Indeed, bioprospecting has been defended as a way of protecting the biodiversity. Collecting specimens from rainforests or coral reefs and bringing them into centralized gardens, zoos, laboratories, repositories, and databases allows them to be protected when their specialized habitats come under threat.

Of course, not all such scientific spaces are open equally to everyone. And sometimes such collection can result in patents issued on biological products derived from these samples. In some such cases, critics have labeled this *biopiracy*: the biological resources of one region have been extracted and exploited by a foreign party (often European or North American scientific institutions or corporations).

One of the best examples of biotech's dependence on biological resources can be found in the use of an enzyme called *Taq polymerase*. Polymerase is an enzyme that is needed for the copying of DNA—it attaches to a single strand of DNA and replicates it into a new double strand (see box 3.1). Taq polymerase is the polymerase extracted from an organism called *Thermus aquaticus* that lives only in high-temperature underwater geothermal vents. Because it lives in this extreme environment, its polymerase has adapted to function at high temperatures (unlike the polymerases found in humans or other animals). This feature makes Taq polymerase the preferred polymerase for use in PCR (the polymerase chain reaction), perhaps the most ubiquitous tool in modern biochemistry, used for amplifying DNA for forensics, sequencing, and disease diagnostics. *Thermus aquaticus* was discovered by researchers in Yellowstone National Park (Wyoming) and samples deposited in a public repository. In the 1980s, the inventors of PCR at the Cetus Corporation were able to acquire Taq and put it to commercial use without paying any royalties to the discoverers or the US National Parks Service. Who has the right to benefit from discoveries arising from such natural products?

In 1992, the United Nations sought to aid in the conservation of biodiversity by regulating the sustainable use of biological resources. The resulting Convention on Biological Diversity, signed at the Earth Summit in Rio de Janeiro in 1992, recognized the need to share the benefits arising from the commercialization of genetic and other biological resources in an equitable way. The Convention paid special attention to the role of traditional knowledge in bioprospecting. Many pharmaceutical companies, in particular, realized that indigenous knowledge of plants and animals, often enshrined in traditional remedies or rituals, was an important way to discover potential drugs. *Ethnobotanists*, often with training in both biology or medicine and anthropology, have ventured into indigenous societies in the hope of tapping into their valuable knowledge about local plants and animals. In such cases, the Convention on Biological Diversity described rules for how traditional knowledge could be used (including provisions for informed consent), for benefit sharing with indigenous people, and for transfer of technology and resources back to local governments and communities.

Nevertheless, biologists, biotechnology companies, and pharmaceutical companies have continued to provoke accusations of biopiracy from indigenous people and developing countries (for a partial list of such incidents see box 22.1). In 2003, the biotech entrepreneur Craig Venter set sail on his yacht, Sorcerer II, sampling marine microorganisms from the world's oceans. Venter sent samples of seawater back to his labs at Institute for Biological Energy Alternatives in Maryland where biologists searched for marine genes that might be used as biofuels or other alternative energy sources. Venter saw such bioprospecting as a way to contribute to saving the planet. Although Venter had signed memoranda of understanding with several nations in South and Central America, groups representing local and indigenous people (especially those assembled at the first Americas Social Forum in July 2004) objected to Venter's sampling in national territorial waters of Ecuador, Bermuda, Costa Rica, Panama, and Chile. Noting that the United States had not ratified the Convention on Biological Diversity, these groups felt that Venter's negotiations with their governments had been underhanded and violated the peoples' rights to biological resources. "Venter's microbe hunting expedition raises serious unanswered questions about sovereignty over genetic resources and resource privatization through patenting," one group wrote.[4]

In another example, between 1993 and 2000 the US National Institutes supported a series of bioprospecting projects in Mexico. These fell under the auspices of a project for drug discovery and sustainable development called the International Cooperative Biodiversity Groups program. Plant extracts were sent from Mexico to the University of Arizona and to the US-based life sciences firm Cyanamid. Under agreements between the United States and Mexico, if a successful drug emerged from this collection and research, royalties of between 2%–5% would be paid to the University of Arizona. About half of this amount would be remitted to Mexico, where it would be distributed amongst contributing institutions and participating communities.

In practice, the agreement proved immensely impractical. The main problems concerned the appropriate definition and boundaries of communities: in the field, people, territories, and plants overlapped in complicated ways.

4. The US has signed but not ratified the Convention on Biological Diversity. See http://www.etcgroup.org/fr/node/91 for quote.

Box 22.1 Some Documented Cases of Bioprospecting

Basmati rice
- Approximate date: Attempted patent in 2000
- Origin: India
- Uses: RiceTec attempted to patent hybrids of basmati rice

Brazzein berries
- Approximate date: 1994
- Origin: Gabon, West Africa
- Uses: University of Wisconsin obtained three patents on proteins isolated from berry for use as sweetener

Hoodia
- Approximate date: 1996
- Origin: San people, Kalahari desert, South Africa
- Uses: appetite suppressant; developed by South Africa's Council for Scientific and Industrial Research for dietary supplements

Neem tree
- Approximate date: Patent granted 1995
- Origin: India and Nepal
- Uses: Patented by US Department of Agriculture and W. R. Grace as insecticide; patent overturned in 2005

Pozol (drink)
- Approximate date: 1999
- Origin: Mayans in Mexico
- Uses: developed by Quest International and University of Minnesota as antimicrobial

Quinoa
- Approximate date: 1994
- Origin: Bolivia
- Uses: Colorado State University obtained patent as food crop; patent later abandoned under pressure from indigenous groups

Turmeric
- Approximate date: 1995
- Origin: India
- Uses: University of Mississippi Medical Center obtained patent for use for wound healing; patent was later canceled

Source: The United Kingdom Select Committee on Environmental Audit 1999; Appendices to the Minutes of Evidence, Appendix 7: Trade Related Intellectual Property Rights (TRIPs) and Farmers' Rights.

What if two or more different communities used the same plant? Should both be compensated? What if the plant in question was utilized by one community but its habitat extended into the land owned or occupied by another? Did the second group have some claim over the plant too? And who had the right to speak for or negotiate on behalf of communities anyway? Ethnobotanists soon gave up trying to solve these problems and began purchasing plants from vendors in cities. This was then considered a straightforward commercial transaction for which the plant seller was not entitled to any further compensation. Negotiations with community organizations were conducted separately.

In a similar set of incidents in 2002, the Diversa Corporation (from San Diego, California) began searching the ocean near Hawaii for microbes of commercial value. Diversa was one of the most prominent bioprospecting companies, with most of their business directed towards seeking biological resources for biotech and pharmaceutical uses.[5] Once again, groups representing indigenous Hawaiians objected to this "biopiracy" and argued that Hawaiians should be entitled to 20% of the profits from any material collected. A bill to this effect was put before the Hawaiian Congress. Although it failed to pass, these examples show the growing political and economic stakes for biotechnology.

Behind these examples is a set of thorny problems concerning the relationship between indigenous rights, sovereignty, scientific knowledge, and biotech innovation. If, for example, bioprospectors are compelled to share their profits, this reduces the incentive for investment, innovation, and exploration; fewer drugs and products may end up being developed. And if rainforests and other delicate ecosystems continue to disappear, the opportunity for discovery may be lost forever. Even in cases where the profits are shared, determining an appropriate balance may be tricky. A mushroom or a flower from a forest is not itself a biotech product: years of research are often required to perform extraction, purification, testing, and (for drugs) clinical trials. Investors expect a handsome return for such work.

Even more problematic are cases (like Taq polymerase) where samples from bioprospecting are first placed into public repositories or databases for general scientific use. Many such samples are collected for the purposes of monitoring biodiversity and global warming. They may be developed into commercial products only at a later time by a third party. Are product developers still obliged to share their profits with indigenous owners in these cases? Should they also share with the bioprospectors responsible for the ini-

5. In 2007, Diversa merged with Celunol to form Verenium, which works on biofuels.

tial extraction? These situations give rise to complex and tangled claims about rights and ownership.

Many other aspects of present-day biology and biotechnology rely on the extraction of biological resources. Whether it is the blood of indigenous people or an enzyme from a bacteria or a drug made from a flower found in the Amazon jungle, the raw materials of biotechnology are the flora and fauna of lands and oceans. Although there are many differences between the story of Malaya's rubber industry and the story of PCR, a few similarities are worth highlighting. First, both involve the transplantation of a biological resource from one context into another. Second, both resulted in massive commercial success (Cetus sold the rights to PCR for $300 million). Third, no financial benefits were received by those closest to the "origins" of the product. Such similarities at least raise the possibility that biotechnology runs the risk of replicating or falling back into old colonial modes of exploitation. If biological samples are considered "free"—to be collected anywhere, by anyone, without restriction—and if most or all of the benefits (commercial and medical) of the subsequent inventions remain available only to small groups (Western drug companies, for instance), then the risk of a kind of neocolonialism is increased. All this is certainly not to say that biotechnology is inherently or inevitably colonial. But biotechnology is dependent on global power relationships—between weak and strong nations, and between wealthy corporations and poor farmers. The forms that biotech takes, the objects or drugs that it is able to produce, depend on these relationships. Whether biotech ultimately becomes neocolonial depends on our ability to understand these power relationships and take steps to ensure they do not lead to exploitation.

CONCLUSIONS: BIOTECHNOLOGY AND GLOBALIZATION

Biotechnology is not just something that takes place in labs, but is immersed in the world. Biotechnology is not just science, but a global enterprise that intersects with crucial political, social, and economic questions and problems. Scrutinizing the collection of DNA samples or the extraction of drugs from plants reveals that these activities are dependent upon particular global distributions of knowledge, power, and wealth. In other words, biotechnology is closely bound to the phenomenon known as globalization. Biotech can be understood both as a product (effect) and vector (cause) of globalization. It would not be possible without global flows of information and capital and it is also contributing to these flows (not just with respect to bioprospecting, but also in the organization of global clinical trials, the sharing of biological information in databases, and so on). It is these global networks of power—the ties between biotech, governments, and corporations—that allow scientists like Spencer Wells to pursue their work and to make their stories compelling. Through biotech, economic and political power translates into the power to make histories.

Globalization, like biotech, has the potential to bring a great many benefits to a great many people. It promises to provide more people with access to markets, education, news media, and ultimately, wealth. But it also entails tremendous risks—the homogenization of culture, the increasing empowerment of multinational corporations, and a growing divide between rich and poor. Biotechnology shares in these risks: it is not value-free or culture-free and brings with it the values and politics of Western science and technology; it is often driven by corporate interests; and many of its benefits are likely to be disproportionately enjoyed by wealthier people and nations. For the Navajo, accepting Wells' stories not only might mean an erosion of their own cultural narratives, but also could entail losses of rights to land and resources.

The question of whether collecting plants in Mexico or blood samples in Africa is good or bad—whether it is "bioprospecting" or "biopiracy"—depends on our ability to find ways of equitably distributing the benefits of biotechnologies to different communities around the world. This is not an easy task: it will mean finding ways of recognizing different kinds of work and different kinds of knowledge (donation of blood, participation in a clinical trail, or knowledge of local plants) in the production of biotechnologies.

FURTHER READING

The documentary film based on the work of Spencer Wells referred to in the introduction is *Journey of Man* [documentary] Dir. Clive Maltby (PBS, 2003). Wells' book (on which the documentary is based) is Spencer Wells, *The Journey of Man: A Genetic Odyssey* (New York: Random House, 2002). The work of Wells, as well as other attempts to categorize indigenous groups according to DNA, has been criticized by Kim Tallbear, *Native American DNA: Tribal Belonging and the False Promise of Genetic Science* (Minneapolis: University of Minnesota Press, 2013).

The most comprehensive effort to catalogue human genetic diversity was undertaken by Luca Cavalli-Sforza and compiled as L. Luca Cavalli-Sforza, Paolo Menozzi, and Alberto Piazza, *The History and Geography of Human Genes* (Princeton, NJ: Princeton University Press, 1994). Cavalli-Sforza went on to organize the Human Genome Diversity Project, the history of which is comprehensively detailed by Jenny Reardon, *Race to the Finish: Identity and Governance in an Age of Genomics* (Princeton, NJ: Princeton University Press,

2004). On this project see also Amade M'Charek, *The Human Genome Diversity Project: An Ethnography of Scientific Practice* (Cambridge: Cambridge University Press, 2005).

There is a diverse historical literature on the importance of plants, agriculture, and botanical gardens in the colonial world. One of the best accounts is Lorna Schiebinger, *Plants and Empire: Colonial Bioprospecting in the Atlantic World* (Cambridge, MA: Harvard University Press, 2007). For more on Henry Ridley and his work at the Singapore botanical gardens see E. J. Salisbury, "Henry Nicolas Ridley 1855–1956," *Biographical Memoirs of Fellows of the Royal Society* 3: 141–159.

For various recent examples of bioprospecting see the work of anthropologists Cori Hayden and Stefan Helmreich: Cori Hayden, "Taking As Giving: Bioscience, Exchange, and the Rise of an Ethic of Benefit Sharing," *Social Studies of Science* 37, no. 5 (2007): 729–775; Cori Hayden, *When Nature Goes Public: The Making and Unmaking of Bioprospecting in Mexico* (Princeton, NJ: Princeton University Press, 2003); and Stefan Helmreich, *Alien Ocean: Anthropological Voyages in Microbial Seas* (Berkeley: University of California Press, 2009). This last discusses the work of Diversa in Hawaii as well as the voyages of Craig Venter. For more on Diversa see Cormac Sheridan, "Diversa Restructures," *Nature Biotechnology* 24 (2006): 229. On the relationship between bioprospecting and synthetic biology see Alain Pottage, "Too Much Ownership: Bio-prospecting in the Age of Synthetic Biology," *Biosocieties* 1, no. 2 (2006): 137–158.

PART XII
BIOLOGICAL FUTURES

SYNTHETIC BIOLOGY AND BIOTERRORISM

INTRODUCTION: THE MECHANISTIC CONCEPTION OF LIFE

The final chapters of this book are about biotechnology and the future. Where is biotech going? And where is it likely to take humanity? These questions will be discussed explicitly in the conclusion. In this chapter, we will examine some recent developments in biotechnology in order to get some insight into what might happen next. Chapter 24 examines the use of biotech in art—these creative uses suggest alternative possibilities for the future.

Many of the biotechnologies in this book entail a very specific view or understanding of life. We have already seen that much of biotechnology is based on *reductionist* ideas about life: that is, it is premised on the notion that life is fundamentally *molecular* and that understanding life requires understanding it on a submicroscopic level. But many biotechnologies also entail a *mechanistic* view of life. As the word itself suggests, this is the idea that life works something like a machine: it has moving parts that can be taken apart, modified, and put back together again.

This might also be called an engineering view of life. Genetic engineering is the process of taking organisms, modifying their parts, and setting them going again. Cloning, stem cells, and pharmaceutical technologies all intervene in organic systems in mechanical ways—they block, interlock, turn, open, close, and rearrange molecular parts in order to stop, start, and control. The notions of "life as a machine" or "engineering life" work as a *metaphor*. Even if biologists don't literally believe that a frog works in the same way as car or a computer, this metaphor suggests particular ways of tinkering with and thinking about organisms. Synthetic biologists, in particular, use the machine metaphor as a powerful way of explaining and making sense of what they are doing.

This chapter explores the possible consequences of what happens when we take this mechanistic or engineering metaphor seriously. What happens when we try to build biological parts that can be manipulated as easily as Lego blocks?

Mechanistic biology now seems so obvious to us that it is almost impossible to imagine a sensible alternative. What else could life be except a kind of gooey machine? How else could it work except by having systems of moving parts? But this view is, in fact, a rather new one, peculiar to the twentieth and twenty-first centuries. Before the nineteenth century, many natural philosophers believed that biological things were governed by special sorts of laws. Machines — steam engines, pulley systems, pumps — obeyed the laws of physics and chemistry. Living things seemed very different from such machines. For one, they seemed much more complicated. For another, it seemed hard to explain how you would get conscious and self-aware behavior out of a mere set of gears and levers.[1] Living matter, therefore, was seen as governed by special *vital* forces, distinct from the laws of physics and chemistry. This set of ideas was known as *vitalism*.

During the nineteenth century, the vitalist view of life came under increasing criticism and pressure. In 1828, Friedrich Wöhler (1800–1882) synthesized the organic substance urea from nonorganic components. This undermined, but did not destroy, the vitalist idea that there was a clear distinction between the organic and the nonorganic.

Indeed, vitalist ideas proved remarkably persistent in the twentieth century. One individual who was violently opposed to vitalism, however, was Jacques Loeb (1859–1924). Loeb was born in the German-speaking lands and was trained as a medical doctor in Strasburg. In 1892, he emigrated to the United States, joining the faculty of the University of Chicago as a professor of physiology and experimental biology. Over the next three decades, Loeb conducted a series of remarkable biological experiments that made him one of the most famous scientists in America.[2]

In 1910, he was appointed to the prestigious Rockefeller Institute for Medical Research in New York (where he worked alongside Alexis Carrel). While based in New York, Loeb spent his summers at the Marine Biological Laboratory in Woods Hole, Massachusetts. Here, he conducted experiments on sea urchin eggs. Loeb took unfertilized eggs and treated them with various concentrations of acid and salt water. Eventually, he managed to induce the eggs

1. This is a problem that we still have not solved. Even though we now think of the brain as an electrical system, philosophers, biologists, and neuroscientists still do not understand how it is possible to explain conscious states or self-awareness in terms of circuits and currents.

2. Loeb was the inspiration for the protagonist in Sinclair Lewis' 1925 novel *Arrowsmith*.

to begin to divide, just as if they had been fertilized by sea urchin sperm. For Loeb, this "artificial parthenogenesis" (as he called it) showed that fertilization was essentially a chemical process. There was no vital force or property transferred to the egg by the sperm. "Life," Loeb wrote in 1911, "can be unequivocally explained in physico-chemical terms."[3]

This was the larger theme of Loeb's work: to show that there was no such thing as a vital force and that all of biology could be explained with physics and chemistry. But, if life could be understood as physics and chemistry, it could also be manipulated by physics and chemistry: Loeb's work also aimed to *control* living things. Loeb sought to build what he called "technologies of living substance." He wanted to be able to take life apart, modifying it at will, and put it back together again. Many of his experiments did just this: his techniques of "parabiosis" involved cutting up invertebrate animals, modifying and adding parts, and stitching them back together. His two-headed worms, hydras with mouth and anus reversed, and reengineered slime molds attracted wide public attention.

Loeb thought that his physical and chemical explanations even extended as far as human nature. In a famous paper, Loeb explained the flight of a moth towards a flame: the moth could be said to fly towards the flame because it "desires" to do so. But, Loeb argued that the impact of light on one side of the body of the moth caused chemical changes that induced the moth to change direction and fly toward the flame. "Desire" could be explained in physicochemical terms. Various human desires and actions could be explained in similar—although perhaps more complicated—terms, Loeb believed. And if the biology and behavior of worms and moths could be controlled, so could the biology and behavior of human beings in society.

WHAT IS SYNTHETIC BIOLOGY?

Loeb's mechanistic attitude, in various guises, has pervaded much of twentieth-century biotechnology. In the early twenty-first century, however, the engineering ideal has emerged in a particularly strong and explicit form. The aim of synthetic biology is to be able to redesign and reengineer life forms for human purposes. This engineering has two aims. First, biologists hope that the process of designing and engineering organisms will teach them something about how biology works. Just as an aspiring mechanic might learn about an internal combustion engine by taking it apart and putting it back together again (or trying to), biologists hope to learn how life works by

---

3. Jacques Loeb, *The Mechanistic Conception of Life: Biological Essays* (Chicago: University of Chicago Press, 1912). Quotation p. 3.

deconstructing and reconstructing it. Second, bioengineers hope to create economically, socially, environmentally, and medically useful organisms. The boosters of synthetic biology imagine its engineered organisms put to work synthesizing drugs, cleaning up toxic spills, making fuels, and manufacturing chemicals or even foods.

The ideals of synthetic biology have much in common with Loeb's notion of "technologies of living substance"—both Loeb and synthetic biologists aim at control over living things. Of course, synthetic biology boasts a more sophisticated toolkit—it puts to use many of the techniques that we have seen in this book: recombinant DNA, DNA sequencing, and DNA synthesis. What distinguishes synthetic biology from these other biotechnologies is not the tools, but rather the *attitude* that it takes towards life. For synthetic biologists, life *is* technology, not just to be tinkered with, but also to be designed and built from scratch. Adding, removing, or altering one or two genes does not go far enough—synthetic biology aims to design whole biological systems made of DNA, RNA, and protein.

In 2007, for instance, Craig Venter and Hamilton Smith set out to build an organism from the ground up. Beginning with the bacteria *Mycoplasma genitalium*, Venter and Smith's team systemically removed each of the organism's 475 genes one by one in order to determine which of them were absolutely essential for life. The team called the resulting set of 382 "essential" genes the "minimal genome." Venter and Smith then aimed to recreate this minimal genome from scratch using DNA synthesis and implant it into a bacterial cell. This would, the researchers claim, represent a new species that they have dubbed *Mycoplasma laboratorium*.[4] This will be a substantial step towards building whole organisms from scratch in the laboratory. Venter commented that this would represent the first species "to have a computer as a parent."[5]

Building whole organisms is the most ambitious aim. But in the shorter term, one of the goals of synthetic biology is to build a library of working biological parts. Based on the engineering principles of standardization, ab-

---

4. Venter and Smith's team has already demonstrated the possibility of building such a large genome by DNA synthesis. And they have demonstrated that they can successfully implant a synthetic genome in a host cell (they achieved this working with the related bacteria species *Mycoplasma mycoides* and *Mycoplasma capricolum*). The team, however, has encountered some difficulties working with *Mycoplasma genitalium* and so have not yet succeeded in implanting their minimal genome.

5. J. Craig Venter, *Life at the Speed of Light: From the Double Helix to the Dawn of Digital Life* (New York: Viking, 2013), 125.

Box 23.1 A Few Biobricks

pLux promoter (BBa_K091156)
- Designer: Erin Feeney
- If this part is placed next to a gene, that gene will be activated if a protein called LuxR is present.

Lead promoter (BBa_I721001)
- Designer: Jeffrey Hoffman
- Allows lead detection inside a cell; this promoter will turn on a gene if lead is present in the cell.

Odorant synthesis (BBa_J45004)
- Designer: Boyuan Zhu
- Makes a protein that converts benzoic acid to methyl benzoate (floral smell) and converts salicylic acid to methyl salicylate (wintergreen smell).

Cellulose degradation (BBa_K118022)
- Designer: Andrew Hall
- Makes a protein that degrades cellulose into a form of glucose.

Cyan fluorescent protein (BBa_E0020)
- Designer: Caitlyn Conboy and Jennifer Braff
- Makes a protein derived from *A. victoria* that fluoresces cyan (blue).

Bacteriophage lysis cassette (BBa_K124017)
- Designer: Neil J. Parikh
- When activated, produces a set of proteins that degrade the cell membrane and kill the cell (via lysis).

Dumbbell hairpin (BBa_J35002)
- Designer: Andrey Kuznetsov
- Produces double-stranded DNA in the shape of a dumbbell (DNA origami).

Source: Registry of Standard Biological Parts: http://partsregistry.org/Main_Page

straction, and decoupling, biologists and bioengineers imagine the parts as being interchangeable, plug-and-play pieces that can be snapped together to construct different kinds of organisms. These *biobricks*, as they are sometimes known, are stored in an online repository called the Registry of Standard Biological Parts (http://partsregistry.org). Anyone with an Internet connection can go to the Registry and browse the parts: there are parts that produce odiferous proteins, parts that produce plastics, parts that kill cells, parts that make DNA origami, parts that turn genes on and off, and lots of parts that help other parts to work properly (see box 23.1).

Many of these parts have been designed by undergraduate students who compete at the annual iGEM (International Genetically Engineered Machine) conference held at the Massachusetts Institute of Technology. In 2004, the Registry contained approximately one hundred parts and in 2013 it contained over twenty thousand parts. Each part is essentially a DNA sequence. To design an organism, you can pick your parts from the Registry and connect them together by joining the pieces of DNA sequence in the right order (by cutting and pasting them on a computer). Once you have a complete sequence built from the various parts you can synthesize the DNA (this can be done by uploading your sequence to specialized DNA-synthesis companies and paying a fee) and implant it into a cell (usually a bacterium).

Bioengineers often compare the Registry to Lego bricks. But another analogy that is often used compares biobricks to programming a computer. In the early days of electronic computing, it was very difficult to write a program. You had to understand the language of the machine and basically input your instructions in zeros and ones. Eventually, however, programmers invented programming languages and compilers—you could write your instructions (or program) in a more straightforward way and let the compiler translate that language into the machine language. Over time, programming languages got easier and easier to use and almost no one needed to understand how to write instructions in machine language. Bioengineers see the Registry as a kind of programming language for biology: using the biobricks will make it much easier to design and build organisms, without needing to speak the language of DNA.

Making working biological parts turned out to be far harder than many synthetic biologists expected. Building organisms from these parts is complicated by the fact there are multiple biobrick standards. On the other hand, advances in DNA synthesis technology (such as Gibson assembly) have made it easier and cheaper to build up long strands of DNA from individual parts. Bioengineers remain hopeful that designing organisms will one day be like software engineering: collaborative, relatively easy to learn, inexpensive, and ubiquitous.

DIY BIOLOGY

The challenges of building new organisms have already been taken on by nonbiologists around the world. Since about 2005, a do-it-yourself (DIY) biology movement has spread from San Francisco and Cambridge to Europe, Australia, and Asia. Students and amateurs with an interest in biology have purchased secondhand lab equipment, built their own instruments, shared ideas online, and rented spaces to set up makeshift labs. DIY biologists make use of whatever they can get their hands on to do their experiments. Although not all DIY biology is aimed at building organisms (some is just for fun, some for education, some aimed at specific medical ends), synthetic biology has played a significant role in encouraging and enabling these activities.

Here too, developments in the world of software have been taken as a model. The growth of the World Wide Web in the 1990s allowed the spread of a worldwide "open source software" movement; programmers (both amateurs and professionals in their spare time) contributed to building large-scale software projects for free (the Linux operating system is the most famous example). The software that resulted was also "open": free to use and free to change. Along with synthetic biologists (who also drew on open source ideas in building their biobrick registries), the DIY biology community has attempted to extend this idea of openness to biology. Just as the open source software movement was attempting to provide an alternative to the "closed" software of Microsoft and other big corporations, open source biology aims to resist the "closed" biology of the biotech industry. The collaborative, open ethos of DIY biology is premised on the idea that biological information, knowledge, and techniques should be free, unlimited, and without ownership. Some DIY biologists even call themselves "biohackers" in analogy with computer hackers who attempt to force open computer systems and software.

BIOTERRORISM AND RISK

Such activities have made some onlookers very nervous. The same factors that make biology more open to amateurs and DIY-ers also make it more open to abuse. For instance, websites like the Registry make sophisticated biological parts available to almost anyone. These parts might be synthesized by simply ordering them online via a DNA-synthesis company (the sequences can usually built within a few days and shipped as a DNA plasmid). A marketplace for cheap instruments and the circulation of DIY knowledge online brings the capabilities for building organisms within the reach of more and more people.

This raises three distinct kinds of risks. First, there is the possibility of accidental release. This was exactly the kind of problem considered in 1975 at the Asilomar Conference (then in response to recombinant DNA experiments—see chapter 4). However, in the 1970s, genetic engineering was confined to corporate and academic laboratories. DIY biology made possible by synthetic parts means that potentially harmful organisms may be created outside traditional lab spaces. Such amateur labs may exercise fewer safety precautions—DIY-ers may not know the risks and their backyard facilities are unlikely to include sophisticated containment technologies (moreover, these labs are more likely to be in urban and suburban buildings where people live). In any case, garage labs will be far more difficult to monitor and regulate.

The second risk emerges from a scenario in which a synthetic organism is deliberately released into the environment with good intentions. For instance, synthetic bacteria might be utilized to clean up a toxic spill. Such bacteria may have unintended and unforeseen side effects when placed into a complex ecosystem. Such risks might be mitigated through extensive testing of a synthetic organism, but without actually observing its behavior in the open, it may be difficult to predict all of its possible effects. Moreover, the risks with wholly synthetic organisms may be greater than for other genetically engineered life forms. Recombinantly modified *E. coli* or genetically modified corn or soy beans are still *E. coli* or corn or soy beans; they may have one or two genes altered but they remain largely similar to their wild ancestors, whose behavior we can mostly understand and predict. Synthetic organisms, as wholly new creations, may be more unpredictable in their interaction with the environment.

Finally, there is the possibility of deliberate and malicious misuse. Governments and antiterrorism agencies worry that rogue individuals or groups may be able to fabricate synthetic organisms that are pathogenic to humans or to animals or plants that we rely on for food. The concern is that synthetic biology has dramatically lowered the levels of knowledge and sophistication that might be required to instigate such acts of bioterrorism. Moreover, there is the possibility that a bioterrorist could engineer a pathogen *more* infectious or deadly than those found in nature. Nevertheless, there is significant disagreement amongst both biologists and terrorism experts as to how difficult it would be to actually weaponize and use a pathogenic synthetic organism. The biologist Jef Boeke, for instance, has argued that the risks are "rather low"—that actually building and growing a substantial quantity of an organism would still require substantial infrastructure. Drew Endy, on the other hand, despite being a big supporter of synthetic biology, claims that it would be surprising if someone *didn't* manage to build a dangerous virus within the next twenty years.

Despite these disagreements, it is apparent that the more successful synthetic biology becomes in making biology open, inexpensive, and ubiquitous, the easier it will become for a single individual or a small group to engineer a potentially dangerous organism.

BIOSECURITY

Just a week after the September 2001 terrorist attacks on the World Trade Center, a series of letters containing spores of the anthrax bacteria were sent to the offices of media and political leaders in the United States. This incident raised fears about a possible bioterrorist attack. In response, the US Patriot Act (2001) made it a crime to possess harmful biological agents, toxins, or delivery systems. In 2002, the Agriculture Bioterrorism Protection Act identified a list of "Select Agent" organisms that required background checks and registration for use.

Such laws fall under the rubric of "biosecurity." Biosecurity aims both to prevent harm from overt acts of bioterrorism, as well as to mitigate the dangers posed by the spread of contagious diseases. Outbreaks of avian influenza and swine influenza in the 2000s have also increased awareness of the need for protective measures against human and animal pathogens. Biosecurity encompasses quarantine and inspection; the creation and enforcement of guidelines for agriculture, transport, and biological research; planning and preparedness for epidemic outbreaks; as well as research into countermeasures such as vaccines and antiviral drugs. The large-scale global flows of people and goods (especially foods) pose a challenging set of biosecurity risks and problems. Some of these can be mitigated by physical security measures—guards, gates, and guns—but others require more subtle and sophisticated methods.

For instance, advocates of synthetic biology have pointed out that at the moment it would be far easier to steal a human pathogen from a biological lab than to bother synthesizing it; a potential bioterrorist could walk into an unsecured academic lab and simply take a virus out of the freezer. So biosecurity is impossible without physical security. Nevertheless, as DIY labs and synthetic biology proliferate, putting locks on freezers will not be sufficient: a potential pathogen could exist as a DNA sequence created on a computer and be shared over the Internet. Biosecurity increasingly requires the regulation and monitoring of *information*.

This has led to an intricate set of debates about how to regulate synthetic biology. For the most part, synthetic biologists and DNA-synthesis companies have taken the initiative to implement biosecurity measures. No doubt they realized that the specter of bioterrorism could threaten their entire enterprise; if they were seen to be doing nothing to address the risks, researchers could find their funds cut off or companies could be forced out of business. Although synthetic biologists have spent a lot of time discussing the risks, no formal regulations or guidelines for research have emerged. Likewise, journal editors have proved reluctant to take action to enforce standards (for example, they could refuse to publish results from certain types of dangerous experiments).

For DNA-synthesis companies, the debate has focused on "screening" of customer DNA orders. It is possible to automatically screen orders against a database of known pathogens; if a match is found, the software will trigger an alarm. Many argue that computerized screening is insufficient and that a *human* expert should be employed. In 2008, the International Association for Synthetic Biology (IASB) developed a standard that did require human experts to screen incoming orders of synthetic DNA for potential threats. Although this standard met with some resistance, by the following year a large majority of DNA synthesis companies (including those in the United States, Europe, and China) had adopted the IASB code.

Governments have been slower to adopt strong standards or other biosecurity measures (such as licensing DNA synthesis companies). The US federal government, for instance, has adopted a standard that is weaker than the IASB and requires only computer screening against its Select Agent list. Debates about screening are further complicated by the fact that even the most sophisticated methods could overlook a dangerous pathogen—in the case of uncharacterized sequences, just looking at a series of As, Gs, Ts, and Cs does not allow a biologist to predict what the sequence might do when put into a cell.

These debates about synthetic biology suggest two important possibilities for the management of risk and the regulation of biotechnology in the future. First, they suggest the increasing entanglement of biotechnologies with state interests, especially national security and public health. All levels of government are going to need to be increasingly aware of the risks posed by biotechnologies and be increasingly prepared to intervene to mitigate and control those risks. Second, the debates suggest the entanglement of biotechnologies with global exchanges. Increasingly, biosecurity is a problem that depends on the worldwide management of the flows of people, goods, and information. Effectively regulating biotechnology in the future will depend on situating it in a global context.

CONCLUSIONS

The debates about biosecurity and synthetic biology also demonstrate that these issues have immense potential to affect all of us. As such, decisions about risk and regulation should not be left in the hands of scientists alone. These are public issues; decisions that are based only on scientific considerations or are driven purely by scientific curiosity are not likely to serve society's best interests. We need to find ways of assessing not just scientific costs, benefits, and risks, but also social, economic, and political costs, benefits, and risks. This need for expansive debate necessitates that philosophers, historians, sociologists, anthropologists, and economists all participate. Most importantly, however, the "open" ideals of DIY biology should inspire a wider debate: not only should the technical aspects of biology be opened up, but also its political and social implications should be communicated and debated within a broader public sphere. Although DIY biology may pose a risk, it also generates a rich set of opportunities for bringing a diverse set of interested and informed voices into the conversations about biotechnology.

FURTHER READING

On the history of vitalism and mechanism see William Bechtel and Robert C. Richardson, "Vitalism," in *Routledge Encyclopedia of Philosophy*, ed. E. Craig (London: Routledge, 1998). On the significance of urea to these debates see Peter J. Ramberg, "The Death of Vitalism and the Birth of Organic Chemistry," *Ambix* 47, no. 3 (2000): 170–195; and John Brooke, "Wöhler's Urea, and Its Vital Force?—A Verdict from the Chemists," *Ambix* 15 (1968): 84–114. On the persistence of vitalism into the twentieth century see Roger Smith, "Biology and Values in Interwar Britain: C. S. Sherrington, Julian Huxley and the Vision of Progress," *Past and Present* 178 (2003): 210–242.

Jacques Loeb's own ideas are contained in his work of 1912: Jacques Loeb, *The Mechanistic Conception of Life: Biological Essays* (Chicago: University of Chicago Press, 1912). For a biography of Loeb see Philip J. Pauly, *Controlling Life: Jacques Loeb and the Engineering Ideal in Biology* (New York: Oxford University Press, 1987). A history that places Loeb in the context of thinking about the plasticity of human life in the twentieth century is Rebecca Lemov, *World as Laboratory: Experiments with Mice, Mazes, and Men* (New York: Farrar, Straus and Giroux, 2006)—Loeb is discussed in chapter 1. The fictional account based on Loeb is Sinclair Lewis, *Arrowsmith* (New York: Signet Classics, 2008 [1926]).

For technical accounts of the creation of an artificial genome see John I. Glass et al., "Essential Genes of a Minimal Bacterium," *Proceedings of the National Academy of Sciences* 103, no. 2 (2006): 425–430; and B. Gibson, "Complete Chemical Synthesis, Assembly, and Cloning of a *Mycoplasma genitalium* Genome," *Science* 319, no. 5867 (2008): 1215–1220. Venter's more reflective account of his quest to recreate an organism from scratch can be found in J. Craig Venter, *Life at the Speed of Light: From the Double Helix to the Dawn of Digital Life* (New York: Penguin, 2013).

An important debate involving natural scientists and social scientists about the possible consequences of synthetic biology is reprinted here: J. Craig Venter, Sarah Franklin, Peter Lipton, Chris Mason, "Debate: Beyond the Genome: The Challenge of Synthetic Biology" *Biosocieties* 3, no. 1 (2008): 3–20. For a historian's account of synthetic biology see Luis Campos, "That Was the Synthetic Biology That Was," in *Synthetic Biology: The Technoscience and Its Societal Consequences*, ed. M. Schmidt et al. (Dordrecht, Netherlands: Springer, 2010), 5–21. For an account of synthetic biology by an anthropologist see Sophia Roosth, "Biobricks and Crocheted Coral: Dispatches from the Life Sciences in the Age of Fabrication," *Science in Context* 26 (2013): 153–171.

There is a lot of information about DIY Bio online. For example, see the DIY Bio website: http://diybio.org/. There are also several good press accounts of what is going on in this field. See Jon Mooallem, "Do-It-Yourself Genetic Engineering," *New York Times*, February 10, 2010, http://www.nytimes.com/2010/02/14/magazine/14Biology-t.html; Ritchie S. King, "When Breakthroughs Begin at Home," *New York Times*, January 16, 2012, http://www.nytimes.com/2012/01/17/science/for-bio-hackers-lab-work-often-begins-at-home.html; Rob Carlson, "Splice It Yourself," *Wired*, May 13, 2005, http://www.wired.com/wired/archive/13.05/view.html?pg=2; and Marcus Wohlson, *Biopunk: Solving Biotech's Biggest Problems in Kitchens and Garages* (New York: Penguin, 2011). On the possibilities of synthetic biology see George M. Church and Ed Regis, *Regenesis: How Synthetic Biology Will Reinvent Nature and Ourselves* (New York: Basic Books, 2012). On the use of open source ideas in biology see Janet Hope, *Biobazaar* (Cambridge: Harvard University Press, 2009). The possibilities of DIY, open, and citizen biology for instituting new regimes of scientific governance are explored in the work of Denisa Kera, "Shenzhen and the Republic of Tinkerers: Open Source Hardware (OSHW) as Tools of Global Governance in the Hackerspaces and DIYBio Labs," http://www.academia.edu/6206001/Shenzhen_and_the_Republic_of_Tinkerers_Open_Source_Hardware_OSHW_as_Tools_of_Global_Governance_in_the_Hackerspaces_and_DIYbio_labs.

The discussions of bioterrorism and biosecurity include biologists, social scientists, and terrorism and security experts. The issues raised in this chapter are discussed in greater detail in Jonathan B. Tucker and Raymond A. Zilinskas, "The Promise and Perils of Synthetic Biology," *New Atlantis* 12 (Spring 2006), http://www.thenewatlantis.com/publications/the-promise-and-perils-of-synthetic-biology; Stephen M. Maurer, "End of the Beginning or Beginning of the End? Synthetic Biology's Stalled Security Agenda and the Prospects for Restarting It," *Valparaiso University Law Review* 45, no. 4 (2011): 73–132; Stephen M. Maurer, "Taking Self-Governance Seriously: Synthetic

Biology's Last, Best Chance to Improve Security," Goldman School of Public Policy, Working Paper No. GSPP12–003 (2012), http://papers.ssrn.com/sol3 /papers.cfm?abstract_id=2183306; and Michael Specter, "A Life of Its Own: Where Will Synthetic Biology Lead Us?" *New Yorker* September 28, 2009, http://www.newyorker.com/reporting/2009/09/28/090928fa_fact_specter.

INTRODUCTION: WHY BIOTECH ART?

Since the 1990s, a handful of artists have begun to engage with biotechnology in their work. For these artists, biotech has become not merely the subject of their work, but also the medium of expression. They utilize the techniques of genetic engineering and tissue culture to produce sculptures, installations, performances, videos, and "living" artworks.

Why include a chapter about art in a book about biotechnology? Throughout these chapters we have explored the connections between biotechnology and imagination. We have seen how science fiction—in the form of novels, movies, and plays—has shaped our hopes and fears about technology. More than this, in some cases these responses have even influenced the way in which technologies are developed and used. Aldous Huxley's *Brave New World*, for instance, had a significant influence on scientists working on the technologies of human reproduction in the 1930s. In the 1980s, cyberpunk literature stimulated disparate groups of software engineers and hardware designers to produce the first virtual reality engines.

Art, of course, is a form of imaginative expression. It too can help us to understand some of the social and cultural responses to biotechnology. Artists express our hopes and fears about technology. Analyzing bioart can reveal what biotechnology *means* for our society and culture. More than this, art has the potential—like science fiction—to shape what biotechnology might become. By placing biotech in different contexts, by deploying it in unusual ways, bioartists challenge our assumptions about what biotech is and open up new possibilities for its use.

This chapter describes and analyzes a cross-section of bioart projects. Much of this book has demonstrated the almost inextricable connections between biotech, capitalism, ownership, and consumerism. The examples here suggest a range of possible alternative biotech futures.

SCIENCE AND ART

Science and art are often assumed to be opposite sorts of activities. Science is said to be about collecting and organizing facts in supposedly objective or disinterested ways. Art is supposed to be about representing ephemeral or partial impressions in subjective ways. Despite these contrasts, the histories of art and science run parallel and are even connected to one another at critical points. After all, art and science might both be said to be involved in the larger project of representing and understanding the world around us.

Before the scientific revolution in the seventeenth century, the boundary between art and science was less clear. Making knowledge about the natural world included attempts to describe and draw that world. To take a famous example, Leonardo da Vinci's *Vitruvian Man* (ca. 1487 — this is a figure of a man, arms and legs outstretched, and inscribed into a square and a circle) has become a symbol of both art and science. For art, it was a remarkable achievement in the use of perspective and proportion. For science, it signified a move towards the realistic and detailed description of human bodies. Likewise, the anatomical drawings that illustrated Andreas Vesalius' (1514–1564) *De Humani Corporis Fabrica* (*On the Fabric of the Human Body* [1543]) were simultaneously a scientific and artistic achievement. In fact, it would be more historically correct to say that, during medieval and early modern periods, there was no meaningful distinction between these two kinds of work.

Even during the scientific revolution, artists and artisans played a central role. Historians of science have shown how certain kinds of artistic work (for example, the prints of Albrecht Dürer, 1471–1528) required detailed observations and depictions. Such work contributed to a "mechanical philosophy" that was crucial to the scientific revolution. Such overlaps between scientific and artistic concerns have not disappeared. In the nineteenth century, both artists and scientists worried about the representational challenges posed by the new technology of the photographic camera.

In the twentieth century, artistic expression has continually been challenged, inspired, and reshaped by science and technology. The expressionist movement, for instance, developed in response to industrialization and urbanization in turn-of-the-century Europe. Later, Salvador Dali's (1904–1989) famous "melting clocks" (for instance, "The Disintegration of the Persistence of Memory," 1952–1954) were artistic responses to Albert Einstein's theories of relativity (in which time and space could become stretched, compressed, and distorted).

This brief overview suggests that art and science are connected by a set of mutual influences. Artistic expression has influenced the development of knowledge about the natural world. Conversely, scientific knowledge (and the technologies produced with it) has continued to rework both the media and the subjects of art.

EXPLORING BIOART

For thousands of years, humans have used plants and animals for decorative and artistic purposes. Gardens, hedgerows, bonsai trees, and ornamental breeds of dogs and cats all involve human manipulation of living things for aesthetic purposes. Chapter 2 described the work of the Californian horticulturalist Luther Burbank. Over his lifetime, Burbank bred a range of new kinds of fruits and flowers. These included hybrid fruits such as the plumcot (plum + apricot) and ornamental daisies and roses. Burbank's work was followed by that of many other plant breeders. Their creations can be seen as a precursor to more recent forms of bioart. In 1936, the plant breeder and photographer Edward Steichen (1879–1973) even exhibited delphiniums that he had bred at the Museum of Modern Art in New York City.

Since the biotech revolution of the 1970s, artists have become increasingly attracted to biotechnology as both a subject and a medium for their work. There are at least two kinds of reasons for this interest. First, artists are aware of the significant cultural implications of biotechnologies. They see that biotechnologies seem to be raising political, ethical, and philosophical questions. They believe that art has an important role to play in exploring these questions in provocative ways and in helping to find answers to them. Some artists, then, see engaging with biotech as responsibility: as biotech becomes more culturally important, it poses a set of challenges to which artists must respond.

Second, some bioartists argue that the new media presented by biotechnologies provide new ways for artists to engage with long-standing cultural problems and questions. For instance, using techniques of genetic engineering or tissue culture as part of artworks allows artists to raise questions about the relationship between humans and technology, between nature and culture, in particularly provocative ways. For these artists, engaging with biotech is an opportunity: it offers new modes of artistic expression and new ways of capturing interest.

Bioart is deliberately pushing and extending the boundaries of artistic expression. This makes it difficult to offer a full characterization of all bioart projects—there are many possible axes along which it could be described (for a partial list see box 24.1). In some cases, works are closely tied to specific programs in experimental biology, while other projects more closely resemble computer hacking, subversion, or political activism. We could categorize bioart according to the types of technologies it utilizes (genetic engineering, tissue culture, etc.). Or, we could describe its chronological development. Here, I have chosen to describe bioart in terms of some of the main themes that it addresses: the relationship between biology and information, the bound-

Box 24.1 A Few Other Bioart Projects

Microvenus
- Year: 1986
- Artist: Joe Davis and Dana Boyd
- A graphic icon (in the shaped of a "Y" and "I" superimposed) were encoded into DNA and transfected into *Escherichia coli*.

Robert Smithson
- Year: 1986–1988
- Artist: George Gessert
- Hybridized iris flower (*Pacifica iris*)

Good and Evil on the Long Voyage
- Year: 1997
- Artist: Paul Perry
- A "hybridoma" cell was created by fusing one of the artist's white blood cells with a mouse cancer cell; during exhibition the cell was placed in a bioreactor in an aluminum canoe.

The Eighth Day
- Year: 2001
- Artist: Eduardo Kac
- Living transgenic bioluminescent life forms are contained in a dome with a biobot (a robot that contains an active biological part, in this case a colony of amoebas). Changes in the colony cause the biobot to move.

Proteic Portrait (study)
- Year: 2002
- Artist: Marta de Menezes
- Specially engineered "marta" protein is compared with known proteins using bioinformatic tools; these predict possible 3-dimensional conformations of the protein that are portrayed on printed canvas.

Source: Eduardo Kac, ed., *Signs of Life: Bioart and Beyond* (MIT Press, 2007).

---

aries between the living and the nonliving, and attempts at criticism and re-sistance.

LIFE AND INFORMATION

One large group of bioartworks experiments with the connections between biology, information, and language. Artists who create these works are drawing attention to the consequences of the digitization of life that is taking place through the technologies of sequencing, DNA chips, bioinformatics, genomics, and computational biology.

24.1 Eduardo Kac—Genesis. Linz, Austria, 1999. Kac's genetically engineered bacteria are under a UV light and webcam in the middle of the exhibition room. The camera also projects an image of the bacterial colonies on the far wall. The left-hand wall shows a video projection of the DNA letters of the synthetic gene produced from the text from the book of Genesis. Source: Eduardo Kac, Genesis, 1999. Transgenic work with artist-created bacteria, ultraviolet light, Internet, video (detail), edition of 2, dimensions variable. Collection Instituto Valenciano de Arte Moderno (IVAM), Valencia, Spain. Used with permission.

Genesis

Eduardo Kac is one of the most active and well-known bioartists. In 1999, Kac created an art installation that he called "Genesis" (figure 24.1). Kac began with a verse from the first book of the Christian Bible (Genesis): "Let man have dominion over the fish of the sea and over the fowl of the air and over every living thing that moves upon the earth." Kac translated this first into Morse code, and then into DNA base pairs (using a special encoding Kac designed himself). Then in the lab, Kac then synthesized the DNA corresponding to his quotation, making a synthetic gene that he inserted into plasmids and transfected into *E. coli* bacteria. These engineered bacteria also contained a yellow fluorescent protein, allowing the bacteria to fluoresce when exposed to UV radiation.

For his installation, Kac placed colonies of his synthetic bacteria in petri dishes with colonies of other bacteria containing blue fluorescent protein. As the bacteria swapped genes with one another, various color combinations emerged. A micro-video camera projected a real-time image of the colored colonies in the petri dish onto the wall of the gallery. The images were also captured and streamed over the World Wide Web. Remote viewers could activate the UV light to stimulate further mutation within the bacterial colonies. The gallery space also contained wall-sized projections of the verse from Genesis and the DNA code of the synthetic gene; "DNA synthesized" music was also played.

This arrangement was on display at the OK Center for Contemporary Art in Linz, Austria, for two weeks. Through it, Kac hoped to explore the relationship between biology, religion, information technology, and ethics. The choice of the Biblical texts draws attention to the potential dangers of using genetic engineering to "play God." But the use of multiple forms of writing or encoding (English text, Morse code, DNA code, binary computer code) also draws attention to the erasing of the boundaries between living and nonliving objects: "Genesis explores the notion that biological processes are now writerly and programmable, as well as capable of storing and processing data in ways not unlike digital computers," Kac wrote.[1]

Biopoetry

Kac and other bioartists have explored a range of other ways to connect biology and language. These projects all draw attention to biotech's ability to "write" and "edit" life. Such activities include *atomic writing* (shaping molecules to spell out words and expressing these words as plant genes), *amoebal scripting* (forming letters and words in agar gel and allowing colonies of microorganism to grow around these shapes), *luciferase signaling* (genetically modifying fireflies to perform patterned displays), *bacterial poetics* (encoding two poems into two different colonies of bacteria and allowing them to compete for resources), *proteopoetics* (coding words directly into three-dimensional proteins to create protein poems), and *agroverbalia* (using an electron beam to write words on seeds and observe which words yield robust plants).

Biopoetry shows how texts can be made into life and how life can be read as texts. But as well as highlighting biotech's power in manipulating living "scripts," biopoetry suggests that these technologies can also become means of *expression*. By displacing and reusing biotechnologies in these strange ways, biopoetry suggests the open-ended, creative potential of biotechnologies. It reminds us that these technologies are not just new ways of doing

---

1. Eduardo Kac, "Genesis," first published in catalogue of *Genesis*, Linz, Austria: OK Center for Contemporary Art, 1999), 45–55, http://www.ekac.org/geninfo.html.

biology, but can also become new ways of speaking, communicating, and interacting. Biopoetry suggests that biotechnologies always have *cultural* as well as technological meanings.

LIVING AND NONLIVING

Tissue Culture and Art Project

Rather than using genetic engineering, other artists have deployed the technologies of tissue culture. One important set of bioartworks emerged from the Tissue Culture and Art Project based at the School of Anatomy and Human Biology at the University of Western Australia. This project has developed a range of techniques for growing "semi-living sculptures" using stem cells and cell culture techniques.

For instance, one piece—by the artists Oron Catts and Ionat Zurr—grew "Guatemalan worry dolls" from living cells (figure 24.2). In Guatemala, children are given small dolls that they are instructed to put under their pillows when they sleep; the dolls are supposed to help alleviate worries. Catts and Zurr built scaffolds out of polymers and then introduced mouse endothelial cells onto the scaffolds. These were then placed in a slowly rotating bioreactor, allowing the cells to grow over the scaffold. This process makes use of techniques similar to those that are being used to grow replacement tissues and organs for therapeutic purposes. The worry dolls were exhibited in 2000—viewers could peer through magnifying classes to observe the growing tissues.

The Tissue Culture and Art project has also grown pig bone cells into the shape of wings (called "Pig Wings"), human and animal cells into a human ear (a collaboration with the performance artist Stelarc), a "blob" of cells taken from many different organisms (called "NoArk"), and mouse muscle and fish nerve into the shape of miniature prehistoric stone artifacts (called "The Stone Age of Biology").

These creations exhibit a playful, even humorous, attitude towards bioart. But the artists also hope to provoke serious public discussion of these objects. In particular, Catts and Zurr aim to raise questions about the status of such "semi-living" things. Are they alive or dead? How should they be treated? "These entities," they write, "blur the boundaries between what is born and what is manufactured, what is animate and what is inanimate, and further challenge our perceptions and our relationships with our bodies and our constructed environment."[2] In other words, these artworks are seeking to bring

2. Oron Catts and Ionat Zurr, "Growing Semi-Living Sculptures: The Tissue Culture and Art Project," *Leonardo* 35, no. 4 (2002): 365–370. Quotation p. 366.

24.2 Tissue Culture and Art Project—Worry Dolls. 2000. Semi-living "dolls" made by growing endothelial, muscle, and osteoblast cells over degradable polymers in a bioreactor. For more details see http://tcaproject.org/projects/worry-dolls. Source: Title: A Semi-Living Worry Doll H. Artists: The Tissue Culture and Art Project (hosted at SymbioticA, School of Anatomy, Physiology and Human Biology, University of Western Australia). Medium: McCoy cell line, biodegradable/bioabsorbable polymers and surgical sutures. Dimension of original: 2 cm x 1.5 cm x 1 cm. Date: from the Tissue Culture and Art(ificial) Wombs Installation, Ars Electronica 2000. Used with permission.

into focus some of the cultural and ethical dilemmas that biotechnologies such as xenotransplantation, tissue banking, tissue engineering, and genomics are forcing us to confront.

REMIX AND RESISTANCE

Superweed

Beginning in 1999, Heath Bunting began to distribute what he called "low tech" do-it-yourself biotechnology kits. These kits, called "Superweed 1.0," contained a variety of genetically modified seeds from plants including canola, wild radish, yellow mustard, and shepherd's purse. Bunting claimed that, if planted, these seeds would generate a "superweed" that would be capable of resisting commercial herbicides such as Roundup (glyphosate). Since governments seemed to be doing little to protect the public from the dangers of genetically modified foods, Bunting reasoned, individuals and small groups could take matters into their own hands. By growing amongst GM crops, superweeds would render "Roundup Ready" and other herbicide-resistant GMOs useless. Ultimately, agricultural biotech companies world be forced to stop selling GM seeds. "By releasing Superweed 1.0 into the environment long before biotech companies have a suitable fix," Bunting wrote, "people can contribute to large losses in their profitability, thus causing them to reassess their future strategies and investments."[3]

Superweed 1.0 is a mixture of political activism and art. It is a potential tool of resistance, but also a means of awareness raising about the dangers of GM crops. The kit also makes use of biotechnology itself, redeploying it as a tool of resistance, rather than a vector of capitalist profit-making. This inversion draws attention to the multiple possibilities and diversity of uses that are immanent in biotechnologies.

i-Biology

In 2004, Diane Ludin created an online system called "i-Biology: a multi-user patent filter system." The software, accessible via the World Wide Web, was conceived as a patent mash-up system. Downloading patents from the US Patents and Trademarks Office online database, i-Biology allowed users to modify and remix patents to make new patents. These new patents would serve similar purposes to the original patents from which they were taken, but would exploit loopholes in the patent system to ensure that they

---

3. "Natural Reality SuperWeed kit 1.0," http://www.irational.org/cta/superweed/kit .html.

didn't violate any existing patents. Finally, the system allowed users to register the patents with the Patent Office and to put them in the public domain.

i-Biology was a "hack" on the patent system—a back door that allowed users to side-step patents held by corporations. Ludin's aim was explicitly subversive, but it was also artistic. She hoped that i-Biology would serve to draw attention to the problems of the patent system as it applied to the life sciences and to raise awareness of the growing corporatization of biotechnologies in general. By displacing and remixing elements of biotechnology (in this case, patents), this bioart provokes its (online) audience to think more carefully about some of the consequences of ownership and commercialization in biology.

CROSSING BOUNDARIES

The examples given here do not completely represent the range of bioart. This dynamic field is innovating to recombine technology, art, performance, and political activism in unexpected ways. However, it is possible to make some general observations about what these artworks are trying to achieve.

First, they are attempting to help society come to terms with biotechnology and its consequences. By reusing and repurposing biotechnologies they pose (and sometimes offer answers to) the kinds of questions we have been addressing throughout this book: What limits should we impose on our powers to "rewrite" life? Should growing human tissues be considered alive or dead? What are the consequences of the patenting of life? Of course, some artworks may be more successful than others, but bioart has become a forum for grappling with the pressing social, political, and ethical issues raised by biotechnologies.

Second, many of these artworks draw special attention to *boundaries*. Biotech involves all kinds of boundary crossing: between alive and dead, between digital and biological, between human and nonhuman, between nature and culture, and between public and private. Many of our fears and hopes about biotechnology arise from precisely these tensions. Like other art forms, bioart derives much of its potency from pushing at, exposing, and calling into question these boundaries. It provokes us to see these binaries in new and different ways and therefore to reassess our views of biotechnology.

CONCLUSIONS: WHAT IS ART FOR?

One of the functions of art in society is to provoke critique and public debate about important political, ethical, and cultural issues. Bioart seems to be playing this role with respect to biotechnology. Some critics of bio-art have argued that bioartists' dependence on biotechnologies have blunted their critical edge; bioart sometimes seems to be merely celebrating and encouraging biotechnology, rather than critiquing it. But celebrating and critiquing should not necessarily be considered mutually exclusive activities. Bioart offers possibilities simultaneously encouraging new uses of biotechnology while critiquing biotechnologies in their current form. At the very least, biotech art fosters an ongoing dialogue between scientists, the public, politicians, and corporations, about the proper role and development of biotechnology.

Perhaps most importantly, even the least critical forms of bioart place biotech in alternative contexts (e.g., art galleries), redeploying the tools of genetic and tissue engineering in new and unexpected ways. This, in itself, may be of great importance for the future of biotechnology. Studies of the history of technology have shown that the meaning and importance of technologies is often determined by their *use*. The Internet, for instance, was designed largely as a network for sharing computer resources (storage and processing power). However, its users reconstituted the Internet as a powerful system for *human* communication (for example, through the invention and use of email).

Most of the uses of biotechnologies that we see around us are embedded in the academic-industrial-medical-commercial spaces and regimes that have been described in this book. They are licensed, commercialized, patented, and marketed. Much bioart manages to take biotech outside of these contexts. Even the act of putting biotechnology to *aesthetic* use, challenges us to see it in ways that go beyond its economic or medical use. Redeploying bioart as anti-commercial (as in Superweed 1.0) or playful (as in the Tissue Culture and Art's "worry dolls") not only opens up biotech's subversive possibilities, but also allows us to see that biotech does not *necessarily* belong to the commercial and industrial regimes in which it mostly now exists. By showing how biotech can be utilized as a means of *expression*, we come to see more clearly that biotech entails multiple meanings and possibilities, and that these meanings and possibilities are under our control: we can choose to use biotech differently and therefore to *remake* it something different. In short, bioart opens up a range of alternative futures for biotechnology.

FURTHER READING

On the history of the relationship between art and science see Pamela H. Smith, *The Body of the Artisan: Art and Experience in the Scientific Revolution* (Chicago: University of Chicago Press, 2004) and Peter Galison and Caroline Jones, eds., *Picturing Science, Producing Art* (London: Routledge, 1998).

There is a body of work by sociologists, bioartists, and others that has begun to reflect on bioart and its significance. See George Gessert, *Green Light: Toward an Art of Evolution* (Cambridge, MA: MIT Press, 2010); Robert Mitchell, *Bioart and the Vitality of Media* (Seattle: University of Washington Press, 2010); Eduardo Kac, *Telepresence and Bio Art: Networking Humans, Rabbits, and Robots* (Ann Arbor: University of Michigan Press, 2005); Eduardo Kac, ed., *Signs of Life: Bioart and Beyond* (Cambridge, MA: MIT Press, 2009); and Beatriz da Costa and Kavita Phillip, eds., *Tactical Biopolitics: Art, Activism, and Technoscience* (Cambridge, MA: MIT Press, 2010).

For the descriptions of the specific projects referred to here, the best sources are online. For Catts and Zurr's work see Oron Catts and Ionat Zurr, "Growing Semi-Living Sculptures: The Tissue Culture and Art Project," *Leonardo* 35, no. 4 (2002): 365–370; and also the Tissue Culture and Art Project website: http://tcaproject.org/. For Kac's "Genesis" see http://www .ekac.org/geninfo.html; for his Biopoetry see: http://www.neme.org/387/bio poetry. Heath Bunting's Superweed can be found (and ordered) at: http:// www.irational.org/cta/superweed/. And Diane Ludin's i-Biology is described at: http://www.thing.net/~diane/i-BPE/index.html.

CONCLUSION: ETERNAL LIFE
AND THE POSTHUMAN FUTURE

OUR BIOTECH FUTURE

Where might biotechnology be taking us? One of the themes that has been emphasized throughout this book is the idea of *promise*. Biotech is oriented towards the future: it is driven by hype, speculation, the promises of future rewards, and the fears of future catastrophes. Therefore, it is appropriate to end this book by describing various visions for the future of biotech. These imaginaries—our fears and hopes—provide important clues as to how society understands and responds to biotechnology and as to where biotech is likely to take us in the near future.

Some of these visions are certainly science fiction. But science fiction—like the art that we saw in chapter 24—can play an important role in shaping how we think and feel about technologies. A particular technology can be used for many purposes—but how we end up using it depends on what sorts of uses we can *imagine* for it. Science fiction plays an important role in generating new ideas about what technologies might be, how they might be used, and how they might transform our lives.

In this conclusion, I introduce the idea of the "posthuman." This concept is a powerful way of imagining how biotechnologies will come to impact humans in the coming decades.

WHAT IS THE POSTHUMAN?

The posthuman (also sometimes called the transhuman) is a way of understanding the evolving relationship between humans and technology. The invention of the electronic computer in the middle of the twentieth century marked a turning point in this respect: machines too, it seemed, could think. In the 1950s, electronic computers were clunky, slow (compared to today), room-sized machines. Nevertheless, people already thought of them as "electronic brains" or "thinking machines."

In 1950, Alan Turing (1912–1954), one of the inventors of electronic computers, proposed a test to evaluate whether computers could really think. The Turing Test requires two humans (a tester and a subject) and a computer. The aim of the test is for the tester to successfully distinguish between the subject and the computer. Obviously, if the tester could see the subject or the computer it would be easy to tell one from the other and the game would be up. So, for the purposes of the test the subject and the computer should be in a separate room, out of sight of the tester. However, the tester can communicate with both the computer and the subject by text—he or she can send both the subject and the computer questions or messages, to which they can respond (you can imagine this working something like instant messenger). Any kinds of questions are allowed. The aim of the subject is to convince the tester that he or she is indeed the human. The aim of the computer is to fool the tester into thinking that it is the human. If the computer succeeds (or if the tester cannot decide in the allotted time), then it is considered to have passed the test. According to Turing, a computer that passes the test should be considered "intelligent."

What is interesting about the Turing Test is that human intelligence is assessed purely by an act of textual communication and information processing. The fleshiness or liveliness of the human body does not matter—all that counts is the ability to send and receive symbols. This idea—that humans and intelligence can exist as pure information—has had a powerful effect on science fiction. From William Gibson's *Neuromancer* (1984) to the Wachowski brothers' *The Matrix* (1999), science fiction has popularized the notion that is possible to upload consciousness into a computer or into cyberspace. What matters in cyberspace (or in the matrix) is not the body, but only the information that can be seamlessly moved from mind into machine and back again.

This view does not distinguish between minds and machines: they are both just different kinds of information processors. And, this allows us to imagine ourselves as seamlessly integrating with machines, becoming part-human and part-machine. That is, it allows us to imagine ourselves as *cyborgs*. We might attach powerful robot arms to our bodies or upgrade our memory with silicon parts or deploy virtual reality to communicate with our friends. Understanding minds as computers allows us to imagine ourselves not just as human beings, but as augmented, extended, enhanced, and developed by and through machines and computers. We can, through cyborg technologies, become *more than human.*

This is what the posthuman is all about—the idea that we can (or will) go beyond the human. The posthuman suggests that we will soon evolve—with the help of machines—far beyond the capabilities of the human species. So far, in fact, that we should not even be called humans: we will become posthumans.

ARE WE (BECOMING) POSTHUMAN?

Some people believe that we are already posthuman — or at least that we are very close to transcending the human species. People who hold this view point to all the technologies with which we already augment our lives: cellular phones, laptop computers, watches, iPods, and so on. We use these electronic devices to supplement and extend the power of our reasoning, our memories, and our senses (e.g., with night-vision goggles or hearing aids). For some people, medical devices too, play a critical role in their day-to-day lives: people's hearts are kept beating by pacemakers; diabetics are automatically injected with insulin; and increasingly sophisticated electromechanical prosthetics can be wired into bodies and brains. We are already critically dependent on machines, many of which we wear on our bodies for much of our lives. These "wearable technologies" have already made us into cyborgs.

These technologies have become wearable due to the rapid advances in solid-state physics and electronics that took place during the second half of the twentieth century. Computers, once room-sized objects, can now be worn on a watch or implanted under the skin. The decrease in the price of silicon technologies has made them available to almost everyone in the developed world; and these computers have become increasingly easy to program and personalize. Computers have become familiar and ubiquitous.

But, just as the twentieth century was the century of physics, the twenty-first century may be the century of biology. The physicist Freeman Dyson is one person who argues that the same "domestication" that occurred with silicon may occur with biotechnologies. At the moment, biotechnologies are expensive, difficult to use, and largely controlled by large corporations or academic laboratories. In 1965, the same could have been said of computers. Biology too may undergo a revolution of "personalization," bringing biotech within the price range and skill level of ordinary people. As synthetic biologists hope, we may all be able to play around with DNA to create our own organisms.

This kind of opening up of biotech would lead, many imagine, to a range of transformative possibilities. Biotech could be used to create organisms to generate energy, replacing fossil fuels; it could be used to build new kinds of foods, new body parts, new drugs, and new weapons. These possibilities for designing and making organisms would reshape how we learn, work, and play. Computers, especially when coupled to the Internet, have fundamentally reshaped society, culture, politics, and commerce. Biotech, now in its infancy, could be poised to do the same.

Biotechnology is already offering us many forms of enhancement and augmentation. IVF, pharmaceuticals, tissue culture, stem cells, and other bio-technologies all attempt to extend and augment our bodies. Many forms of biotechnology intervene in our bodies at the most basic molecular level. As such, the transformations offered by biotech are perhaps even more funda-mental than those offered by computers.

ARTs have allowed us to fundamentally transform and rewire our repro-ductive capacities. In the future, other biotechnologies may transform our metabolism, our cellular development, or our thought processes (by rewiring neurons, for instance). Increasingly, too, biotechnology and machines are themselves becoming integrated. Fields such as nanobiotechnology are de-veloping ways to use insights from biology, robotics, and artificial intelligence to intervene in our bodies using nano-scale devices. These devices might help to deliver oxygen to our tissues, repair damaged organs, hunt down bacteria in the body, or interface with neurons in our brain. In other words, as the boundaries between biotech and nanotech blur, biological knowledge may help us to interface with machines in even more subtle and pervasive ways. Biotech is likely to play a significant role in profoundly transforming human beings.

As we are able to gain greater control over our bodies, we may be able to dramatically slow the aging process, or even forestall death indefinitely. Im-mortality would represent the ultimate way in which biotechnology could transform the meanings and purposes of human life.

Of course, it is possible that none of these things will ever occur. Many people believe that bodies, let alone brains, are far too complex to be com-pletely understood in the way that these developments would require. Never-theless, the idea of the posthuman suggests what is at stake in our biotech future. The posthuman makes it clear that we, as a species, are being radically transformed by biotech. It suggests that biotech is not just about a set of tech-nological transformations, but about transformations in how we reproduce, how we eat, how we think, and what we are. Whatever biotech is becoming, we are becoming too.

FURTHER READING

There are a number of different perspectives on the posthuman. The approach here draws especially on ideas developed in English litera-ture. See N. Katherine Hayles, *How We Became Posthuman: Virtual Bodies in Cybernetics, Literature, and Informatics* (Chicago: University of Chicago Press, 1999) and Richard Doyle, *On Beyond Living: Rhetorical Transforma-tions in the Life Sciences* (Palo Alto, CA: Stanford University Press, 1997).

Donna Haraway's analysis of cyborg thinking is also closely related to the idea of the posthuman developed here: Donna Haraway, "A Cyborg Manifesto," in *Simians, Cyborgs, and Women* (New York: Routledge, 1991).

For Dyson's view on the domestication of biotech in the twenty-first century see Freeman Dyson, "Our Biotech Future," *New York Review of Books*, July 19, 2007, http://www.nybooks.com/articles/archives/2007/jul/19/our -biotech-future/.

The views of the singulatarians—who believe that dramatic shifts in human experience and consciousness are imminent—can be found in Vernor Vinge, "The Coming Technological Singularity: How to Survive in the Post-Human Era," originally presented at VISION-21 symposium, NASA Lewis Research Center and Ohio Aerospace Institute, March 30–31, 1993, http://www.aleph .se/Trans/Global/Singularity/sing.html; and Ray Kurzweil, *The Singularity Is Near: When Humans Transcend Biology* (New York: Penguin, 2006). In this futurist vein see also Robert Pepperell, *The Posthuman Condition: Consciousness beyond the Brain* (Bristol, UK: Intellect, 1995). For a more pessimistic view see Frances Fukuyama, *Our Posthuman Future: Consequences of the Biotechnology Revolution* (New York: Picador, 2003).

ACKNOWLEDGMENTS

A work of this breadth necessarily draws deeply on the work of others. My primary debt is to all those scholars whose works are listed in the "Further Reading" sections of each chapter and without whose research and insights this book could not exist.

Biotechnology and Society developed through my teaching an undergraduate class first at the University of Melbourne, then at Harvard College, and finally at Nanyang Technological University. At Melbourne, I took over the teaching of a class called "Biotechnology and Modern Society" from Rosemary Robbins. At Harvard, I served as a teaching assistant to Hannah Landecker in her class, "Biotechnology, 1900 to Now." My own version of the class emerged as a kind of hybrid of these two. I am deeply grateful to Janet Browne for providing the opportunity to teach this class at Harvard in 2010–2011. There, Sam Schweber encouraged me to adapt my ideas for an audience beyond the classroom. My students in Melbourne, Cambridge, and Singapore also demonstrated to me the importance of the kind of interdisciplinary approach to biotechnology pursued in this book.

Most of the writing of this manuscript was carried out at Nanyang Technological University. There I have been lucky enough to be able to engage with a lively group of scholars in the School of Humanities and Social Sciences, especially through the "Humanities, Science, and Society" Cluster. In particular, I have benefitted greatly from sharing ideas with Shirley Sun, Sulfikar Amir, Lisa Onaga, Lyle Fearnley, and Miles Powell. This work was also supported by a Start-Up Grant from the University. The last parts of this book were completed during a fellowship at the Max Planck Institute for the History of Science in Berlin in 2014.

The original illustrations for this book were produced by Jerry Teo. He has done a remarkable job and has been amazingly patient as I pestered him over the smallest details. Karen Darling, my editor, has supported this work from beginning to end. Her guidance has been invaluable. The anonymous reviewers for the University of Chicago Press have also allowed this to develop into a much better book than would otherwise have been possible.

For reading portions of the manuscript and offering invaluable comments

375

(as well as encouragement) I thank Jenny Bangham, Sarah Blacker, Joanna Radin, Jenny Reardon, Lukas Rieppel, and Sophia Roosth, as well as members of my family Peter, Gill, Rosemary, Lara, and Eddie.

Finally, extreme and extraordinary thanks go to my wife, Yvonne Ruperti. Her work and support sustained me, literally.

INDEX

Page numbers followed by the letter *b*, *f*, or *t* indicate a box, figure, or table, respectively.

categories of disease, 213–14; secularization of, 216

Bioethics (Potter), 210

biological determinism, 171, 189

Biometrika, 162

biopiracy, 336, 337, 338b

biopoetry, 363–64

bioprospecting, 335–36, 337, 338b, 339. *See also* biocolonialism

biosecurity, 353–54

biotech art: basis of artists' interest in biotech, 360; examples of bioart projects, 361b; functions of art with respect to biotech, 367–68; goals of, 367; life and information theme in, 361–64; living and nonliving theme in, 364–66; overlaps in history of science and art, 358–59; relationship between art and biology, 358; remix and resistance theme in, 366–67; use of plants and animals for decorative purposes, 360

biotech industry: basis of, 47; bioethics and, 218; business of genetic testing, 198–99; GMFs and enabling of corporate control over agriculture, 127; growth of pharmaceutical industry and, 227; impact on development of science and technology, 74–76; initial obstacles to biotech commercialization, 71–72; origin of (*see* Silicon Valley); patent law applied to organisms (*see* patent law); relationship between the academy and, 75–76; scientists' hesitancy to pursue commercialization, 70; transfer and exchange of tissues and organs and, 152; transformation from pure to applied biology, 76–77; worldwide earnings, 65

biotechnology: approach used to examine, 4–5; audience and scope of this book, 3–4; defined in relation to active biological processes, 17–18;

definition of technology and, 15–16; domains not examined, 4–5; drugs as biotechnologies (*see* pharmaceutical drugs); history of (*see* history of biotechnology); molecular control component of, 18–20; relationship to technology, 19–20; significance in our society, 1; sociotechnical system concept and, 2, 4, 15, 16–18; susceptibility to ideological agendas, 2–3; theme of beyond controversy, 9–10; theme of biotech is not new, 6–7; theme of plasticity, 7–8; theme of promise, 8–9; theme of risk, 10–11

Biotechnology and Biological Sciences Research Council, 239

bioterrorism, 352, 353

birth control. *See* contraception

blood donation, 148–49

Blood Music (Bear), 266

Blumberg, Baruch, 141

Blumenbach, Johann Friedrich, 314

Boeke, Jef, 352

Borlaug, Norman, 99

boundaries in biology. *See* hybrids and hybridity; organ and tissue ownership

Boveri, Theodor, 255

bovine somatotropin (BST), 102–3

Boyer, Herbert, 41, 45f, 70, 71, 72, 84

Boyle, Robert, 145

Brattain, Walter, 68

Brave New World (Huxley), 35, 358

BRCA cancer genes, 89, 197

Brenner, Sydney, 179–80

Briggs, Robert, 244

Brown, Louise, 231

Brown-Séquard, Charles-Édouard, 27

BST (bovine somatotropin), 102–3

Bt (*Bacillus thuringiensis*), 104

Buck v. Bell, 166

Bunting, Heath, 366

Burbank, Luther, 26, 360

Bush, George W., 211, 262, 305

Cooperative Wheat Research Production Program, 98–99
coproduction, 2
corn, 120–22, 121f, 122n5
cortisone, 226, 227
cosmetic psychopharmacology, 289–90
cotton, 103
Coulson, Alan, 180
counterculture, 50–51, 209, 228
Crassostrea gigas (Pacific oysters), 85
Crichton, Michael, 90
Crick, Francis, 37, 182
crosses. *See* hybrids and hybridity
cryobiology, 146
cryopreservation, 145–47
cultured meat, 153
Cusenza, Paul, 301
Cyanamid, 337
cyborgs, 1, 21, 371–72. *See also* hybrids and hybridity
cystic fibrosis (CF), 182, 188

Dali, Salvador, 359
Darwin, Charles, 160, 315
Davenport, Charles, 163
da Vinci, Leonardo, 359
DeCODE Genetics, 300–301
De Humani Corporis Fabrica (Vesalius), 359
DeKalb Genetics, 104
De Lisi, Charles, 181, 182
Department of Energy (DOE), 181, 182
depression, 284–85
designer babies: absence of defined property rights in surrogacy cases, 274; argument against regulation of ARTs, 271; critique of genetic engineering offered in *Gattaca*, 269; danger of the commodification of babies, 275; egg donations, 272–73; ethical questions raised by ARTs, 271; genetic tests available for pregnant women, 199b, 269–70; growing use of PGD, 270, 273; market forces influence on ARTs, 273; need for regulation, 274–75; notion of property rights in a market, 273–74; restriction on use of PGD, 270; sperm banks, 272; surrogacy and, 272–73; use of PGD to select for desired trait, 270–71. *See also* reproduction
designer bodies. *See* pharmaceutical drugs
Diamond v. Chakrabarty, 73, 82–84, 124
diazepam (Valium), 281
dieldrin, 57n7
diosgenin, 226
Dishley Ram sheep, 237, 238f
Diversa Corporation, 339
DIY biology movement, 350–51
DNA: attempts to patent DNA sequences, 88; cloning of, 43–45; "code" view of, 174, 176; discoveries about its structure and function, 175; explanation of function, 38f, 39b; "junk" designation, 184; postulation of its existence, 36–37; process used to insert human DNA into animal cells, 239–40; recombinant (*see* recombinant DNA); registry of designed DNA sequences, 348–50; sequencing process, 176–79, 293–96, 294b (*see also* Human Genome Project)
Dolly the sheep, 18, 152, 239, 242, 243f, 245–47
Drosophila melanogaster (fruit fly), 179
drugs. *See* pharmaceutical drugs
DuPont, 85–86, 104, 126
Dürer, Albrecht, 359
Dyson, Freeman, 372

economic meanings of biotech: absence of defined property rights in surrogacy cases, 274; advances in hormone synthesis, 226; attempts to create economic meanings of biotech (*cont'd*) drug-making pharm animals, 239; biocolonialism and, 336–37, 338b, 339–40, 341; business and biotechnology (*see* biotech industry); costs and scale of the HGP, 182, 187; first commercially grown GMF for human consumption, 103; genetic engineering and, 47–48; genetic testing and, 198–99; GMFs and impact on how people eat (*see* economics of eating); hybrids and hybridity and, 153, 154; impact of granting of patents for living things, 84; legislative reforms to promote commercialization, 82; link to interests of government and private enterprise, 30–31; market forces and influence on ARTs, 271, 273; organ and tissue ownership and, 136, 139, 140; pharmaceutical drugs as marketed products, 287, 288, 289; purpose of patents, 80; reproduction seen as a market (*see* designer babies); side effects of the Green Revolution, 100

economics of eating: agricultural transformations due to GMFs, 117–19; consolidation of economic power in farming, 126; field to table path of corn and soybean, 121f; food consumption transformations due to GMFs, 127; food production transformations due to GMFs, 120–23; food supply control afforded to owners of plants, 125; GMFs and enabling of corporate control over agriculture, 127; impacts of industrialization of livestock farming, 119; nonfood uses of corn and soybeans, 122n5; ownership of the worldwide seed supply, 123, 125, 126; ownership transformations due to GMFs, 123–25; scope of the PPA and PVPA, 123–24; socio-

technical system of GMFs, 116–17; technology use agreements required of farmers, 124–25, 126

ED (erectile dysfunction), 285

Edwards, Robert, 230, 255

Efficiency of Our Race and the Protection of the Weak, The (Ploetz), 167

EFSA (European Food Safety Authority), 108

egg donations, 272–73

Egyptians and beer, 22

Einstein, Albert, 359

Eli Lilly, 72–73, 282

Elliott, Carl, 287

ELSI (ethical, legal, and social issues), 211

embryonic stem cell research, 260–63

endocrine glands, 28n2

endocrinology. *See* hormone research

Endy, Drew, 352

England. *See* United Kingdom

Enlightenment, the, 314

Enovid, 228

environment: bioethics and, 211; impact of agricultural practices, 50, 53, 119; impact of the Green Revolution, 100; impact on differences between social groups, 324; patent law and, 91; risk assessment limitations, 55, 109–10; risks from GMFs and, 103–4, 106, 107, 108, 127; risks from modified genes, 9, 10, 87, 107, 352, 366

EPA (US Environmental Protection Agency), 103

erectile dysfunction (ED), 285

Ereky, Karl, 118

Escherichia coli (*E. coli*), 44, 53, 179, 352, 361

Essential Biologicals, 141

estradiol, 226

ESTs (expressed sequence tags), 88, 186

ethical, legal, and social issues (ELSI), 211

ethnobotany, 336–37

eugenics: after World War II, 168–70; biological determinism and, 171; development of modern statistics and, 161; differences with genetic testing, 204; field of behavior genetics and, 169; genetic testing and, 203–5; in Germany, 166–68; implementation of policies by governments, 162–63; investigation of the heritability of traits, 161; legacy of, 168–69, 170, 204–5; level of control achieved by scientists, 170; natural selection applied to society as a whole, 160–61; in other countries, 162–63; premise of, 159; racial superiority focus of, 162; shift away from, 169; similarities and differences with biotechnologies, 170–71; status as a science, 162; sterilization laws and, 166, 167; supporting institutions, 160; support of contraception, 225–26; in the United Kingdom, 160–62; in the United States (*see* eugenics in the United States)

eugenics in the United States: adoption of restrictive marriage laws, 166; adoption of sterilization laws, 166; cooperation with other countries, 167; Davenport's research on trait heredity, 163–64; efforts to encourage certain marriages, 165; extension to immigration policies, 164–65; Mendelian inheritance and, 163; race hygiene movement and, 165

European Food Safety Authority (EFSA), 108

European Patent Convention, 86–87, 91

European Union, 108

Evans, Martin, 257

Existence Genetics, 301

Ex Parte Allen, 85

Ex Parte Hibberd, 124

Ex Parte Latimer, 80

expressed sequence tags (ESTs), 88, 186

expressionist movement, 359

Farley, Peter, 69, 71

FDA (US Food and Drug Administration), 103, 280, 306

Federal Food, Drug, and Cosmetic Act (1938), 280

Federation of Spirit Producers (*Verein der Spiritus Fabrikanten*), 25

fermentation process, 22, 23f, 24

fiduciary duty, 138

Fischer, Eugen, 167

Flavr Savr, 103

fluoxetine (Prozac), 281–85, 287, 289

food production: consideration of the origins of food, 120; decoupling of food from plants and animals, 122–23; field to table path of corn and soybean, 120–22, 121f; impacts of industrialization of livestock farming, 119; nonfood uses of corn and soybeans, 122n5. *See also* agriculture; economics of eating

Ford, Henry, 117

Ford Foundation, 99

Foucault, Michel, 232

454 Life Sciences, 293, 301

Fox Chase Cancer Center, 141

France, 162–63, 185

Franklin, Sarah, 246

freezing tissues, 145–47

French Eugenics Society, 162–63

"Frosty One" and "Frosty Two," 146

fruit fly (*Drosophila melanogaster*), 179

Fukuyama, Francis, 47

Funk, Casimir, 29

Galton, Francis, 160–61, 315

Gattaca, 203, 269

Gaud, W. S., 98

G. D. Searle, 227, 228

genetic disease, 197–98, 199b; measuring value of different lives and, 201–2, 203, 204; precursors to, 196–97; welfare of the individual and, 204

Genographic Project, 293, 332

Genome-Wide Association Studies (GWAS): common variants analysis, 297, 300; explanation of, 298b, 299f; heritability of traits accountability and, 300; process used, 296–97; questions being addressed by, 296

Germany: beer brewing and, 25; eugenics and, 163; regulation of PGD, 270

germ plasm theory, 254–55

gestational surrogacy, 272

Gey, George, 134

Gibson, William, 371

Giddens, Anthony, 108

Gilbert, Walter, 176

GINA (Genetic Information Nondiscrimination Act), 305

Glaser, Donald, 69, 71

Glaxo, 29

globalization, 340–41

glyphosate (Roundup), 104, 366

Golde, David, 137–40

Golden Rice, 104, 127

Google, 301

grapefruit, 101

Green Revolution: connection to stopping the spread of communism, 98; efforts to improve wheat yields in Mexico, 98–99; increase in rice yields in Asia and, 99–100; negative side effects of, 100; relationship to GMFs, 100–101

grey goo, 265–67

grey market, 273–74

Gross, Alan, 60

Guatemalan worry dolls, 364, 365f

Gurdon, John, 244–45

GWAS (Genome-Wide Association

Studies): common variants analysis, 297–300; explanation of, 298b, 299f; heritability of traits accountability and, 300; process used, 296–97; questions being addressed by, 296

Häcker, Valentin, 255

HapMap (International Haplotype Map Project), 188, 293, 331–32

Harrison, Ross, 133

Harvard Medical School, 85–86

Hastings Center, 209

Hastings Center Report, 209

Hawking, Stephen, 202

Hayflick limit, 134, 258

Heape, Walter, 229

Hegel, Friedrich, 315

HeLa cell line, 134–36, 135f

Helicos Biosciences, 293

HelloGene, 301

hemophilia, 141

hepatitis B virus, 141

herbicides, 104

Hereditary Genius (Galton), 161

Heredity in Relation to Eugenics (Davenport), 163

Herrnstein, Richard J., 318

Hewlett, William, 66

HGDP (Human Genome Diversity Project), 330–32, 334

HGP. *See* Human Genome Project

Hibberd, Kenneth A., 124

Hi-Bred Corn, 27

Hilgartner, Stephen, 91

Hippocrates, 208

Hippocratic Oath, 208

history of biotechnology: agriculture and, 25–27; beer brewing and (*see* zymotechnology); beginnings of biotech, 21–22; biotechnology's link to interests of government and private enterprise, 30–31; biotech revolution and, 76–77; genetic engineering

history of biotechnology (*cont'd*)
and (*see* genetic engineering); hormone research, 27–29; horticulture advances, 26; importance of studying, 6–7; interest in manipulation of higher animals, 27; protests against science, 50–52; reasons to examine, 21–22; scientists' hesitancy to pursue commercialization, 70; transformation from pure to applied biology, 76–77; vitamins and diet and, 29–30

Hoffman-La Roche, 74

Homo faber, 15–16

Hood, Leroy, 179, 182

Hooper, Martin, 241

horizontal gene transfer, 107

hormone research, 27–29, 226

horticulture, 26

HR977, 90

Hubbard, Pauline, 124

Hubbard, Ruth, 57

Hughes, Sally Smith, 70

Human Fertilization and Embryology Authority, UK, 211

Human Genome Diversity Project (HGDP), 330–32, 334

Human Genome Project (HGP): argument of Eurocentrism in, 330; bioethics and, 211; building of support for mapping the human genome, 181–82; case made for sequencing more genes, 183; *C. elegans* mapping project, 180; claims of commonness and uniqueness in the genome, 328–29; completion of the sequencing, 186–87; concerns about the implications of, 174–75; costs and scale of the projects, 182, 187; criticisms and concerns about the project, 183–84; genetic code and DNA, 174, 176; "genome" term use, 175–76; impact on the way biology is organized, 184–85; international collaboration rising from, 185; knowledge gained from the completed genome, 187–88; limitations of the sequence, 188; model organisms study, 179–81; opposition to, 183–84; patents and, 87–88; results of, 292–93; rethinking of biological determinism due to, 189; sequencing approach developed by Venter, 87–88, 186; sequencing process, 176–79, 185–86; sociotechnical system concept and, 175; subdisciplines emerging from, 188–89; technology used for, 176–79, 185–86

Huxley, Aldous, 35, 48, 358

Huxley, Julian, 35

hybrid corn, 26–27

hybrids and hybridity: anxiety over mixing perceived categories, 144; assigning value to life outside the body and, 154; cryopreservation and, 145–47; as an economic issue, 153, 154; ethical consequences of tissues crossing bodily boundaries, 153–54; in vitro meat research, 153; issues raised by some medical technologies and devices, 145, 153–54; organ and tissue ownership and, 147–49; relationship between biology and business and, 152; societal problems inherent in the exchange of body parts, 149, 154; successful use of a frozen human embryo, 146; tissue banks and, 147–49; types of tissues stored by freezing, 147; uses of pigs, 152–53; xenotransplantation, 149–52

IASB (International Association for Synthetic Biology), 354

i-Biology, 366–67

iGEM (International Genetically Engineered Machine) conference, 350

Illumina, 293, 295n1, 301

Immigration Act (1924), 164

PKU (phenylketonuria), 196, 203
Planned Parenthood Federation of America, 225. *See also* American Birth Control League
plant breeding, 25–27, 99, 360. *See also* Burbank, Luther; genetically modified foods; Green Revolution
Plant Patent Act (PPA, 1930), 79, 83, 123–24
Plant Variety Protection Act (PVPA, 1970), 83, 123
plasmids, 43
plasticity, 7–8, 31, 290
Ploetz, Alfred, 167
Polge, Christopher, 146
political meanings of biotech: argument that politics and science should be separate, 59; biocolonialism and, 328, 332–34, 335–36, 339–40; European opposition to patents on living organisms, 86–87; experimentation with inclusive forms of policy making, 113; framing of debates over ARTs, 232–33; genetic engineering and, 46–47; hormone research in post-WWI Europe, 28–30, 226; impetus for support of research in the United States, 81–82; legislation allowing patents on genetically engineered organisms, 73; legislation designed to protect individuals, 305; link to interests of government and private enterprise, 30–31; organ and tissue ownership and, 139, 141; stem cells and, 260–63; women's rights movement and contraception, 224–25
Pollan, Michael, 120, 122
Polly (sheep), 244
polymerase chain reaction (PCR), 336
postgenomics, 187–89, 292
posthuman: implications of understanding minds as computers, 371; prevalence of machines in our lives,

372; revolution of personalization in biology, 372; test designed to determine if a computer is intelligent, 370–71; transformative possibilities of biotech, 372–73
potatoes, 103
Potter, Van Rensselaer, II, 210
Poultry Research Centre, 239
PPL Therapeutics, 152, 239, 240n1
pramipexole (Mirapex), 288
preimplantation genetic diagnosis (PGD), 198; growing use of, 270, 273; market forces influence on, 273; to select for desired trait, 270–71
prenatal testing, 195, 197, 198, 199
Presidential Commission for the Study of Bioethical Issues, 212
President's Council on Bioethics, 211–12
product of nature doctrine, 80, 84
progesterone, 226
promissory science, 8–9, 31, 47, 174, 245, 263–64, 370
pronuclear microinjection, 240
prosthetic limbs, 145
proteopoetics, 363
Prozac (fluoxetine), 281–85, 287, 289
Prozac Nation (Wurtzel), 284
Ptashne, Mark, 58
Pure Food and Drug Act (1906), 280

Quetelet, Adolphe, 315
Quinlan, Karen Ann, 216

Rabinow, Paul, 70
race and biology: attempt to measure cognitive characteristics between races, 317–18; classification of plants and animals, 313–14; consideration of nongenetic influences on racial differences, 324–25; designation of races among humans, 314; discrediting of biologically based race theories, 316–17; the Enlightenment and

race and biology (*cont'd*)
human rights, 314; eugenicists' focus
on racial superiority, 162; history of
the expansion of the science of race,
315–16, 318–19; human differences
and, 162, 318, 320, 323–24; hypothesis of a genetic basis for differences in
drug reactions, 319–20; Lahn's study
of the MCPH1 gene, 321; proposed
genomic basis for differences in intelligence, 321–23; questions about importance of human differences, 323;
race-based categories used to market
a drug, 320–21
race hygiene movement, 165, 168f
Radiation Laboratory, 66–67
radium, 57n7
Rajan, Kaushik Sunder, 47–48
recombinant bovine somatotropin
(rBST), 102–3
recombinant DNA: activists' efforts
against recombinant DNA research,
56–59; arguments that politics and
science should be separate, 59; Asilomar conference, 54–56, 351; commercialization of, 70, 73–74; debates
over risks, 54–58, 60–61; economic
meanings of, 47–48; initial obstacles
to commercializing, 71–72; invention
of, 41, 43, 53–54, 70; patent award, 84;
in plant breeding, 103; political meanings of, 46–47; possible research-
outcome scenarios, 54–55; process of
making, 42f, 43–45, 54; recommendations for experimentation with,
55–56; setting of experimental-work
guidelines, 53–54; social meanings of,
46; use of recombinant techniques
to alter plant genes, 101–2; voluntary
moratorium on research, 54, 55n5
Reemtsma, Keith, 150
Reeve, Christopher, 1, 263
regenerative therapies, 263, 264b

Registry of Standard Biological Parts,
349–50
Reimers, Niels, 70
reproduction: cloning's implications for,
246; contraception and (*see* contraception); as an industrial complex,
270–71; in-vitro fertilization (*see* in-
vitro fertilization); as a market (*see*
designer babies); perception of as
fundamental right, 231; PGD and, 198,
270–71, 273; reality of freedom of reproductive choice, 232–33; successful
implantation of a frozen embryo, 146.
See also assisted reproductive technologies
reproductive cloning, 36n1, 223, 246
restless legs syndrome, 288
restriction enzymes, 41–43, 42f
rice, 99, 101. *See also* Golden Rice
Ridley, Henry, 335
Rinfret, Arthur, 146
risk: bioethical considerations of, 210,
270, 332, 340, 341; of bioterrorism,
351–52, 353; cancer and biotechnologies and, 53, 54, 55, 57, 87, 100, 108,
197, 306; debates over recombinant
DNA, 54–58, 60–61; framework for
assessment of for GMFs, 108–13, 109f,
127; framework for assessment of in
biotechnology, 10–11; improving risk
assessment approaches, 111; information from personal genomics and,
292, 301–6, 302f, 307–8; information
provided by genetic tests, 198, 204,
269–70; possibility of "unknown
unknowns" and, 110; of synthetic
biology, 5, 9, 351–52
risk society, 114
Ritvo, Harriet, 237
Rock, John, 227–28, 230
Rockefeller Foundation, 98–99, 167
Rockefeller Institute for Medical Research, 346

Roe v. Wade, 261
Rosenberg, Charles, 212
Roslin Institute, 239
Roszak, Theodore, 51
Roundup (glyphosate), 104, 366
rubber, 335
Ruby Red Grapefruit, 101

sacrifice siblings, 270–71
salmon, AquaAdvantage, 104
Sandoz Pharmaceuticals, 137
Sanger, Frederick, 176, 181
Sanger, Margaret, 224–25, 227
Sanger Sequencing, 177b, 178f
savior siblings, 270–71
Saxenian, Anna Lee, 69
Schallmayer, Wilhelm, 167
Schrödinger, Erwin, 36
Science, 53
SCOT (Social Construction of Technology), 20
Second Creation, The (Wilmut and Campbell), 245
Select Agent list, 354
sequencing DNA, 176, 177b, 178f, 179, 293–96, 294b. *See also* DNA
Shannon, Claude, 37
sheep breeding: advances in, 237, 238f; animal research institutes in the United Kingdom, 238–39; creation of Dolly, 239, 242, 243f; economic importance of sheep, 237; symbolic roles of sheep, 236–37. *See also* animal husbandry
Shiva, Vandana, 112
shmeat (in vitro meat), 153
Shockley, William, 68
Shull, George H., 27
sildenafil (Viagra), 285–86, 287
Silent Spring (Carson), 50
Silicon Valley: creation of, 65–68; Genentech and (*see* Genentech); impact of World War II on science and

engineering, 67–68; location in the United States, 65, 66f; network effect and social climate in, 68–69; prevalence of defense contracts in, 68; transistor invention and, 68
Singapore Botanical Gardens, 335
Singer, Maxine, 53
single nucleotide polymorphism (SNP), 329
Sinsheimer, Robert, 181
Skloot, Rebecca, 135n2
Slavin, Ted, 140–41
Smith, Audrey, 146
Smith, Hamilton, 348
Smith, Lloyd, 179
SNP (single nucleotide polymorphism), 329
Social Construction of Technology (SCOT), 20
social meanings of biotech: artists' interest in biotech, 360; ARTs and, 232–33, 271; claims of commonness and uniqueness in the genome, 328–29; cloning and, 246–47, 248; complications when considering biomedical inventions, 90–91; construction of "human family tree," 330, 331f; counterculture and, 50–51, 209, 228; cultural impact of oral contraceptives, 228, 280–81; DNA testing's significance to the histories of groups, 328, 333–34; Dolly the sheep's implications for the science of cloning, 245–46; ethical issues (*see* bioethics); eugenics and (*see* eugenics); European opposition to patents on living organisms, 86–87; fears of the commodification of babies, 275; genetic engineering and, 46, 60–61; GMFs and (*see* genetically modified foods); growth of personal genomics (*see* personal genomics); HeLa cell line, 134–36; issues raised by

things theme, 348; debates about screening sequences, 354; debates on how to regulate, 353–54; DIY biology movement, 350–51; implications of a physico-chemical basis for life, 347; mechanistic view of life and, 345, 346; need for expansive debate, 354–55; parallels to software development, 350, 351; registry of designed DNA sequences, 348–50; risk management possibilities, 354; risks in public availability of, 351–52; vitalist view of life and, 346

syphilis, 209

Systema Naturae, 314

tacrolimus, 150

Taq polymerase, 336

Taylor, Frederick Winslow, 117

Tay-Sachs disease, 197, 201, 202

technology use agreements, 124–25, 126

Terman, Frederick, 65, 67

Terman, Lewis, 318

testosterone, 226

thalidomide, 57n7

therapeutic cloning, 36n1, 246

Thermus aquaticus, 336

Third World Network, 330

Thomson, James, 257, 260

Till, James, 257

tissue banks, 147–49

tissue culture: in bioart, 360, 364, 365t, 366; defined, 133; HeLa cell line story, 134–36; implications for the boundaries of life, 134, 136; skepticism about viability of, 133

Tissue Culture and Art Project, 364, 365f, 366

tissue ownership. *See* organ and tissue ownership

tomatoes, 103

Tracy (sheep), 240n1

transgenic modification, 102, 244

transhuman. *See* posthuman

transistors, 68

triticale, 102n2

Turing, Alan, 370

Turing Test, 370

Tuskegee Institute, 209, 210

23andMe (company): argument that they provide a beneficial service, 302–4; concerns raised by services, 305–6; FDA-forced suspension of services, 306; founding of, 300, 301; information presented by, 302–4; process for collecting and analyzing genomes, 301–2

Unger, Ernst, 150

United Kingdom: animal research in, 238–39; beer brewing in England, 25; bioethics and, 211; HGP and, 185; regulation of PGD, 270, 272

United Nations, 336

United States: activists' efforts against genetic research, 56–58; agriculture in, 116–19; countercultural attitudes towards science in, 50–51; eugenics movement in (*see* eugenics in the United States); GMFs approved for use in, 105b; Green Revolution and, 98–99; instances of ethical abuses of patients, 209; legality of organ sales in, 152; legal protections for biotechnology (*see* patent law); legislation regarding genetic information, 305; number of successful pregnancies by IVF per year, 231; ownership of food processing facilities in, 126; ownership of reproductive material, 148; politics and biotech (*see* political meanings of biotech); post-WWI hormone research in, 28–30, 226;

.